AF765323

NEUROMETHODS

Series Editor
Wolfgang Walz
University of Saskatchewan
Saskatoon, Canada

For further volumes:
http://www.springer.com/series/7657

Neuromethods publishes cutting-edge methods and protocols in all areas of neuroscience as well as translational neurological and mental research. Each volume in the series offers tested laboratory protocols research. Each volume in the series offers tested laboratory protocols step-by-step methods for reproducible lab experiments and addresses methodological controversies and pitfalls in order to aid neuroscientists in experimentation. *Neuromethods* focuses on traditional and emerging topics with wide-ranging implications to brain function, such as electrophysiology, neuroimaging, behavioral analysis, genomics, neurodegeneration, translational research, and clinical trials. *Neuromethods* provides investigators and trainees with highly useful compendiums of key strategies and approaches for successful research in animal and human brain function, including translational "bench to bedside" approaches to mental and neurological diseases.

Intrathecal Drug Delivery

Edited by

Manuel Menéndez-González

Instituto de Investigación Sanitaria del Principado de Asturias (ISPA), Oviedo, Spain;
Departamento de Medicina, Universidad de Oviedo, Oviedo, Asturias, Spain;
Hospital Universitario Central de Asturias, Servicio de Neurología, Oviedo, Spain

Humana Press

Editor
Manuel Menéndez-González
Instituto de Investigación Sanitaria del
Principado de Asturias (ISPA)
Oviedo, Spain

Departamento de Medicina
Universidad de Oviedo
Oviedo, Asturias, Spain

Hospital Universitario Central de Asturias
Servicio de Neurología
Oviedo, Spain

ISSN 0893-2336 ISSN 1940-6045 (electronic)
Neuromethods
ISBN 978-1-0716-5240-4 ISBN 978-1-0716-5241-1 (eBook)
https://doi.org/10.1007/978-1-0716-5241-1

© The Editor(s) (if applicable) and The Author(s), under exclusive license to Springer Science+Business Media, LLC, part of Springer Nature 2026

This work is subject to copyright. All rights are solely and exclusively licensed by the Publisher, whether the whole or part of the material is concerned, specifically the rights of translation, reprinting, reuse of illustrations, recitation, broadcasting, reproduction on microfilms or in any other physical way, and transmission or information storage and retrieval, electronic adaptation, computer software, or by similar or dissimilar methodology now known or hereafter developed.
The use of general descriptive names, registered names, trademarks, service marks, etc. in this publication does not imply, even in the absence of a specific statement, that such names are exempt from the relevant protective laws and regulations and therefore free for general use.
The publisher, the authors and the editors are safe to assume that the advice and information in this book are believed to be true and accurate at the date of publication. Neither the publisher nor the authors or the editors give a warranty, expressed or implied, with respect to the material contained herein or for any errors or omissions that may have been made. The publisher remains neutral with regard to jurisdictional claims in published maps and institutional affiliations.

This Humana imprint is published by the registered company Springer Science+Business Media, LLC, part of Springer Nature.
The registered company address is: 1 New York Plaza, New York, NY 10004, U.S.A.

If disposing of this product, please recycle the paper.

Preface to the Series

Experimental life sciences have two basic foundations: concepts and tools. The *Neuromethods* series focuses on the tools and techniques unique to the investigation of the nervous system and excitable cells. It will not, however, shortchange the concept side of things as care has been taken to integrate these tools within the context of the concepts and questions under investigation. In this way, the series is unique in that it not only collects protocols but also includes theoretical background information and critiques which led to the methods and their development. Thus it gives the reader a better understanding of the origin of the techniques and their potential future development. The *Neuromethods* publishing program strikes a balance between recent and exciting developments like those concerning new animal models of disease, imaging, in vivo methods, and more established techniques, including, for example, immunocytochemistry and electrophysiological technologies. New trainees in neurosciences still need a sound footing in these older methods in order to apply a critical approach to their results.

Under the guidance of its founders, Alan Boulton and Glen Baker, the *Neuromethods* series has been a success since its first volume published through Humana Press in 1985. The series continues to flourish through many changes over the years. It is now published under the umbrella of Springer Protocols. While methods involving brain research have changed a lot since the series started, the publishing environment and technology have changed even more radically. Neuromethods has the distinct layout and style of the Springer Protocols program, designed specifically for readability and ease of reference in a laboratory setting.

The careful application of methods is potentially the most important step in the process of scientific inquiry. In the past, new methodologies led the way in developing new disciplines in the biological and medical sciences. For example, Physiology emerged out of Anatomy in the 19th century by harnessing new methods based on the newly discovered phenomenon of electricity. Nowadays, the relationships between disciplines and methods are more complex. Methods are now widely shared between disciplines and research areas. New developments in electronic publishing make it possible for scientists that encounter new methods to quickly find sources of information electronically. The design of individual volumes and chapters in this series takes this new access technology into account. Springer Protocols makes it possible to download single protocols separately. In addition, Springer makes its print-on-demand technology available globally. A print copy can therefore be acquired quickly and for a competitive price anywhere in the world.

Saskatoon, Canada ***Wolfgang Walz***

Preface

Intrathecal drug delivery (ITDD) has evolved into a cornerstone technique for the management of a wide range of neurological conditions, offering targeted therapy to the central nervous system while minimizing systemic exposure. The clinical relevance of ITDD spans chronic pain, spasticity, central nervous system infections, oncology, and emerging neurodegenerative conditions. Despite decades of development, the field continues to face challenges in drug distribution, catheter design, device biocompatibility, and patient safety, necessitating ongoing research and innovation.

This volume is structured into four interconnected parts. Part I, "Introduction," provides foundational knowledge on the history of ITDD, key neuroanatomical considerations, cerebrospinal fluid physiology, and pharmacokinetics. This background equips the reader with the essential principles needed to understand subsequent clinical applications. Part II, "Marketed Intrathecal Drug Delivery Systems and Applications," focuses on currently available implantable systems, surgical techniques, and therapeutic strategies for chronic pain, spasticity, infections, and oncology. Here, the reader will find detailed insights into device mechanics, infusion strategies, and complication management.

Part III, "New Therapeutics and Formulations," explores novel intrathecal approaches, including targeted drug delivery for neurodegenerative and leptomeningeal diseases, and advanced formulation strategies such as mRNA nanoemulsions. These chapters highlight translational research bridging bench to bedside, emphasizing mechanistic understanding, formulation optimization, and preclinical evaluation. Finally, Part IV, "The Need for New Intrathecal Drug Delivery Systems," addresses the emerging technological and clinical challenges in ITDD, including sustained-release devices, nanoporous membranes, and pumpless implantable prototypes. These chapters provide a forward-looking perspective on innovation, regulatory considerations, and the design principles for next-generation intrathecal therapeutics.

This volume is intended for a broad audience, from medical students, graduate students, and residents to postdoctoral researchers and clinicians involved in neurology, anesthesiology, and translational neuroscience. By combining comprehensive theoretical foundations, detailed methodological instructions, and forward-looking analyses, this book aims to be a definitive resource for both academic research and clinical practice in ITDD.

Oviedo, Asturias, Spain *Manuel Menéndez-González*

Contents

Contributors

MARÍA JOSÉ ALONSO • *Center for Research in Molecular Medicine and Chronic Diseases (CiMUS), Campus Vida, Universidade de Santiago de Compostela, Santiago de Compostela, Spain; Department of Pharmacology, Pharmacy and Pharmaceutical Technology, Campus Vida, Universidade de Santiago de Compostela, Santiago de Compostela, Spain; Health Research Institute of Santiago de Compostela (IDIS), Santiago de Compostela, Spain*

GABRIEL ÁLVAREZ • *Neuroscience Innovative Technologies, Llanera, Spain*

PHILIPP LAPUHS • *Center for Research in Molecular Medicine and Chronic Diseases (CiMUS), Campus Vida, Universidade de Santiago de Compostela, Santiago de Compostela, Spain; Department of Pharmacology, Pharmacy and Pharmaceutical Technology, Campus Vida, Universidade de Santiago de Compostela, Santiago de Compostela, Spain; Health Research Institute of Santiago de Compostela (IDIS), Santiago de Compostela, Spain*

VICTOR M. LU • *Department of Neurological Surgery, University of Miami, Miami, FL, USA*

ÁLVARO JOSÉ VÁZQUEZ MARÍN • *Pain Unit, Anesthesiology and Resuscitation Department, Virgen de las Nieves University Hospital, Granada, Spain*

MANUEL MENÉNDEZ-GONZÁLEZ • *Instituto de Investigación Sanitaria del Principado de Asturias (ISPA), Oviedo, Spain; Departamento de Medicina, Universidad de Oviedo, Oviedo, Spain; Hospital Universitario Central de Asturias, Servicio de Neurología, Oviedo, Spain*

ESTER PÉREZ-MARTÍN • *Neuroscience Innovative Technologies, Llanera, Spain; Instituto de Investigación Sanitaria del Principado de Asturias (ISPA), Oviedo, Spain*

LAURA PINEIRO-ALONSO • *Center for Research in Molecular Medicine and Chronic Diseases (CiMUS), Campus Vida, Universidade de Santiago de Compostela, Santiago de Compostela, Spain; Department of Pharmacology, Pharmacy and Pharmaceutical Technology, Campus Vida, Universidade de Santiago de Compostela, Santiago de Compostela, Spain; Health Research Institute of Santiago de Compostela (IDIS), Santiago de Compostela, Spain*

CATUXA PRADO • *Neuroscience Innovative Technologies, Llanera, Spain*

CLAUDIA ROCES-LLORIÁN • *Neuroscience Innovative Technologies, Llanera, Spain*

LAURA ÁLVAREZ RODRÍGUEZ • *Pain Unit, Anesthesiology and Resuscitation Department, Virgen de las Nieves University Hospital, Granada, Spain*

MARÍA RODRÍGUEZ-CAÑÓN • *Neuroscience Innovative Technologies, Llanera, Spain*

INÉS RUBIO-PREGO • *Center for Research in Molecular Medicine and Chronic Diseases (CiMUS), Campus Vida, Universidade de Santiago de Compostela, Santiago de Compostela, Spain; Department of Pharmacology, Pharmacy and Pharmaceutical Technology, Campus Vida, Universidade de Santiago de Compostela, Santiago de Compostela, Spain; Health Research Institute of Santiago de Compostela (IDIS), Santiago de Compostela, Spain*

MANUEL ALEJANDRO SÁNCHEZ-GARCÍA • *Pain Unit, Anesthesiology and Resuscitation Department, Virgen de las Nieves University Hospital, Granada, Spain*
ASHISH H. SHAH • *Department of Neurological Surgery, University of Miami, Miami, FL, USA*
CRISTINA TOMÁS-ZAPICO • *Instituto de Investigación Sanitaria del Principado de Asturias (ISPA), Oviedo, Spain; Universidad de Oviedo, Oviedo, Spain*

Part I

Introduction

Check for updates

Chapter 1

Foundations and History of Intrathecal Drug Delivery

Manuel Menéndez-González

Abstract

The effective treatment of central nervous system (CNS) disorders is profoundly hindered by the blood–brain and blood–cerebrospinal fluid barriers, which prevent approximately 98% of small-molecule drugs and nearly all large-molecule biologics from reaching their intended targets. Intrathecal (IT) drug delivery, the direct administration of therapeutic agents into the cerebrospinal fluid (CSF), offers a powerful strategy to bypass these restrictive barriers and achieve therapeutic drug concentrations within the CNS. This chapter provides a comprehensive overview of the principles, history, and clinical application of IT therapy. First, it details the fundamental neuroanatomical and physiological landscape governing IT drug delivery, including the structure of the meninges, the dynamics of CSF production, circulation, and clearance pathways such as the glymphatic system. The chapter explains how the unique physical and biochemical properties of the CSF, combined with drug-specific characteristics and patient-related factors, determine the pharmacokinetics and biodistribution of intrathecally administered agents. Next, it traces the historical evolution of neuraxial therapy, from its nineteenth-century origins in spinal anesthesia to the modern era of programmable implantable pumps. This section highlights the key scientific discoveries, such as the identification of spinal opioid receptors, and the technological innovations that enabled the transition to continuous infusion for chronic conditions. Finally, the chapter reviews the current clinical applications of IT delivery. It covers its well-established role as a standard of care for severe chronic pain and spasticity, its use in neuro-oncology for treating leptomeningeal disease, and its growing importance for delivering novel therapeutics like antisense oligonucleotides for neurodegenerative diseases. By integrating principles of neurobiology, pharmacology, and biomedical engineering, this chapter establishes a foundational understanding of the science and practice of intrathecal drug delivery.

Keywords Intrathecal drug delivery, Cerebrospinal fluid, Neuroanatomy, Pharmacokinetics, Blood–brain barrier, History of neuraxial therapy

1 Introduction: The Concept and Need for Intrathecal Drug Delivery

The CNS is afforded remarkable protection by two highly regulated physiological barriers: the blood–brain barrier (BBB) and the blood–cerebrospinal fluid barrier (BCSFB). The BBB, formed by intricate tight junctions between brain microvascular endothelial cells, rigorously controls the passage of substances from the bloodstream

Manuel Menéndez-González (ed.), *Intrathecal Drug Delivery*, Neuromethods, vol. 234,
https://doi.org/10.1007/978-1-0716-5241-1_1,
© The Author(s), under exclusive license to Springer Science+Business Media, LLC, part of Springer Nature 2026

into the brain parenchyma. This barrier severely restricts the entry of most drugs, especially large molecules (those exceeding 400–500 Da in molecular weight) and hydrophilic compounds. Indeed, it is estimated that approximately 98% of small molecules and nearly all large therapeutic molecules, such as monoclonal antibodies or viral vectors, are unable to traverse this barrier effectively [1].

The BCSFB, primarily composed of the choroid plexus epithelium, also regulates drug entry into the CSF. While often considered comparatively "leakier" than the BBB, it still imposes selective permeability. Molecules that typically cannot cross the BBB, such as sucrose, inulin, and albumin, can enter the CSF, with their penetration inversely proportional to their molecular weight. These protective barriers, while essential for safeguarding neural tissue from pathogens and toxins, present a substantial impediment to CNS drug development, leading to high failure rates in drug translation and limiting the clinical availability of many promising therapeutics for neurological disorders.

The intrathecal (IT) route of drug administration involves delivering therapeutic agents directly into the CSF, bypassing the restrictive blood–brain barrier (BBB). This enables therapeutics, particularly large molecules such as biologics, ASOs, and stem cells, to reach target tissues in the CNS at therapeutic concentrations. Although the clinical application of IT delivery is still evolving, the great potential of this method depends primarily on understanding the anatomical and physiological principles that underlie the intrathecal route and the improvement of first-generation devices.

The relevant anatomical and physiological aspects from a therapeutic perspective are related to the meninges surrounding the CNS and the role of the CSF in providing mechanical cushioning, immunological protection, nutritional support, and waste exchange. The CSF, summing approximately 150 mL, circulates through the ventricular system and subarachnoid space, undergoing complete turnover approximately four to five times per day via absorption at the arachnoid villi into the venous sinuses. Moreover, when considering peripherally administered medication, the BBB acts as a highly selective barrier for substances entering the CNS from the systemic circulation. Although this structure plays a crucial role in protecting the brain from toxins and maintaining homeostasis, it severely restricts the entry of potentially therapeutic macromolecules.

Intrathecal administration bypasses the BBB by directly accessing the CSF. From the CSF, therapeutic agents can distribute along the neuraxis, particularly targeting the periventricular, cortical, and spinal cord regions. IT administration follows distinct pharmacokinetic principles compared to systemic delivery. In this chapter, we will understand the principles of IT drug delivery. We will see how upon injection into the lumbar subarachnoid space, therapeutics distribute in a rostral direction due to pulsatile CSF flow driven by cardiac and respiratory cycles. The distribution and clearance of drugs within the CSF are influenced by several factors, including drug-related factors such as molecular size, charge, or lipophilicity, as well as patient-related factors, including CSF flow dynamics. Larger molecules (e.g., ASOs, proteins, viral vectors) exhibit limited diffusion and are more reliant on convective CSF flow. Charged and hydrophilic molecules tend to remain within the CSF compartment, whereas lipophilic compounds have an increased biodisponibility. Factors related to administration protocols, such as injection volume, drug concentration, and excipients, can also affect CNS spread and penetration. When considering CSF dynamics, pathological conditions such as hydrocephalus or spinal stenosis can alter mechanically drug dispersion and lead to localized drug accumulation or reduced efficacy. Clearance of IT-administered drugs occurs primarily via bulk flow into systemic circulation through arachnoid granulations, glymphatic drainage pathways and, to a lesser extent, perivascular routes. Or, in NDDs, particularly AD, there is a reduced rate of substance clearance from the CNS, influencing the drug concentration balance. The half-life of IT-delivered substances in the CSF is typically longer than plasma administered drugs, enhancing the potential for sustained CNS exposure with appropriately designed delivery systems. To understand all these facts in detail, we need to review the fundamentals of CNS drug delivery and CSF dynamics first.

2 The Neuroanatomical and Physiological Landscape of the Cerebrospinal Fluid Compartment

The success of any IT therapy is contingent upon a sophisticated understanding of the target environment: the CSF system. This is not a passive conduit but a dynamic, biochemically unique, and anatomically complex compartment that defines the physical and chemical milieu of the CNS. Its anatomical boundaries, fluid composition, and the physiological barriers that enclose it are the primary determinants of the pharmacokinetics and pharmacodynamics of any agent delivered via this route.

The BBB is a highly specialized and dynamic cellular barrier that meticulously regulates the passage of molecules between the systemic circulation and the delicate environment of the CNS. It is composed of a continuous endothelium of cerebral microvessels, characterized by complex tight junctions, a near-absence of fenestrations, and minimal pinocytotic activity. This endothelial layer is ensheathed by pericytes and the end-feet of astrocytes, which collectively form the "neurovascular unit."

While this barrier is essential for protecting the CNS from toxins, pathogens, and peripheral inflammation, it is also exceptionally effective at excluding potential therapeutics. It is estimated that over

98% of small-molecule drugs and virtually 100% of large-molecule biologics—such as monoclonal antibodies, enzymes, gene therapy vectors, and antisense oligonucleotides—do not cross the BBB in therapeutically relevant concentrations when administered systemically [1]. This fundamental obstacle has rendered countless promising drug candidates ineffective for CNS diseases.

To overcome the challenge of the BBB, researchers and clinicians have turned to strategies that deliver therapies directly to the CNS. Intrathecal (IT) delivery, which involves the administration of a therapeutic agent directly into the subarachnoid space and, thus, into the CSF, stands out as the most direct and clinically advanced of these strategies. The CSF, which bathes the brain and spinal cord, acts as a natural distribution system, allowing drugs to bypass the BBB entirely and access their CNS targets.

This approach is not merely theoretical; it has been validated by the transformative success of Nusinersen (Spinraza®), an intrathecally delivered ASO for the treatment of spinal muscular atrophy (SMA). The approval and profound clinical efficacy of Nusinersen demonstrated that direct delivery to the CSF could fundamentally alter the course of a devastating neurological disease [2, 3]. This success has catalyzed a surge of interest and investment in developing IT therapies for a wide range of NDDs, fueling the rise of advanced therapeutic modalities specifically engineered for CNS targets. However, as these powerful and complex new therapies advance through the clinical pipeline, they are exposing the profound limitations of the decades-old devices used to deliver them. The future success of CNS medicine is, therefore, inextricably linked not only to the discovery of new drugs but also to the engineering of advanced delivery systems capable of realizing their full potential.

2.1 The Meninges: A Multilayered Protective and Functional Envelope

The CNS is encased by three classical meningeal membranes: the dura mater, the arachnoid mater, and the pia mater. The dura mater, or "tough mother," is the outermost layer, a thick, fibrous connective tissue sheath that provides robust mechanical protection and anchors the brain to the skull and the spinal cord within the vertebral canal [4]. The arachnoid mater, named for its delicate, web-like appearance, is an avascular layer that forms a watertight enclosure, creating a ceiling for the CSF-filled space beneath it. Adhering intimately to the entire surface of the brain and spinal cord, following every sulcus and gyrus, is the pia mater, or "tender mother." This highly vascularized membrane serves as a critical scaffold for the superficial blood vessels that nourish the CNS [4].

The space between the arachnoid and pia mater is the subarachnoid space (SAS), filled with CSF and traversed by a delicate network of connective tissue filaments known as arachnoid trabeculae. For centuries, this tripartite model was considered complete. However, in a landmark 2023 study, Møllgård and colleagues unveiled a fourth functional meningeal layer, which they termed the Subarachnoid LYmphatic-like Membrane (SLYM). Using advanced 3D imaging and tracer experiments, they identified a mesothelium-like, single-cell-thick membrane that partitions the SAS into an outer, superficial compartment and an inner, deeper compartment adjacent to the pia mater. The SLYM was found to be a significant diffusion barrier, selectively restricting the passage of solutes larger than 3 kDa. Furthermore, it houses a rich population of myeloid cells, including macrophages and dendritic cells, which actively survey the CSF and expand during inflammation and aging. This discovery redefines the SAS as a functionally compartmentalized space, with the SLYM acting as a critical checkpoint for both hydraulic control and innate immune surveillance [5]. The implications for IT drug delivery are profound, as the SLYM may represent a previously unappreciated barrier to the parenchymal penetration of large-molecule drugs like antisense oligonucleotides (ASOs), monoclonal antibodies, and viral vectors, while also serving as a site of immune-mediated drug clearance.

2.2 CNS Fluid Compartments and Protective Barriers

A sophisticated understanding of the environment into which intrathecal drugs are delivered is paramount for optimizing their therapeutic effect. The CSF space is not a static reservoir but a dynamic, complex compartment with unique physiological properties that govern drug distribution, stability, and clearance.

Three principal extracellular fluids sustain CNS homeostasis: blood, CSF, and the interstitial fluid (ISF). The ISF is the microenvironment directly bathing the neurons and glia, occupying approximately 20% of the total brain volume (~250 mL). It is not a static pool but engages in convective mixing with the CSF, a process vital for distributing essential nutrients like glucose, folate, and thiamine, and for clearing metabolic byproducts such as peroxidized lipids, lactate, and excess neurotransmitters [6].

The CSF is a clear, colorless fluid that provides essential functions for the CNS, including mechanical cushioning against trauma, maintenance of a stable biochemical environment, immunological protection, and removal of metabolic waste products. In an adult, the total CSF volume is approximately 150 mL, with about 25 mL contained within the ventricles and the remainder filling the subarachnoid space surrounding the brain and spinal cord. The CSF is actively produced, primarily by the choroid plexus within the ventricles, at a rate of about 0.3–0.4 mL/min, leading to a complete turnover of the entire CSF volume three to four times per day.

The classical model of CSF circulation describes a unidirectional bulk flow: from its production in the lateral ventricles, CSF flows through the third and fourth ventricles, exits into the subarachnoid space via the foramina of Luschka and Magendie, and then circulates over the cerebral convexities and down the spinal canal. It is ultimately absorbed back into the systemic circulation, primarily through arachnoid granulations that project into the major dural venous sinuses. More recent evidence has highlighted the crucial role of the glymphatic system, a perivascular network that facilitates the exchange of CSF with interstitial fluid, playing a key role in clearing solutes like Aβ and tau from the brain parenchyma, particularly during sleep. Impairment of this clearance system is increasingly implicated in the pathophysiology of NDDs like Alzheimer's disease.

The CSF itself is produced at a brisk rate of 18–24 mL/h, resulting in a total daily production of 400–600 mL. This allows for the complete turnover of the adult CSF volume (~150 mL) three to four times per day. The majority of this production (~70%) occurs in the choroid plexuses, specialized vascular structures located within the brain's ventricles. Here, a complex process of active secretion transforms a plasma filtrate into CSF. This process is driven by an osmotic gradient maintained by ion transporters like the NaK2Cl symporter and facilitated by aquaporin-1 (AQP-1) water channels [7]. The remaining 30% of CSF is thought to originate from extrachoroidal sources, primarily as a net flow of fluid across the capillary walls of the brain parenchyma.

This delicate fluidic system is guarded by two principal barriers. The blood–brain barrier (BBB) is an exceptionally restrictive interface formed by the endothelial cells of cerebral capillaries. These cells are sealed together by complex protein structures known as tight junctions (composed of claudins, occludins, and junctional adhesion molecules), are supported by pericytes and astrocytic end-feet, and are lined by a luminal glycocalyx. The BBB's combination of physical restriction and a panoply of active efflux transporters (e.g., P-glycoprotein, Breast Cancer Resistance Protein) effectively excludes over 98% of small-molecule drugs and virtually all large molecules from the CNS parenchyma. It is this barrier that necessitates direct CNS delivery routes for a vast array of neurotherapeutics [1]. The second barrier, the blood–CSF barrier (BCSFB), is located at the choroid plexus epithelium. While architecturally "leakier" than the BBB, its tight junctions strictly regulate ion and protein flux, meticulously crafting the unique chemical composition of the CSF that is essential for normal neuronal function [8].

2.3 Physical Properties and Biochemical Composition of CSF

Physically, CSF is an incompressible Newtonian fluid with a density (ρ) of approximately $1.0 \times 10^3\,kg/m^3$ and a kinematic viscosity (ν) of around $0.7 \times 10^{-6}\,m^2/s$, properties closely resembling those of water [9]. Its biochemical composition, however, is markedly different from that of blood plasma, reflecting its role as a highly regulated "ultrafiltrate." As detailed in Table 1, CSF is characterized by higher concentrations of sodium (Na^+), chloride (Cl^-), and magnesium (Mg^{2+}) ions, but significantly lower concentrations of potassium (K^+), calcium (Ca^{2+}), glucose, and, most notably, protein. The extremely low potassium concentration is critical for maintaining the neuronal resting membrane potential and preventing hyperexcitability.

Table 1
Key physiological and biochemical parameters of human cerebrospinal fluid (CSF) compared to plasma

Parameter	Cerebrospinal fluid (CSF)	Blood plasma	Clinical significance
Physical properties			
Total volume (Adult)	~150 mL (25 mL ventricular)	~3 L	Defines drug dilution space
Production rate	18–24 mL/h (400–600 mL/day)	N/A	High turnover necessitates continuous or repeated dosing
Opening pressure	70–180 mm H_2O (lateral decubitus)	N/A	Key diagnostic for intracranial pressure disorders
Specific gravity	1.003–1.008	1.025–1.029	Benchmark for determining drug baricity
Biochemical composition			
pH	7.31–7.34	7.35–7.45	Tightly regulated for enzymatic function
Protein	15–45 mg/dL	6000–8000 mg/dL	Low protein minimizes drug binding in CSF
Glucose	50–80 mg/dL (~60% of plasma)	70–110 mg/dL	Primary energy source for CNS
Sodium (Na^+)	138–150 mmol/L	135–145 mmol/L	Main caution, drives fluid balance
Potassium (K^+)	2.7–3.9 mmol/L	3.5–5.0 mmol/L	Low level critical for neuronal excitability
Chloride (Cl^-)	118–132 mmol/L	98–107 mmol/L	Main anion, balances sodium
Magnesium (Mg^{2+})	1.0–1.4 mmol/L	0.7–1.0 mmol/L	Higher concentration modulates NMDA receptors
Cells	<5 leukocytes/μL	4500–11,000/μL	Low count indicates immune quiescence

The low protein content (<0.5 g/L, compared to ~70 g/L in plasma) minimizes buffering capacity and oncotic pressure, while the acellular nature (<5 cells/μL) reflects an immunologically quiescent state under physiological conditions [10]. These properties are of paramount clinical importance. For instance, the specific gravity of CSF (typically 1.003–1.008) serves as the benchmark for defining a drug formulation's baricity—a key determinant of its gravity-driven distribution within the SAS.

2.4 CSF Production, Circulation, and Clearance Pathways

The cerebrospinal fluid (CSF) is an indispensable biological medium that envelops the brain and spinal cord, performing critical functions in mechanical protection, waste removal, and maintaining the homeostatic balance of the CNS microenvironment. A comprehensive understanding of its intricate dynamics is paramount for optimizing the delivery of intrathecal drugs.

CSF is primarily generated by the choroid plexuses located within the brain's ventricular system. This process occurs at an approximate rate of 500 mL/day, or about 20 mL/h, with the total CSF volume maintained at a dynamic equilibrium of 125–150 mL at any given moment. CSF production involves a two-step mechanism: initial passive filtration of plasma from fenestrated capillaries into an interstitial space, followed by the active secretion of ions (sodium, potassium, chloride) across choroid plexus epithelial cells. This active transport creates an osmotic gradient that draws water into the CSF, a process facilitated by aquaporins. The epithelial cells of the choroid plexus possess tight junctions, forming the blood-CSF barrier, which selectively restricts the free flow of most substances into the CSF.

The circulation of CSF follows a well-defined, yet, dynamic pathway throughout the CNS:

1. **Production:** The majority of CSF originates from the choroid plexus within the two lateral ventricles.
2. **Ventricular Flow:** From the lateral ventricles, CSF traverses the interventricular foramina to enter the third ventricle and subsequently flows through the cerebral aqueduct to reach the fourth ventricle.
3. **Subarachnoid Space Entry:** Upon exiting the fourth ventricle, CSF flows into the subarachnoid space (SAS) through four distinct apertures: the central canal of the spinal cord, the median aperture, and the two lateral apertures.
4. **Distribution:** The CSF then fills the entire SAS, surrounding both the brain and spinal cord, extending into cisterns and sulci. A connection also exists from the subarachnoid space to the bony labyrinth of the inner ear via the perilymphatic duct.
5. **Absorption:** The primary mechanism for CSF reabsorption back into the vascular system involves its entry into the dural venous sinuses via arachnoid granulations. This process is driven by a pressure differential between the arachnoid mater and the venous sinuses, with one-way valves ensuring efficient drainage. Additionally, CSF has been observed to drain into lymphatic vessels, particularly those associated with the olfactory nerve via the cribriform plate, and through the sheaths of cranial and spinal nerves, as well as the ependyma. The entire

CSF volume undergoes turnover approximately three to four times daily.

The classical model of CSF circulation describes a ventricular-cisternal-cortical pathway. CSF originates in the lateral ventricles, flows through the foramen of Monro into the third ventricle, descends via the cerebral aqueduct of Sylvius into the fourth ventricle, and then exits into the SAS through the foramina of Magendie and Luschka. From there, it bathes the surfaces of the brain and spinal cord [11]. While the net bulk flow is slow (<0.5 mL/min), the CSF is in a state of constant, pulsatile motion, driven primarily by the systolic expansion and diastolic recoil of the cerebral arteries and, to a lesser extent, by respiratory pressure changes.

Experimental data strongly support CSF outflow at the cribriform plate, where lymphatic vessels transport fluid and solutes to the deep cervical lymph nodes. Additional outflow pathways have been identified along the optic nerves and at the skull base, potentially passing through various foramina via perineural routes or dural lymphatics. There is also evidence for CSF drainage from the spinal region into lymphatic vessels, particularly in the lumbar and sacral areas. In mice, a few studies have observed CSF outflow into dural lymphatic vessels on the superior aspect of the skull. By contrast, outflow through arachnoid villi or granulations is poorly supported under normal physiological conditions, though this pathway may become active when intracranial pressure rises [14] (Fig. 1).

A paradigm-shifting advance in our understanding of CSF dynamics has been the characterization of the glymphatic system. This model posits that CSF from the SAS enters the brain parenchyma along perivascular spaces surrounding penetrating arteries. Within the brain tissue, it exchanges with the ISF in a convective flow process facilitated by AQP-4 water channels densely expressed on astrocytic end-feet. This "glymphatic" flow collects metabolic solutes, such as amyloid-β and tau, and directs them toward perivenous spaces for clearance from the brain [12]. This system is most active during deep, slow-wave sleep, and its dysfunction is increasingly implicated in the pathophysiology of neurodegenerative diseases [13]. For IT therapy, the glymphatic system represents a critical pathway for drug penetration from the CSF into deep brain structures, as well as a potential route for drug clearance.

The final clearance of CSF from the SAS occurs via multiple egress routes. The arachnoid granulations (or villi), which protrude into the major dural venous sinuses, have long been considered the primary site of CSF absorption. Recent evidence suggests they function less like one-way valves and more like porous lymphatic conduits, channeling CSF into the dural stroma and adjacent bone marrow spaces [14]. Parallel drainage occurs along the sheaths of cranial and spinal nerves. Notably, CSF drains along the olfactory nerves, through the cribriform plate of the ethmoid bone, and into

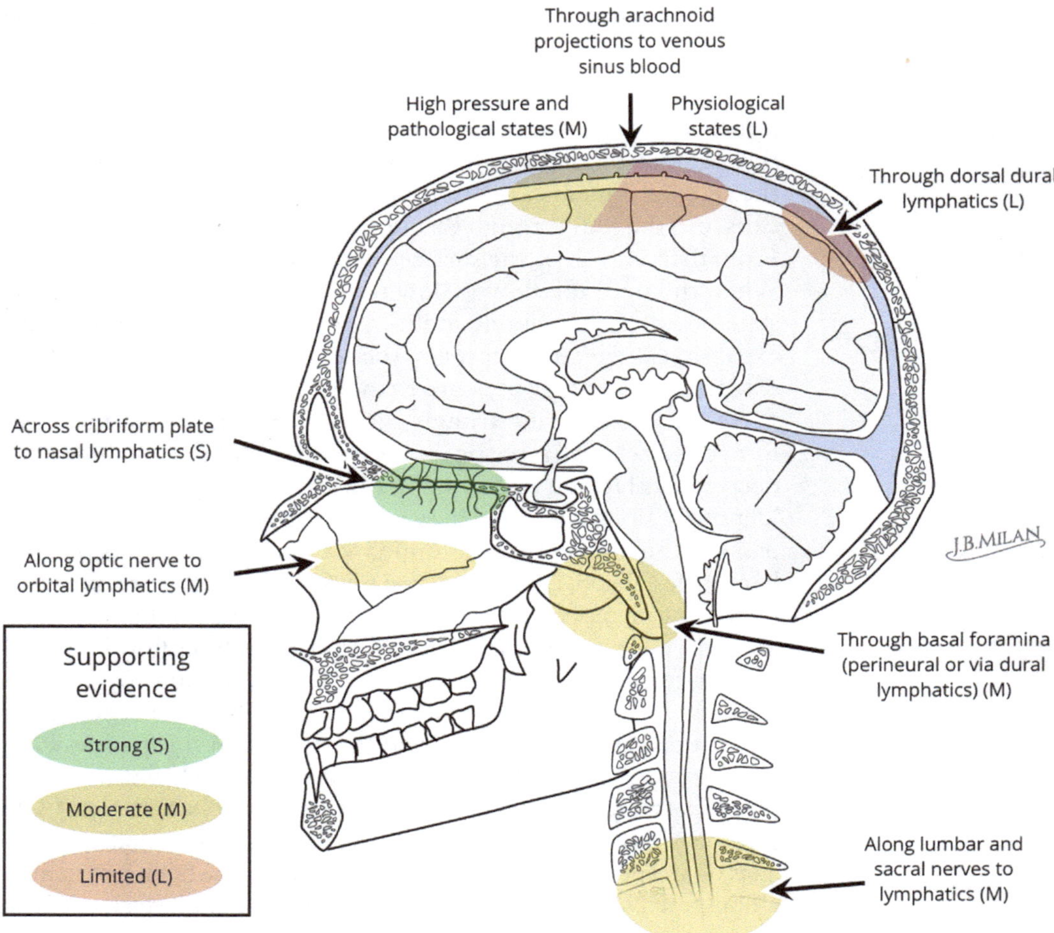

Fig. 1 An overview of the evidence for routes of CSF outflow. A qualitative assessment of the level of supporting evidence for various CSF outflow pathways was made based on the number of published studies, uniformity of the results across different species and plausibility of the anatomical and physiological mechanisms [scheme by Joachim Birch Milan]

the deep cervical lymph nodes [15, 16]. Additional efflux occurs along spinal nerve roots into sacral lymphatics [17]. While the relative contribution of these pathways likely varies with age and species, there is a clear consensus that the lymphatic system is a major player in CSF clearance.

CSF flow is not static; rather, it is inherently pulsatile, primarily propelled by the cardiac cycle. The rhythmic pulsations of arterial walls generate pulse waves that drive CSF flow through the perivascular spaces, which surround cerebral arteries. This intrinsic pulsatility plays a significant role in enhancing drug dispersion within the spinal canal, facilitating the distribution of intrathecally administered therapeutics to cell populations in both the spinal cord and the brain at rates considerably faster than can be accounted for by molecular diffusion alone. Studies indicate that both the heart rate and the CSF stroke volume of a patient profoundly influence drug distribution; higher heart rates and larger CSF stroke volumes generally lead to lower peak drug concentrations, suggesting a more widespread and dispersed drug profile.

Beyond cardiac pulsatility, respiratory cycles also contribute to CSF movement. Recent investigations highlight inspiration as a particularly significant driving force for CSF flow in humans. The reduction in thoracic pressure during inspiration is believed to modulate the hydrostatic pressure conditions, thereby influencing CSF clearance routes. This illustrates a complex interplay of physiological forces that govern the dynamic movement of CSF throughout the CNS. The inherent pulsatility of CSF flow, driven by cardiac and respiratory cycles, along with the bulk flow mechanism of the glymphatic system, demonstrates that the CNS naturally employs active fluid dynamics for solute transport and waste clearance. This "natural active flow" has been shown to substantially enhance drug dispersion compared to purely diffusive processes. This physiological reality serves as a foundational blueprint for developing engineered active fluid delivery systems. By understanding and potentially augmenting these intrinsic processes, new technologies can achieve more effective and widespread drug distribution, moving beyond the limitations of passive diffusion. This also suggests that therapeutic strategies could be designed to modulate intrinsic CSF dynamics, for example, through sleep optimization or targeted respiratory interventions, to improve drug delivery efficiency.

2.5 The Glymphatic System: Role in Waste Clearance and Solute Distribution

The glymphatic system, a macroscopic waste clearance pathway in the CNS discovered relatively recently, functions as the brain's analog to the peripheral lymphatic system. It employs a unique network of perivascular channels, formed by astroglial cells, to facilitate the efficient elimination of soluble proteins and metabolites, such as amyloid-β and tau proteins, from the brain.

The operational mechanism of the glymphatic system involves a bulk movement of fluid: CSF from the subarachnoid space enters the brain along periarterial spaces, where it mixes with interstitial fluid (ISF) within the brain parenchyma. This mixed fluid, now laden with metabolic waste, subsequently exits the parenchyma via perivenous spaces, ultimately draining into the peripheral lymphatic system. This fluid movement is a synergistic combination of advection (convective flow) and diffusion, resulting in a net directional flow towards the venous system. The aquaporin-4 (AQP4) water channels situated on astrocytic end-feet are critical facilitators of this directed fluid transport.

A crucial characteristic of the glymphatic system is its dramatic enhancement during sleep and suppression during wakefulness. This sleep–wake differential correlates with an expansion of the interstitial space during sleep, which creates conditions more conducive to convective fluid fluxes and efficient metabolite clearance. Beyond its primary role in waste removal, the glymphatic system also contributes to the distribution of nonwaste compounds, including essential nutrients, signaling molecules, and potentially therapeutic agents, throughout the brain.

The variability in CSF pulsatility, encompassing both frequency (heart rate) and stroke volume, is significant among patients, leading to highly variable drug distribution even when identical dosing regimens are employed. This variability directly impacts peak drug concentrations and the potential for toxicity. This inherent interpatient variability presents a considerable challenge for standardized ITDD protocols but simultaneously offers a clear opportunity for personalized medicine. Patient-specific modeling, such as medical image-based computational fluid dynamics (miCFD), derived from individual CSF flow measurements, becomes essential for predicting drug distribution and optimizing infusion parameters. This approach aims to enhance therapeutic efficacy while mitigating toxicity for each unique patient, pointing towards a future where ITDD is highly individualized.

Furthermore, the glymphatic system's critical role in waste clearance means that its dysfunction is implicated in neurodegenerative diseases and traumatic brain injury. Impaired glymphatic function leads to the accumulation of neurotoxic proteins and reduced CSF influx. Given that this system also distributes nonwaste compounds and potentially drugs, enhancing glymphatic function, whether pharmacologically or through lifestyle interventions (e.g., optimizing sleep), could serve as a novel strategy to augment intrathecal drug distribution, particularly for large molecules or in diffuse pathologies. This opens avenues for combination therapies that address both disease pathophysiology and drug delivery efficiency.

3 The Historical Trajectory and Technological Evolution of Neuraxial Therapy

Today's IT therapies are the result of nearly 150 years of iterative discovery in neuroanatomy, pharmacology, and biomedical engineering. Yet, we are still in the down of IT therapy as systems and compounds are becoming more and more sophisticated. This history is a compelling narrative of how fundamental science, clinical need, and technological innovation have converged to transform patient care. From a historical perspective, the concept of IT administration dates back to the early 20th century, when it was originally employed for spinal anesthesia and the treatment of infectious meningitis. The use of the IT route for chronic neurological conditions, however, gained in importance in the latter half of the 20th century. Among the key milestones worth mentioning are the 1970s and 1980s, when implantable pumps for the continuous IT administration of analgesics (e.g., morphine) and muscle relaxants (e.g., baclofen) were developed. During the 1990s and 2000s, IT delivery was proposed for chemotherapy in CNS malignancies and leptomeningeal metastases.

3.1 Early Diagnostic Puncture and the Dawn of Therapeutics (1820–1910)

The anatomical rationale for subarachnoid access was firmly established in 1822 when François Magendie experimentally demonstrated the distinct sensory function of the dorsal spinal roots and the motor function of the ventral roots [18]. By the late 19th century, improvements in syringe and needle manufacturing made diagnostic lumbar puncture a routine procedure. The momentous leap from diagnosis to therapy occurred in 1898 when German surgeon August Bier intentionally injected cocaine intrathecally to induce "spinal narcosis" for surgery, performing the procedure on himself and his assistant [19]. The era of spinal anesthesia had begun. Early investigators were already grappling with pharmacokinetics; Barker, in 1909, documented the rapid renal elimination of the anesthetic stovaine after spinal injection [20]. However, safety concerns arose almost immediately, with seminal reports by Wossidlo and Spielmeyer in 1908 describing neurohistological damage following intrathecal stovaine administration. These early observations presaged the modern imperative for rigorous preclinical safety evaluation [21, 22].

3.2 From Regional Anesthesia to the Receptor Revolution (1910–1976)

For the first half of the twentieth century, spinal blocks were almost exclusively the domain of surgical anesthesia. Continuous techniques using ureteral catheters were described by Love as early as 1935 [23] and refined by Tuohy with his specialized needle in 1944 [24], yet, their application to chronic pain was rare. The management of intractable pain was dominated by neuroablative procedures, exemplified by the prefrontal lobotomy reported by Freeman and Watts in 1948 [25]. The paradigm shifted dramatically in 1965 with the publication of Melzack and Wall's gate-control theory of pain [26]. This theory proposed that nociceptive signals could be modulated within the dorsal horn of the spinal cord by non-nociceptive inputs, framing the spinal cord as an active, processing "gate" rather than a passive relay.

This conceptual framework ignited a firestorm of research into spinal neurochemistry. Between 1967 and 1973, a series of landmark discoveries using radioligand binding techniques by research groups led by Martin, Goldstein, Pert, Snyder, Terenius, and Simon definitively identified and localized stereospecific μ-, δ-, and κ-opioid receptors in mammalian CNS tissue, particularly in the substantia gelatinosa of the dorsal horn [27–32]. Electrophysiological studies ran in parallel, with Kitahata et al. showing that morphine selectively suppressed nociceptive neurons in the cat dorsal horn [33]. The final pieces of the puzzle were put in place by Tony Yaksh and Thomas

Rudy. They first developed a method for chronic catheterization of the rodent spinal subarachnoid space [34] and then used it to demonstrate conclusively that morphine delivered directly to the spine produced profound, dose-dependent analgesia without the sedation, confusion, or respiratory depression associated with systemic administration [35]. These discoveries established the spinal cord as an autonomous, drug-responsive analgesic target.

3.3 First-in-Human Trials and the Engineering of Continuous Intrathecal Infusion (1977–1995)

The translation from the laboratory to human patients was remarkably swift. In 1979, Wang et al. at the Mayo Clinic reported successful treatment of intractable cancer pain with lumbar morphine injections [36]. The same year, Behar and colleagues in Israel reported effective analgesia with epidural morphine [37]. The technique was rapidly adapted for postoperative and obstetric pain [38–41]. Concurrently, preclinical work demonstrated the antinociceptive effects of intrathecally administered adrenergic agonists, baclofen, and benzodiazepines [42–44], paving the way for their eventual clinical evaluation [45–47].

The success of intermittent boluses highlighted the need for sustained, continuous delivery. This required implantable hardware that could maintain sterility, provide accurate dosing, and allow patient mobility. Perry Blackshear's 1979 description of an implantable, battery-free osmotic pump, which used an osmotic pressure differential to drive drug flow, spurred commercial development [48]. This led to devices like the gas-driven, constant-rate INFUSAID system [49]. In 1981, Onofrio and Yaksh published their seminal paper showing that continuous low-dose morphine delivered by these pumps could control malignant pain for months [50]. A major breakthrough came with the development of the programmable, battery-powered, peristaltic pump, pioneered at Johns Hopkins and the Mayo Clinic [51]. These devices offered unprecedented flexibility. A landmark 1989 trial by Richard Penn and colleagues confirmed the profound efficacy of continuous intrathecal baclofen for severe spasticity [52]. By the mid-1990s, fully programmable pumps allowing ambulatory titration, complex dosing schedules, and download of the patient's dosing history had become the standard of care, dramatically broadening the indications for IT therapy [53].

3.4 Safety Lessons: Neurotoxicity, Granuloma, and Regulatory Guidance (1986–Present)

Wider application and longer-term treatment in larger patient cohorts inevitably revealed new challenges. The most significant of these was the emergence of inflammatory granulomas at the catheter tip, first characterized by North et al. in 1991 [54]. These noninfectious masses, capable of causing devastating spinal cord compression, prompted urgent preclinical toxicology research. Meticulous studies in dog and sheep models by Yaksh and others delineated the dose-dependent and concentration-dependent parenchymal neurotoxicity of high-concentration morphine, clonidine, and baclofen [43–55]. The histopathology revealed an inflammatory mass rich in lymphocytes, plasma cells, macrophages, and eosinophils, often surrounding precipitated drug crystals, confirming that high local drug concentration was the primary culprit [56].

This body of safety data formed the scientific backbone of the Polyanalgesic Consensus Conference (PACC) guidelines, first published in 2000 and updated periodically since. These international expert guidelines standardized best practices, establishing concentration limits for common IT drugs, creating algorithmic approaches to therapy, and recommending surveillance imaging protocols [57]. The foundational principle, elegantly articulated by Yaksh and Collins in 1989—that rigorous animal studies must precede the intrathecal use of any new drug in humans [58]—remains codified in all current regulatory frameworks and serves as a constant reminder of the field's commitment to patient safety.

4 Present Clinical Applications

The modern IT delivery platform, built on a rich foundation of basic science and refined through decades of clinical experience, is now being applied to an expanding array of CNS disorders, pushing the boundaries of neurotherapeutics. A retrospective analysis of 32,784 patients reveals significant off-label use in ITDD, with only 7.9% of prescriptions adhering to FDA-approved indications for morphine, ziconotide, and baclofen. Monotherapies, primarily opioids, comprised 55% of prescriptions, while combination therapies, often including opioids with bupivacaine or clonidine, accounted for 45%. Off-label formulations dominated, reflecting the need for tailored approaches to manage complex neurological conditions. The findings highlight the prevalence and therapeutic potential of customized ITDD regimens, emphasizing their role in addressing refractory pain and spasticity despite deviations from regulatory guidelines.

The fundamental rationale for IT drug administration lies in its unique ability to circumvent the BBB. This highly selective interface effectively excludes large molecules, polar compounds, and substrates of potent efflux transporters like P-glycoprotein from the CNS. Consequently, for entire classes of therapeutic agents, direct administration into the CSF is the only viable route to achieve and sustain therapeutic concentrations at the desired neural targets. This principle has given rise to a diverse and expanding portfolio of clinical applications, each tailored to specific disease states, from the management of chronic symptoms to the deployment of disease-modifying biologics. Detailed examples of current therapeutic applications can be found in Table 2.

Two of the most well-established pillars of IT therapy are the management of severe spasticity and chronic refractory pain. For spasticity of either cerebral or spinal origin (e.g., from multiple

Table 2
Approved intrathecal therapies

Therapy name	Indication	Molecule type	Mechanism	Delivery	Dosing regimen	Stage
Nusinersen (Spinraza)	Spinal muscular atrophy	Antisense oligonucleotide	Modulates SMN2 splicing to increase SMN protein levels	Free form	Loading phase (four doses), maintenance every 4 months	Approved
Methotrexate	CNS lymphomas, leukemias	Antimetabolite chemotherapy	Inhibits DNA synthesis in proliferating cells	Free form	Weekly dosing during treatment cycles	Approved
Liposomal Cytarabine (DepoCyt)	Lymphomatous meningitis	Chemotherapy (Liposomal)	Sustained-release of cytarabine for prolonged exposure	Liposomal	Every 2 weeks during induction, maintenance dosing	Approved
Morphine	Intractable pain	Opioid analgesic	Binds μ-opioid receptors for pain relief	Free form	Continuous infusion or repeated dosing	Approved
Baclofen	Spasticity	GABA-B receptor agonist	Reduces muscle spasticity	Free form	Continuous or repeated intrathecal dosing	Approved
Ziconotide	Severe chronic pain	Synthetic peptide	Blocks N-type calcium channels	Free form	Continuous infusion	Approved
Rituximab	CNS B-cell lymphomas	Monoclonal antibody	Targets CD20 on B cells	Free form	Weekly or biweekly injections	Approved

sclerosis, cerebral palsy, or spinal cord injury) that is refractory to oral medications, intrathecal baclofen (ITB) therapy is the undisputed gold standard. Baclofen, a gamma-aminobutyric acid (GABA-B) receptor agonist, acts directly on inhibitory pathways within the spinal cord. By delivering it intrathecally, clinicians can achieve profound reductions in muscle tone and spasms with doses that are orders of magnitude lower than systemic equivalents, thereby dramatically improving patient function, comfort, and quality of life while minimizing systemic side effects like sedation.

In parallel, IT therapy has revolutionized the management of severe chronic pain, particularly cancer-related pain and complex regional pain syndrome (CRPS). By delivering analgesics directly to the opioid and other nociceptive receptors in the dorsal horn, the IT route provides potent pain relief with a fraction of the systemic dose, reducing the burden of common opioid side effects such as constipation, nausea, and cognitive impairment. Opioids like morphine and hydromorphone are mainstays for severe, diffuse nociceptive pain. For refractory neuropathic pain, the nonopioid agent ziconotide offers a vital alternative. A synthetic peptide antagonist of N-type voltage-gated calcium channels, ziconotide disrupts pain signaling through a distinct mechanism, though its narrow therapeutic window necessitates careful and expert titration.

Another core application is in neuro-oncology, specifically for the treatment and prophylaxis of leptomeningeal metastases. Cancers that spread to the meninges are shielded from most systemic chemotherapies by the BBB. Intrathecal delivery of agents such as methotrexate and cytarabine directly into the CSF is essential to achieve cytotoxic concentrations throughout the neuraxis, forming a critical component of treatment protocols for leukemia, lymphoma, and certain solid tumors with CNS involvement.

Conflict of Interest

The authors have no conflicts of interest to declare that are relevant to the content of this chapter.

References

1. Pardridge WM (2005) The blood-brain barrier: bottleneck in brain drug development. NeuroRx 2(1):3–14
2. Finkel RS, Mercuri E, Darras BT, Connolly AM, Kuntz NL, Kirschner J et al (2017) Nusinersen versus sham control in infantile-onset spinal muscular atrophy. N Engl J Med 377(18):1723–1732
3. Mercuri E, Darras BT, Chiriboga CA, Day JW, Campbell C, Connolly AM et al (2018) Nusinersen versus sham control in later-onset spinal muscular atrophy. N Engl J Med 378(7):625–635
4. Haines DE, Mihailoff GA (2018) Neuroanatomy: An Atlas of Structures, Sections, and Systems, 10th edn. Wolters Kluwer, Philadelphia
5. Møllgård K, Beinlich FR, Kusk M, Miyakoshi LM, Delle C, Plá V et al (2023) A mesothelium divides the subarachnoid space into functional compartments. Science 379(6627):94–98

6. Nicholson C, Hrabětová S (2017) Brain extracellular space: the final frontier of neuroscience. Biophys J 113(10):2133–2142
7. Spector R, Keep RF, Snodgrass SR (2015) A balanced view of the force of the CSF sink. Exp Neurol 269:54–59
8. Menéndez González M (2023) CNS compartments: the anatomy and physiology of the cerebrospinal fluid. In: Liquorpheresis. Springer, Cham. https://doi.org/10.1007/978-3-031-43482-2_1
9. Menéndez González M (2023) Pathology related to dysfunction of CSF production, flow, and clearance. In: Liquorpheresis. Springer, Cham. https://doi.org/10.1007/978-3-031-43482-2_2
10. Seeley RR, Stephens TD, Tate P (2008) Anatomy & physiology. McGraw-Hill
11. Sakka L, Coll G, Chazal J (2011) Anatomy and physiology of cerebrospinal fluid. Eur Ann Otorhinolaryngol Head Neck Dis 128(6):309–316
12. Iliff JJ, Wang M, Liao Y, Plogg BA, Peng W, Gundersen GA et al (2012) A paravascular pathway facilitates CSF flow through the brain parenchyma and the clearance of interstitial solutes, including amyloid-β. Sci Transl Med 4(147):147ra111
13. Tarasoff-Conway JM, Carare RO, Osorio RS, Glodzik L, Butler T, Fieremans E et al (2015) Clearance systems in the brain—implications for Alzheimer disease. Nat Rev Neurol 11(8):457–470
14. Proulx ST (2021) Cerebrospinal fluid outflow: a review of the historical and contemporary evidence for arachnoid villi, perineural routes, and dural lymphatics. Cell Mol Life Sci 78(6):2429–2457
15. Johnston M, Zakharov A, Papaiconomou C, Salmasi G, Armstrong D (2004) Evidence of connections between cerebrospinal fluid and nasal lymphatic vessels in humans, non-human primates and other mammalian species. Cerebrospinal Fluid Res 1(1):1–13
16. Kida S, Pantazis A, Weller RO (1993) CSF drains from the subarachnoid space into nasal and cervical lymphatics in the rat. J Neurosurg 79(4):583–591
17. Zakharov A, Papaiconomou C, Djenic J, Midha R, Johnston M (2004) Lymphatic drainage of the spinal cord in rats. Cerebrospinal Fluid Res 1(1):1–11
18. Magendie F (1822) Expériences sur les fonctions des racines des nerfs rachidiens. J Physiol Expér Pathol 2:276–279
19. Bier A (1899) Versuche über Cocainisirung des Rückenmarkes. Deutsche Zeitschrift für Chirurgie 51(3–4):361–369
20. Barker AE (1907) A report on clinical experiences with spinal analgesia in 100 cases. Br Med J 1(2412):665
21. Wossidlo E (1908) Ueber die histologischen Veränderungen nach Stovainanästhesie. Arch Klin Chir 86:1017–1054
22. Spielmeyer W (1908) Die Stovainanästhesie in histologischer Beziehung. Münch Med Wochenschr 55:1629–1634
23. Love JG (1935) Continuous spinal anesthesia. Proc Staff Meet Mayo Clin 10:409–411
24. Tuohy EB (1945) Continuous spinal anesthesia: a new method utilizing a ureteral catheter. Surg Clin North Am 25(4):834–840
25. Freeman W, Watts JW (1942) Psychosurgery: intelligence, emotion and social behavior following prefrontal lobotomy for mental disorders. Charles C Thomas Publisher
26. Melzack R, Wall PD (1965) Pain mechanisms: a new theory. Science 150(3699):971–979
27. Martin WR (1967) Opioid antagonists. Pharmacol Rev 19(4):463–521
28. Goldstein A, Lowney LI, Pal BK (1971) Stereospecific and nonspecific interactions of the morphine congener levorphanol in subcellular fractions of mouse brain. Proc Natl Acad Sci USA 68(8):1742–1747
29. Pert CB, Snyder SH (1973) Opiate receptor: demonstration in nervous tissue. Science 179(4077):1011–1014
30. Snyder SH (1975) Opiate receptor in normal and drug altered brain function. Nature 257(5523):185–189
31. Terenius L (1973) Stereospecific interaction between narcotic analgesics and a synaptic plasma membrane fraction of rat cerebral cortex. Acta Pharmacol Toxicol 32(3):317–320
32. Simon EJ, Hiller JM, Edelman I (1973) Stereospecific binding of the potent narcotic analgesic [3H] Etorphine to rat-brain homogenate. Proc Natl Acad Sci USA 70(7):1947–1949
33. Kitahata LM, Kosaka Y, Taub A, Bonikos K, Hoffert M (1974) Lamina-specific suppression of dorsal-horn unit activity by morphine sulfate. Anesthesiology 41(1):39–48
34. Yaksh TL, Rudy TA (1976) Analgesia mediated by a direct spinal action of narcotics. Science 192(4246):1357–1358
35. Yaksh TL, Rudy TA (1977) Studies on the direct spinal action of narcotics in the production of analgesia in the rat. J Pharmacol Exp Ther 202(2):411–428
36. Wang JK, Nauss LA, Thomas JE (1979) Pain relief by intrathecally applied morphine in man. Anesthesiology 50(2):149–151
37. Behar M, Olshwang D, Magora F, Davidson JT (1979) Epidural morphine in treatment of pain. Lancet 313(8115):527–529

38. Bromage PR, Camporesi E, Chestnut D (1980) Epidural narcotics for postoperative analgesia. Anesth Analg 59(6):473–480
39. Bromage PR, Camporesi EM (1981) Epidural narcotics in obstetrics. In: Anesthesia for obstetrics. WB Saunders, pp 215–231
40. Cousins MJ, Mather LE, Glynn CJ, Wilson PR, Graham JR (1979) Selective spinal analgesia. Lancet 313(8126):1141–1142
41. Yaksh TL (1981) Spinal opiate analgesia: characteristics and principles of action. Pain 11(3):293–346
42. Reddy SVR, Maderdrut JL, Yaksh TL (1980) Spinal cord pharmacology of adrenergic agonist-mediated antinociception. J Pharmacol Exp Ther 213(3):525–533
43. Yaksh TL, Reddy SVR (1981) Studies in the primate on the analgetic effects associated with intrathecal actions of opiates, α-adrenergic agonists and baclofen. Anesthesiology 54(6):451–467
44. Yaksh TL (1983) In vivo studies on spinal opiate receptor systems. In: Advances in pain research and therapy, vol 5. Raven Press, pp 411–420
45. Yaksh TL, Howe JR (1982) Opiate-sensitive and opiate-insensitive spinal analgesia systems. In: Spinal cord dysfunction. Springer, Vienna, pp 141–171
46. Yaksh TL (1987) Spinal opiates: a review of their effect on spinal function with an emphasis on pain processing. Acta Anaesthesiol Scand 31(s85):25–37
47. Yaksh TL, Noueihed R (1985) The physiology and pharmacology of spinal opiates. Annu Rev Pharmacol Toxicol 25(1):433–462
48. Blackshear PJ, Rohde TD, Prosl F (1979) The implantable infusion pump: a new concept in drug delivery. Med Prog Technol 6(3):149–161
49. Buchwald H, Grage TB, Vassilopoulos PP, Rohde TD, Varco RL, Blackshear PJ (1980) Intraarterial infusion chemotherapy for hepatic carcinoma using a totally implantable infusion pump. Cancer 45(5):866–869
50. Onofrio BM, Yaksh TL (1983) Intrathecal delta-receptor ligand produces analgesia in man. Lancet 321(8337):1386–1387
51. Penn RD, Paice JA, Kroin JS (1986) The programmable implantable medication pump. Neurosurgery 18(2):215–218
52. Penn RD, Savoy SM, Corcos D, Latash M, Gottlieb G, Parke B (1989) Intrathecal baclofen for severe spinal spasticity. N Engl J Med 320(23):1517–1521
53. Krames ES (1993) Intrathecal infusional therapies for intractable pain: patient management guidelines. J Pain Symptom Manage 8(1):36–46
54. North RB, Cutchis PN, Epstein JA, Long DM (1991) Spinal cord compression from intrathecal morphine-induced inflammatory mass. Neurosurgery 29(5):778–784
55. Yaksh TL, To-Phan V, Ozkan T (1995) The spinal actions of opioids. In: The pharmacology of pain. Springer, Berlin, Heidelberg, pp 153–182
56. Yaksh TL, Collins JG (1989) Studies in animals should precede human use of spinally administered drugs. Anesthesiology 70(1):4–6
57. Hildebrand KR, Elsberry DD, Deer TR (2005) Stability, compatibility, and safety of intrathecal drug admixtures. J Pain Symptom Manage 30(5):407–417
58. Hao D, Bishop J, K-Singh S (2020) Off-label use of intrathecal drug delivery for chronic non-cancer pain: a retrospective analysis of 32,784 patients. Pain Physician 23(4):E351–E360

Chapter 2

Factors Affecting the Distribution of Intrathecally Administered Drugs

Manuel Menéndez-González

Abstract

Intrathecal (IT) drug administration is a critical strategy for treating central nervous system (CNS) diseases by delivering therapeutics directly into the cerebrospinal fluid (CSF), thereby bypassing the blood–brain barrier. However, the clinical efficacy of this approach is contingent upon achieving an adequate and predictable distribution of the drug throughout the subarachnoid space (SAS). This process is far from simple, being governed by a complex interplay between the drug's physicochemical properties, the specific parameters of the injection protocol, and the highly variable anatomical and physiological landscape of the individual patient. This chapter provides a comprehensive review of the fundamental factors that dictate drug fate following IT injection. It examines the dominant role of CSF hydrodynamics, including bulk flow and cardiac-driven pulsatility, and explores how gravitational forces can be harnessed through patient posture and drug baricity. Furthermore, it details the significant impact of injection parameters, the challenges posed by microanatomical structures and biological barriers within the SAS, and the overarching influence of patient-specific pathophysiology. A thorough understanding of these interconnected variables is paramount for the rational design of novel therapeutics and the optimization of delivery strategies to improve drug distribution and patient outcomes in neurology.

Keywords Intrathecal drug delivery, Cerebrospinal fluid (CSF), CSF hydrodynamics, Drug distribution, Pharmacokinetics, Baricity, Blood–brain barrier, Patient-specific factors

1 Principles of Intrathecal Drug Release and Distribution

The delivery of therapeutic agents directly into the intrathecal (IT) space represents a powerful clinical strategy, offering a unique opportunity to circumvent the restrictive blood–brain barrier (BBB) and achieve therapeutic concentrations within the central nervous system (CNS) that would be unattainable with systemic administration [1–43]. Despite its conceptual simplicity, the journey of a drug from the point of injection to its molecular target is fraught with complexity. The final distribution pattern is not a product of simple diffusion but rather the cumulative result of numerous interacting variables. These include the intrinsic physicochemical properties of

Manuel Menéndez-González (ed.), *Intrathecal Drug Delivery*, Neuromethods, vol. 234,
https://doi.org/10.1007/978-1-0716-5241-1_2,
© The Author(s), under exclusive license to Springer Science+Business Media, LLC, part of Springer Nature 2026

the drug molecule itself, such as its size, charge, and lipophilicity; the precise parameters of the administration protocol, including the volume, rate, and anatomical location of the injection; and a host of patient-specific factors, ranging from spinal anatomy to the dynamic state of their cerebrospinal fluid (CSF) circulation [44–100]. A comprehensive appreciation of these intricate hydrodynamics and physiological barriers is therefore essential for the rational design of novel IT therapeutics and the development of personalized delivery strategies aimed at enhancing efficacy while mitigating potential toxicity.

1.1 CSF Hydrodynamics and Parameters Governing Drug Distribution

Once an agent is introduced into the subarachnoid space (SAS), its fate is governed by a complex interplay of forces that collectively determine its spread, target engagement, and eventual clearance. The dominant transport mechanism within the CSF is convection, or bulk flow, which far outweighs the contribution of passive diffusion, especially for large molecules [8, 89]. The CSF itself is in constant motion, driven by two primary forces. First, there is a slow, continuous, caudocranial bulk flow generated by the net production of CSF from the choroid plexuses and its subsequent absorption into the venous circulation at the arachnoid granulations [9, 13]. Superimposed on this slow, directional current is a much more vigorous, oscillatory movement known as CSF pulsatility. This to-and-fro motion, driven primarily by the systolic expansion and diastolic relaxation of the cerebral vasculature during the cardiac cycle and secondarily by respiratory pressures, generates significant local mixing and is a major factor in distributing solutes throughout the SAS [10, 90]. The complex interplay of these flow patterns means that a drug injected into the lumbar sac does not simply diffuse upwards but is actively mixed and transported by these powerful hydrodynamic forces.

1.2 Injection Parameters: Volume, Rate, and Location

The specifics of the injection technique itself profoundly impact the initial and subsequent drug dispersion. A confluence of evidence from in vitro bench models, in vivo animal studies, and computational fluid dynamics (CFD) has established several key principles [23, 72]. Injection volume is a critical driver of spread. Small boluses are rapidly diluted in the local CSF volume with minimal distribution, whereas larger volumes act as a "piston," displacing CSF and promoting a wider, more immediate distribution [91]. A widely cited rule of thumb suggests that an injection volume of approximately 10% of the total CSF pool (~15 mL in an adult) represents an effective compromise between maximizing spread and minimizing the risk of acute pressure-related side effects such as headache [22]. The anatomical location of the injection also matters. CFD modeling and clinical studies have shown that moving from a low lumbar (L4/L5) to a mid-lumbar (L3/L4) interspace can increase cranial delivery, as it places the drug higher up the lumbar lordosis, facilitating its entry into the main rostral flow stream [21]. In contrast, the rate of bolus injection appears less influential than volume, provided it is slow enough (e.g., ≤0.2 mL/s) to prevent high-pressure jetting that could cause nerve root irritation or promote leakage from the dural puncture site [23, 92].

1.3 Convection Vs. Diffusion in CNS Drug Transport

Understanding the fundamental distinction between diffusion and convection is crucial for appreciating the challenges and opportunities in CNS drug delivery. Diffusion is a passive process where molecules move randomly from an area of higher concentration to one of lower concentration. Within the tortuous and crowded extracellular space of the CNS parenchyma, diffusion is severely restricted and effective only over very short distances, typically on the order of millimeters [93]. This results in a steep exponential drop in drug concentration as distance from the CSF-parenchymal interface increases, rendering it highly inefficient for delivering therapeutic agents to deep or diffusely affected brain targets. Convection, or bulk flow, is the movement of molecules carried along by the flow of a fluid, in this case, the CSF. Within the SAS, convection is the dominant force for distributing drugs over large distances. This principle is harnessed in advanced delivery techniques like convection-enhanced delivery (CED), which uses a continuous pressure gradient to drive bulk fluid flow directly through the brain's interstitial spaces, achieving widespread distribution of even large molecules that would otherwise be immobile [94]. For standard IT delivery, it is the convective forces of CSF circulation, not diffusion, that are responsible for transporting a drug from the lumbar spine to the cerebral convexities.

1.4 The Influence of Posture, Gravity, and Baricity

Gravity is arguably the most powerful and clinically manipulable force governing the distribution of IT drugs. The simple act of a patient moving from a supine to an upright posture creates a hydrostatic column within the neuraxis, causing intracranial pressure to fall and lumbar CSF pressure to rise significantly [17, 18]. This establishes a steep pressure gradient that drives a net caudal shift of CSF, a phenomenon clinicians have long exploited to control the spread of spinal anesthetics. This gravitational influence is mediated by the drug formulation's baricity—its density relative to that of the patient's CSF [19]. Hyperbaric solutions, typically made denser than CSF by the addition of an excipient like dextrose, will reliably sink within the SAS, moving in the direction of gravity. This principle allows for highly predictable and controllable sensory blocks, for instance, by positioning a patient to direct the anesthetic toward specific nerve roots [95]. Conversely, hypobaric solutions, which are less dense than CSF, will rise against gravity. Isobaric solutions have a density similar to that of CSF and are therefore less influenced by gravity; their spread tends to be more limited and is driven primarily by the volume of injection and the inherent pulsatile dynamics of the

CSF [20]. Consequently, achieving broad cranial delivery of a therapeutic agent often necessitates the use of a large-volume, isobaric or slightly hypobaric formulation, frequently combined with placing the patient in a head-down (Trendelenburg) position to leverage gravity to assist rostral flow [24].

2 Barriers and Modulators of Distribution

2.1 Microanatomical Structures and Biological Barriers

The SAS should not be conceptualized as a simple, open conduit. It is a complex, three-dimensional space populated by millions of delicate, web-like strands known as arachnoid trabeculae, as well as by spinal nerve roots and blood vessels. These structures create a porous medium that significantly impacts fluid dynamics [96]. Advanced CFD models that incorporate these microanatomical features reveal that they dramatically increase hydraulic resistance and generate chaotic advection—a complex pattern of eddies and vortices that greatly enhances the mixing of CSF and accelerates caudo-cranial drug transport compared to what would occur in a simple, smooth-walled tube [24, 97]. In addition to these physical structures, formidable biological barriers exist within the SAS itself. The entire space is lined by meningeal cells, including a high density of resident macrophages and other phagocytic cells located in the dura, pia, and the recently described subarachnoid lymphatic-like membrane (SLYM) [3, 25]. These immune cells can actively recognize and clear foreign materials, including therapeutic proteins, antisense oligonucleotides (ASOs), and viral vectors, thereby reducing their parenchymal penetration and therapeutic efficacy [98].

2.2 Pharmacokinetics and Clearance from the CSF

The clearance of drugs from the CSF follows principles distinct from systemic pharmacokinetics. For most molecules, the primary route of elimination is not metabolic degradation but rather bulk flow absorption into the venous circulation via the arachnoid granulations and, to a lesser extent, clearance into the lymphatic system along cranial and spinal nerve roots [13–15]. This means that most drugs, regardless of their size or chemical properties, are cleared from the CSF at roughly the same rate as the CSF itself turns over, which is approximately 0.3–0.4 mL/min in a healthy adult [9]. This results in a much longer half-life in the CSF compared to plasma, an attribute that can be advantageous for sustaining therapeutic concentrations. However, for a therapy to be successful, the drug must not only reach its target but also remain chemically and biologically stable within the unique biochemical environment of the CSF long enough to exert its intended effect before being cleared.

2.3 The Overarching Influence of Patient Pathophysiology

A critical consideration that supersedes all generalized principles is that the patient's underlying disease state can profoundly alter CSF dynamics and, consequently, drug distribution. The assumptions of normal CSF flow and anatomy do not hold true for a significant subset of patients with neurological disorders. In communicating hydrocephalus, for example, altered CSF absorption dynamics can change drug residence time, while in obstructive hydrocephalus, drugs may be sequestered within enlarged ventricles, preventing distribution to the spinal cord or cerebral convexities. In patients with a Chiari I malformation, the physical obstruction of CSF flow at the foramen magnum can severely impede the cranial distribution of any drug injected into the lumbar space [72, 99]. Similarly, conditions such as spinal stenosis, arachnoiditis, or the aftermath of spinal surgery can create compartments or obstructions that completely prevent local drug delivery. Therefore, a thorough neuroradiological and clinical assessment of the individual patient's CSF space anatomy and physiology is an absolute prerequisite for planning IT therapy, as these conditions may necessitate altered delivery strategies to ensure the therapeutic agent can reach its intended target.

3 Neurotoxicity and Safety Considerations

The direct administration of substances into the neuraxis carries inherent risks, and a paramount concern is iatrogenic neurotoxicity. This is not a single entity but a spectrum of potential adverse effects ranging from direct neuronal or glial injury to delayed inflammatory reactions. The history of spinal therapeutics is punctuated by episodes of severe neurological injury resulting from the inadvertent administration of neurotoxic agents, most notably preservatives such as benzyl alcohol and sodium bisulfite, which were included in formulations never intended for spinal use [102]. This has led to the strict maxim that only preservative-free solutions specifically formulated for intrathecal administration should ever be used. Beyond excipients, the active pharmaceutical ingredient itself can be neurotoxic, often in a dose- and concentration-dependent manner [103].

A classic example of a device- and drug-related inflammatory reaction is the formation of catheter-tip granulomas in patients receiving chronic IT opioid infusions [63, 67]. These non-infectious inflammatory masses, composed of plasma cells, lymphocytes, and fibroblasts, can cause devastating spinal cord compression. Their formation is thought to be driven by a combination of factors, including a local foreign body reaction to the catheter and, more importantly, the high local concentration of the infused drug, with morphine being most frequently implicated [67]. This underscores the principle that maintaining the lowest effective dose and ensuring adequate drug dispersal away from the catheter tip are critical safety measures.

With the advent of modern genetic therapies, new patterns of neurotoxicity have emerged. A primary safety concern for both ASOs and AAV-based gene therapies is the potential for dorsal root ganglion (DRG) toxicity [76]. The DRGs occupy a unique anatomical niche; they are bathed by CSF, but their capillaries are fenestrated and lack the tight junctions of the BBB, rendering the sensory neurons within them highly susceptible to high local concentrations of therapeutic agents [80, 81]. Off-target effects in these neurons can lead to a sensory neuronopathy, manifesting as pain, paresthesia, or proprioceptive loss. This has been observed in preclinical studies across multiple species and with various ASO chemistries and AAV serotypes [101, 104]. Mitigating this risk is a major focus of current research and involves a multi-pronged approach: designing AAV capsids with reduced DRG tropism, using cell-specific promoters to limit transgene expression, chemically modifying ASOs to reduce their intrinsic toxicity [105], and optimizing delivery protocols (e.g., using slow infusions of large, dilute volumes) to avoid the high peak drug concentrations that are thought to drive this specific form of toxicity.

Conflict of Interest

The authors have no conflicts of interest to declare that are relevant to the content of this chapter.

References

1. Haines DE (1991) On the question of a subdural space. Anat Rec 230(1):3–21. https://doi.org/10.1002/ar.1092300102
2. Bissenas A, Fleeting C, Patel D et al (2022) CSF dynamics: implications for hydrocephalus and glymphatic clearance. Curr Res Med Sci 1:24–42.https://doi.org/10.56397/CRMS.2022.12.04
3. Møllgård K, Beinlich FR, Kusk P et al (2023) A mesothelium divides the sub-arachnoid space into functional compartments. Science 379(6627):84–88. https://doi.org/10.1126/science.adc8810
4. Kylkilahti TM, Berends E, Ramos M et al (2021) Achieving brain clearance and preventing neurodegenerative diseases—a glymphatic perspective. J Cereb Blood Flow Metab 41(9):2137–2149. https://doi.org/10.1177/0271678X211026955
5. Trevisi G, Frassanito P, Di Rocco C (2014) Idiopathic CSF overproduction: case-based review. Croat Med J 55(4):377–387. https://doi.org/10.3325/cmj.2014.55.377
6. Pardridge WM (2005) The blood-brain barrier: bottleneck in brain drug development. NeuroRx 2(1):3–14. https://doi.org/10.1602/neurorx.2.1.3
7. Chen H (2023) The blood-brain barrier: the frontier in aging and neurodegeneration. AHA blogs. https://doi.org/10.1161/blog.20230629.718116
8. Kelley DH, Thomas JH (2023) Cerebrospinal fluid flow. Annu Rev Fluid Mech 55:237–260. https://doi.org/10.1146/annurev-fluid-030322-114421
9. Fishman, RA (1992 Cerebrospinal fluid in diseases of the nervous system, 2nd edn. Saunders, Philadelphia, PA
10. Mehta NH, Suss RA, Dyke JP et al (2022) Quantifying CSF dynamics: a review. Neurobiol Dis 170:105776. https://doi.org/10.1016/j.nbd.2022.105776

11. Jessen NA, Munk ASF, Lundgaard I, Nedergaard M (2015) The glymphatic system: a beginner's guide. Neurochem Res 40(12):2583–2599. https://doi.org/10.1007/s11064-015-1581-6
12. Grubb S, Lauritzen M (2019) Deep sleep drives brain fluid oscillations. Science 366(6465):572–573. https://doi.org/10.1126/science.aaz5450
13. Shah T, Leurgans SE, Mehta RI et al (2023) Arachnoid granulations are lymphatic conduits. J Exp Med 220(3):e20220618. https://doi.org/10.1084/jem.20220618
14. Spera I, Cousin N, Ries M et al (2023) Open pathways for CSF outflow at the cribriform plate. EBioMedicine 91:104558. https://doi.org/10.1016/j.ebiom.2023.104558
15. Mehta NH, Sherbansky J, Kamer AR et al (2022) The brain–nose interface: a potential CSF clearance site. Front Physiol 12:769948. https://doi.org/10.3389/fphys.2021.769948
16. Stokes C, White EF, Toddes S et al (2023) Whole CNS 3D cryo-fluorescence tomography. J Imaging 9(2):45. https://doi.org/10.3390/jimaging9020045
17. Magnaes B (1976) Body position and cerebrospinal fluid pressure. Part 1: clinical studies on the effect of rapid postural changes. J Neurosurg 44(6):687–697. https://doi.org/10.3171/jns.1976.44.6.0687
18. Muccio M, Sun Z, Chu D et al (2024) The impact of body position on neurofluid dynamics. Front Aging Neurosci 16:1454282. https://doi.org/10.3389/fnagi.2024.1454282
19. Hocking G, Wildsmith JA (2004) Intrathecal drug spread. Br J Anaesth 93(4):568–578. https://doi.org/10.1093/bja/aeh204
20. Burla GKR, Shrestha D, Bowen M et al (2021) Effect of injection parameters and CSF dynamics on intrathecal drug distribution: A review. Fluids Barriers CNS 18(1):37. https://doi.org/10.1186/s12987-021-00270-3
21. Ayansiji AO, Gehrke DS, Baralle B et al (2023) Determination of spinal tracer dispersion after intrathecal injection in a deformable CNS model. Front Physiol 14:1244016. https://doi.org/10.3389/fphys.2023.1244016
22. Bernards CM (2004) Recent insights into the pharmacokinetics of spinal opioids and the relevance to opioid-induced respiratory depression. Curr Opin Anaesthesiol 17(1):1–4. https://doi.org/10.1097/00001503-200402000-00001
23. Tangen KM, Leval R, Mehta AI, Linninger AA (2017) Computational and in vitro investigation of IT drug distribution. Anesth Analg 124(5):1686–1696. https://doi.org/10.1213/ANE.0000000000001669
24. Tangen KM, Hsu Y, Zhu DC, Linninger AA (2015) CNS wide drug delivery to the brain and spinal cord by intrathecal injection. J R Soc Interface 12(106):20150022. https://doi.org/10.1098/rsif.2015.0022
25. Ransohoff RM, Engelhardt B (2012) The anatomical and cellular basis of immune surveillance in the central nervous system. Nat Rev Immunol 12(9):623–635. https://doi.org/10.1038/nri3265
26. Magendie F (1822) Experiences sur les fonctions des racines des nerfs rachidiens. J Physiol Exp 2:276–279
27. Van Zundert A, Goerig M, De Andrés J (2006) August Bier 1861–1949. A tribute to a great surgeon who was the first to describe spinal anaesthesia. Rev Esp Anestesiol Reanim 53(4):240–248
28. Barker AE (1909) Elimination of stovaine after spinal analgesia. Br Med J 2(2544):789–791. https://doi.org/10.1136/bmj.2.2544.789
29. Wossidlo E (1908) Experimentelle untersuchungen über veränderungen der nissl'schen granula bei der lumbalanästhesie. Arch Klin Chir 86:1017–1053
30. Spielmeyer W (1908) Veränderungen des nervensystems nach stovain-anästhesie. Münch Med Wochenschr 55:1629–1634
31. Love JG (1935) Continuous subarachnoid drainage for meningitis by means of a ureteral catheter. JAMA 104(18):1595–1597. https://doi.org/10.1001/jama.1935.02760180021006
32. Tuohy EB (1944) Continuous spinal anesthesia: its usefulness and technic involved. Anesthesiology 5(2):142–148. https://doi.org/10.1097/00000542-194403000-00004
33. Freeman W, Watts JW (1948) Pain mechanisms and the frontal lobes; a study of prefrontal lobotomy for intractable pain. Ann Intern Med 28(4):747–754. https://doi.org/10.7326/0003-4819-28-4-747
34. Melzack R, Wall PD (1965) Pain mechanisms: a new theory. Science 150(3699):971–979. https://doi.org/10.1126/science.150.3699.971
35. Martin WR (1967) Opioid antagonists. Pharmacol Rev 19(4):463–521
36. Goldstein A, Lowney LI, Pal BK (1971) Stereospecific and nonspecific interactions of the morphine congener levorphanol in subcellular fractions of mouse brain. Proc Natl Acad Sci USA 68(8):1742–1747. https://doi.org/10.1073/pnas.68.8.1742
37. Pert CB, Snyder SH (1973) Opiate receptor: demonstration in nervous tissue. Science 179(4077):1011–1014. https://doi.org/10.1126/science.179.4077.1011

38. Terenius L (1973) Stereospecific interaction between narcotic analgesics and a synaptic plasma membrane fraction of rat cerebral cortex. Acta Pharmacol Toxicol (Copenh) 32(3):317–320. https://doi.org/10.1111/j.1600-0773.1973.tb01497.x
39. Simon EJ, Hiller JM, Edelman I (1973) Stereospecific binding of the potent narcotic analgesic [3H]Etorphine to rat-brain homogenate. Proc Natl Acad Sci U S A 70(7):1947–1949. https://doi.org/10.1073/pnas.70.7.1947
40. Kuhar MJ, Pert CB, Snyder SH (1973) Regional distribution of opiate receptor binding in monkey and human brain. Nature 245(5426):447–450. https://doi.org/10.1038/245447a0
41. Lamotte C, Pert CB, Snyder SH (1976) Opiate receptor binding in primate spinal cord: distribution and changes after dorsal root section. Brain Res 112(2):407–412. https://doi.org/10.1016/0006-8993(76)90818-6
42. Kitahata LM, Kosaka Y, Taub A, Bonikos K, Hoffert M (1974) Lamina-specific suppression of dorsal-horn unit activity by morphine sulfate. Anesthesiology 41(1):39–48. https://doi.org/10.1097/00000542-197407000-00008
43. Yaksh TL, Rudy TA (1976) Chronic catheterization of the spinal subarachnoid space. Physiol Behav 17(6):1031–1036. https://doi.org/10.1016/0031-9384(76)90027-4
44. Yaksh TL, Rudy TA (1976) Analgesia mediated by a direct spinal action of narcotics. Science 192(4246):1357–1358. https://doi.org/10.1126/science.1273597
45. Wang JK, Nauss LA, Thomas JE (1979) Pain relief by intrathecally applied morphine in man. Anesthesiology 50(2):149–151. https://doi.org/10.1097/00000542-197902000-00011
46. Behar M, Magora F, Olshwang D, Davidson JT (1979) Epidural morphine in treatment of pain. Lancet 1(8115):527–529. https://doi.org/10.1016/s0140-6736(79)91113-6
47. Baraka A, Noueihid R, Hajj S (1981) Intrathecal injection of morphine for obstetric analgesia. Anesthesiology 54(2):136–140. https://doi.org/10.1097/00000542-198102000-00010
48. Kalso E (1983) Effects of intrathecal morphine injected with bupivacaine on pain after orthopaedic surgery. Br J Anaesth 55(5):415–422. https://doi.org/10.1093/bja/55.5.415
49. Dirksen R, Nijhuis GM (1980) Epidural opiate and perioperative analgesia. Acta Anaesthesiol Scand 24(5):367–374. https://doi.org/10.1111/j.1399-6576.1980.tb01563.x
50. Samii K, Chauvin M, Viars P (1981) Postoperative spinal analgesia with morphine. Br J Anaesth 53(8):817–820. https://doi.org/10.1093/bja/53.8.817
51. Reddy SV, Maderdrut JL, Yaksh TL (1980) Spinal cord pharmacology of adrenergic agonist-mediated antinociception. J Pharmacol Exp Ther 213(3):525–33.
52. Wilson PR, Yaksh TL (1978) Baclofen is antinociceptive in the spinal intrathecal space of animals. Eur J Pharmacol 51(4):323–330. https://doi.org/10.1016/0014-2999(78)90159-8
53. Whitwam JG, Niv D, Loh L, Jack RD (1982) Depression of nociceptive reflexes by intrathecal benzodiazepine in dogs. Lancet 2(8313):1465. https://doi.org/10.1016/s0140-6736(82)91689-5
54. Tamsen A, Gordh T (1984) Epidural clonidine produces analgesia. Lancet 2(8396):231–232. https://doi.org/10.1016/s0140-6736(84)90363-2
55. Penn RD, Kroin JS (1985) Continuous intrathecal baclofen for severe spasticity. Lancet 2(8447):125–127. https://doi.org/10.1016/s0140-6736(85)90247-4
56. Goodchild CS, Noble J (1987) The effects of intrathecal midazolam on sympathetic nervous system reflexes in man: a pilot study. Br J Clin Pharmacol 23(3):279–285. https://doi.org/10.1111/j.1365-2125.1987.tb03036.x
57. Blackshear PJ (1979) Implantable drug-delivery systems. Sci Am 241(6):66–73. https://doi.org/10.1038/scientificamerican1279-66
58. Johnston J, Reich S, Bailey A, Sluetz J (1988) Shiley INFUSAID pump technology. Ann N Y Acad Sci 531:57–65. https://doi.org/10.1111/j.1749-6632.1988.tb31317.x
59. Onofrio BM, Yaksh TL, Arnold PG (1981) Continuous low-dose intrathecal morphine administration in the treatment of chronic pain of malignant origin. Mayo Clin Proc 56(8):516–20.
60. Saudek CD, Fischell RE, Swindle MM (1990) The programmable implantable medication system (PIMS): design features and pre-clinical trials. Horm Metab Res 22(4):201–206. https://doi.org/10.1055/s-2007-1004889
61. Penn RD, Savoy SM, Corcos D et al (1989) Intrathecal baclofen for severe spinal spasticity. N Engl J Med 320(23):1517–1521. https://doi.org/10.1056/NEJM198906083202303
62. Wilkes D (2014) Programmable intrathecal pumps for the management of chronic pain: recommendations for improved efficiency. J Pain Res 7:571–577. https://doi.org/10.2147/JPR.S68975
63. North RB, Cutchis PN, Epstein JA, Long DM (1991) Spinal cord compression complicating subarachnoid infusion of morphine: case report and laboratory experience. Neurosurgery 29(5):778–784. https://doi.org/10.1227/00006123-199111000-00021

64. Sabbe MB, Grafe MR, Mjanger E et al (1994) Spinal delivery of sufentanil, alfentanil, and morphine in dogs: physiologic and toxicologic investigations. Anesthesiology 81(4):899–920. https://doi.org/10.1097/00000542-199410000-00014
65. Sabbe MB, Grafe MR, Pfeifer BL, Mirzai TH, Yaksh TL (1993) Toxicology of baclofen continuously infused into the spinal intrathecal space of the dog. Neurotoxicology 14(4):397–410
66. Yaksh TL, Rathbun M, Jage J et al (1994) Pharmacology and toxicology of chronically infused epidural clonidine HCl in dogs. Fundam Appl Toxicol 23(3):319–335. https://doi.org/10.1006/faat.1994.1112
67. Allen JW, Horais KA, Tozier NA et al (2006) Time course and role of morphine dose and concentration in intrathecal granuloma formation in dogs: a combined magnetic resonance imaging and histopathology investigation. Anesthesiology 105(3):581–589. https://doi.org/10.1097/00000542-200609000-00025
68. Deer TR, Pope JE, Hayek SM et al (2017) The polyanalgesic consensus conference (PACC): recommendations on intrathecal drug infusion systems best practices and guidelines. Neuromodulation 20(2):96–132. https://doi.org/10.1111/ner.12538
69. Yaksh TL, Collins JG (1989) Studies in animals should precede human use of spinally administered drugs. Anesthesiology 70(1):4–6. https://doi.org/10.1097/00000542-198901000-00003
70. Yaksh TL, Allen JW (2004) The use of intrathecal midazolam in humans: a case study of process. Anesth Analg 98(6):1536–1545. https://doi.org/10.1213/01.ANE.0000117336.57343.4D
71. Madadi AK, Sohn M (2024) Advances in intrathecal nanoparticle delivery. Pharmaceuticals 17(3):1070. https://doi.org/10.3390/ph17031070
72. Moral-Pulido F, Jiménez-González JI, Gutiérrez-Montes C et al (2023) In vitro characterisation of solute transport in the spinal canal. Phys Fluids 35(5):051905. https://doi.org/10.1063/5.0146059
73. Liu H, Barthélemy NR, Ovod V et al (2023) Acute sleep loss decreases CSF-to-blood clearance of Alzheimer biomarkers. Alzheimers Dement 19(7):3055–3064. https://doi.org/10.1002/alz.13017
74. Flessner MF (2001) The transport of peritoneal dialysis solutions. Semin Dial 14(1):50–60. https://doi.org/10.1046/j.1525-139x.2001.00011.x
75. Salegio EA, Hancock K, Korszen S (2023) Pre-clinical delivery of gene therapy products to CSF. Front Mol Neurosci 16:1248271. https://doi.org/10.3389/fnmol.2023.1248271
76. Hinderer C, Wilson JM (2021) The challenge of translating AAV-mediated gene therapy to the CNS. Trends Mol Med 27(7):679–691. https://doi.org/10.1016/j.molmed.2021.03.007
77. Boster KAS, Cai S, Ladrón-de-Guevara A et al (2023) AI velocimetry reveals in vivo perivascular flows. PNAS 120(13):e2217744120. https://doi.org/10.1073/pnas.2217744120
78. Kurtcuoglu V, Soellinger M, Summers P et al (2018) Computational fluid dynamics for the assessment of cerebrospinal fluid dynamics. J Magn Reson Imaging 47(3):581–596. https://doi.org/10.1002/jmri.25778
79. Williams SD, Setzer B, Fultz NE et al (2023) Sensory-induced neural activity drives CSF flow during wakefulness. PLoS Biol 21(4):e3002035. https://doi.org/10.1371/journal.pbio.3002035
80. Atweh SF, Kuhar MJ (1977) Autoradiographic localization of opiate receptors in rat brain. I. Spinal cord and lower medulla. Brain Res 124(1):53–67. https://doi.org/10.1016/0006-8993(77)90416-8
81. Fields HL, Emson PC, Leigh BK et al (1980) Multiple opiate receptor sites on primary afferent fibres. Nature 284(5754):351–353. https://doi.org/10.1038/284351a0
82. Gamse R, Holzer P, Lembeck F (1979) Indirect evidence for presynaptic location of opiate receptors on chemosensitive primary sensory neurones. Naunyn Schmiedebergs Arch Pharmacol 308(3):281–285. https://doi.org/10.1007/BF00501186
83. Eisenach JC, Yaksh TL (2002) Safety in numbers: how do we study toxicity of spinal analgesics? Anesthesiology 97(5):1047–1049. https://doi.org/10.1097/00000542-200211000-00001
84. Menéndez-González M (2024) Liquorpheresis: CSF filtration to treat CNS conditions. Springer, Cham. https://doi.org/10.1007/978-3-031-49599-1
85. Hornkjøl M, Valnes LM, Ringstad G et al (2022) CSF circulation and dispersion yield rapid clearance. Front Bioeng Biotechnol 10:932469. https://doi.org/10.3389/fbioe.2022.932469
86. Holstein-Rønsbo S, Gan Y, Giannetto MJ et al (2023) Glymphatic influx and clearance accelerated by neurovascular coupling. Nat Neurosci 26(6):1042–1053. https://doi.org/10.1038/s41593-023-01327-z
87. Rangroo Thrane V, Hynnekleiv L, Wang X et al (2021) Ocular glymphatic clearance in glaucoma. Acta Ophthalmol 99(2):e283–e284. https://doi.org/10.1111/aos.14588

88. Lilius TO, Waris M, Koskela A et al (2022) Factors affecting the distribution of intrathecally administered drugs: a review of the literature. Pharmaceutics 14(9):1855. https://doi.org/10.3390/pharmaceutics14091855
89. Martin BA, Lentsch A, Mestre H et al (2022) The state of the art and future of technologies for the study of fluid flow in the central nervous system. J R Soc Interface 19(193):20220247. https://doi.org/10.1098/rsif.2022.0247
90. Mestre H, Tithof J, Du T et al (2018) Flow of cerebrospinal fluid is driven by arterial pulsations and is reduced in hypertension. Nat Commun 9(1):4878. https://doi.org/10.1038/s41467-018-07318-3
91. Papisov MI (2021) The piston effect of the administration bolus in intrathecal drug delivery. Adv Drug Deliv Rev 178:113884. https://doi.org/10.1016/j.addr.2021.113884
92. Kuttler A, Bauknecht V, Schlanderer S et al (2020) The effect of injection speed on intrathecal drug distribution in a new ex-vivo model. Fluids Barriers CNS 17(1):63. https://doi.org/10.1186/s12987-020-00222-3
93. Nicholson C, Hrabětová S (2017) Brain extracellular space: the final frontier of neuroscience. Biophys J 113(10):2133–2142. https://doi.org/10.1016/j.bpj.2017.06.052
94. Bobo RH, Laske DW, Akbasak A et al (1994) Convection-enhanced delivery of macromolecules in the brain. Proc Natl Acad Sci USA 91(6):2076–2080. https://doi.org/10.1073/pnas.91.6.2076
95. Richardson MG, Wissler RN (2000) Densities of dextrose-containing intrathecal local anesthetics and cerebrospinal fluid in pregnant and nonpregnant women. Anesth Analg 90(5):1167–1171. https://doi.org/10.1097/00000539-200005000-00032
96. Schain M, Tithof J, Kelley DH et al (2017) Visualization of solute transport in the spinal canal of the rat. Nat Commun 8:14225. https://doi.org/10.1038/ncomms14225
97. Khani M, Xing T, Linninger A (2021) Hydrodynamics of the spinal subarachnoid space: the effect of idealised trabeculae on CSF mixing. J Fluid Mech 928:A3. https://doi.org/10.1017/jfm.2021.758
98. Butler S, Ghaffari L, Boddapati V et al (2023) Single-cell analysis of the adult human brain and spinal cord meninges. Nat Neurosci 26(6):1107–1119. https://doi.org/10.1038/s41593-023-01343-z
99. Hofmann M, Kuttler A, Bauknecht V et al (2022) Drug distribution in the cerebrospinal fluid of a Chiari I malformation patient. Fluids Barriers CNS 19(1):1. https://doi.org/10.1186/s12987-021-00298-5
100. Bey-doun N, He T, Papacostas M et al (2021) Intrathecal delivery in non-human primates: a translational model for ASO distribution to the central nervous system. Toxicol Appl Pharmacol 411:115372. https://doi.org/10.1016/j.taap.2020.115372
101. Elmallah MK, Falk DJ, Lane M et al (2022) AAV-mediated gene therapy for SMA: the importance of dosing, route of administration, and the blood-brain barrier. Mol Ther 30(2):594–610. https://doi.org/10.1016/j.ymthe.2021.11.015
102. Hildebrand KR, Elsberry DD, Hassenbusch SJ (2008) Stability and compatibility of morphine-sufentanil-bupivacaine mixtures for intrathecal administration. Anesth Analg 106(5):1540–1549. https://doi.org/10.1213/ane.0b013e3181682898
103. Yaksh TL, Hassenbusch SJ, Allen JW et al (2003) Preclinical safety evaluation of intrathecal drugs. Anesthesiology 99(1):189–201. https://doi.org/10.1097/00000542-200307000-00029
104. Hordeaux J, Buza EL, Hinderer C et al (2018) AAV-mediated gene therapy for MPS IIIB in non-human primates. Mol Ther 26(4):1131–1141. https://doi.org/10.1016/j.ymthe.2018.02.013
105. Burel S, Hart C, Fiallos D et al (2016) Platform-wide assessment of the effect of 2'-O-methoxyethyl-modified antisense oligonucleotides on pro-inflammatory cytokines and coagulation factors in monkeys. Nucleic Acid Ther 26(4):215–226. https://doi.org/10.1089/nat.2016.0616

Part II

Marketed Intrathecal Drug Delivery Systems and Applications

Check for updates

Chapter 3

Implantable Systems for CNS Drug Delivery: Passive Reservoirs and Active Infusion Pumps

Manuel Menéndez-González

Abstract

Implantable devices that provide direct and repeated access to the cerebrospinal fluid (CSF) represent a cornerstone of modern neurotherapeutics, enabling the circumvention of the blood–brain barrier for the treatment of oncological, infectious, and functional neurological disorders. This chapter provides a comprehensive review of the two major classes of these technologies: passive subcutaneous reservoirs and active programmable infusion pumps. We trace the parallel evolution of these systems, from the simple, manually operated ventricular reservoirs conceived by Rickham and Ommaya in the mid-twentieth century, to the sophisticated, telemetry-enabled, implantable micropumps that define contemporary practice. For each class of device, we detail the engineering principles, material science, surgical implantation techniques, and the distinct pharmacokinetic profiles they generate—bolus versus continuous infusion. Clinical indications, ranging from intraventricular chemotherapy and management of shunt infections to the delivery of advanced biologics and gene vectors, are discussed in depth. A comparative analysis of their respective advantages, limitations, and characteristic complications, such as reservoir-related infections and pump-associated catheter-tip granulomas, is provided. Finally, we explore the next generation of innovations, including antibiotic-impregnated materials, 3D-printed customized ports, nanofluidic metering, and the development of closed-loop, sensor-integrated smart pumps, highlighting how both passive and active systems continue to evolve as indispensable platforms for central nervous system drug delivery.

Keywords Implantable pumps, Intrathecal infusion systems, Programmable pumps, Fixed-flow pumps, Device engineering, Cerebrospinal drug administration

1 Introduction

Effective pharmacological treatment of central nervous system (CNS) disease is profoundly limited by the blood–brain barrier (BBB), a dynamic, tightly regulated interface that excludes the most small and nearly all large or hydrophilic molecules from the parenchyma. The clinical consequences of this gatekeeping are well-documented across neuro-oncology and neuroimmunology and help explain why otherwise active systemic regimens may underperform once malignant or inflammatory processes reach the neuraxis. These constraints,

Manuel Menéndez-González (ed.), *Intrathecal Drug Delivery*, Neuromethods, vol. 234,
https://doi.org/10.1007/978-1-0716-5241-1_3,
© The Author(s), under exclusive license to Springer Science+Business Media, LLC, part of Springer Nature 2026

summarized in classic critiques of BBB pharmacology and echoed in contemporary practice resources, continue to motivate alternative routes and devices that place drugs into the cerebrospinal fluid (CSF) compartment directly, where exposure can be orders of magnitude higher than with systemic administration alone [1–4].

Two families of implantable technologies have matured around this therapeutic objective. The first comprises passive access systems—principally Ommaya-type and Rickham-type reservoirs—that establish a stable, subcutaneous portal to the ventricular system or subarachnoid space. They have become workhorse tools for repeated CSF sampling and for bolus intrathecal or intraventricular chemotherapy in hematologic malignancies and leptomeningeal disease, sparing patients the cumulative burden of serial lumbar punctures and improving the reliability of ventricular delivery when rostral distribution is essential [1, 5, 6]. The second family includes active, programmable infusion pumps capable of delivering continuous or scheduled intrathecal dosing. Although these devices were refined mainly in pain and spasticity care, their pharmacokinetic logic—flattening peak–trough oscillations, enabling complex schedules, and supporting long-term exposure—maps naturally onto neuro-oncologic needs as agents diversify beyond classic antimetabolites [2, 7].

Taken together, these platforms convert the CSF from a diagnostic fluid into a therapeutic medium. The choice between them is more than procedural preference; it encodes assumptions about pharmacokinetics, disease topology, and care trajectory. Ventricular access through a reservoir leverages physiologic CSF flow from the choroid plexus to the cisternal and convexity surfaces, which can be decisive in leptomeningeal metastasis or CNS lymphoma where uniform craniospinal bathing is desired. Conversely, programmable pumps—typically placed with lumbar catheters—favor spinal targets or scenarios where a time–concentration profile (rather than a bolus) is central to efficacy or tolerability. In practice, programs individualize across these axes, sometimes combining ventricular bolus therapy for induction with maintenance strategies that exploit prolonged exposure or patient-controlled schedules [1, 6, 8].

Guidelines and reviews increasingly reflect this nuance. European and international consensus statements frame intrathecal therapy as one pillar within multimodal care for leptomeningeal disease—alongside focal radiotherapy and systemic drugs with CNS activity—while emphasizing that device selection should follow CSF flow studies, imaging of nodular disease, and the expected duration of treatment [8, 9]. Parallel methodological reviews chart a growing armamentarium of agents and routes, including intraventricular monoclonal antibodies for HER2-positive disease and investigational checkpoint blockade delivered into CSF, developments that depend on reliable, repeatable access to the ventricular space [10–12].

The device landscape is also being reshaped by the biology of what we now deliver. As targeted antibodies, antibody–drug conjugates, RNA therapeutics, and cell-based treatments enter CNS trials, the limitations of simple diffusion in the subarachnoid space become more salient. Formulation work in nanomedicine—spanning liposomes, lipid nanoparticles, and ionizable nanoemulsions—aims to improve persistence and tissue contact after intrathecal or intraventricular dosing, a direction that will likely intensify the role of implantable access to support repeated administrations with consistent pharmacokinetics [13–16]. Early translational signals from intrathecal or intracerebroventricular immunotherapies and pediatric CAR-T programs underscore the same theme: the route and the hardware are inseparable determinants of exposure, distribution, and safety in the CNS [11, 12, 17].

This chapter situates passive reservoirs and active pumps within that broader translational arc. We first trace the design logic, implantation techniques, and clinical use cases for ventricular reservoirs, including their role in diagnostics, drug delivery, and patient-centered logistics. We then examine programmable intrathecal pumps through a pharmacokinetic lens, contrasting continuous infusion with intermittent schedules and mapping those profiles to indications where concentration–time heuristics matter. Throughout, we integrate current guidance and emerging evidence on agent selection, route matching to disease patterns, and multidisciplinary governance of complications and neurotoxicity. By setting the engineering of access alongside the pharmacology of what is delivered, we aim to provide a practical framework for choosing, combining, and future-proofing implantable systems in neurotherapeutics [1, 2, 5, 6, 8, 9, 11, 12, 18, 19].

2 Part I: Passive Reservoir Systems

Passive reservoir systems represent the simplest and the longest-standing solution for chronic CSF access. Their design philosophy prioritizes durability, simplicity, and reliability over programmable complexity, providing a dependable portal for manual drug instillation and fluid sampling.

2.1 Historical Origins and Design Philosophy

The conceptual framework for a subcutaneous ventricular access device emerged from the challenges of managing neonatal hydrocephalus. In 1949, the pediatric surgeon Richard Rickham introduced a device comprising a stainless-steel reservoir connected to a ventricular cannula. This “subcutaneous reservoir” was designed for repeated, sterile CSF taps in infants with posthemorrhagic hydrocephalus, allowing for temporary ventricular decompression while avoiding the high risks of repeated direct ventricular puncture [1]. While functional, the rigid nature of the device posed challenges.

The seminal refinement came in 1963 from the neurosurgeon Ayub Ommaya. Recognizing the need for a more compliant and durable system, Ommaya developed a reservoir with a dome molded from medical-grade silicone (Silastic) and, crucially, integrated a self-sealing silicone septum at its apex. This innovative septum could withstand thousands of needle punctures without leaking, creating a truly reliable long-term access port [20]. The Ommaya reservoir was rapidly adopted as the standard of care for administering intraventricular chemotherapy, particularly methotrexate, for the treatment and prophylaxis of leptomeningeal leukemia and lymphoma. Its utility and safety have been validated over more than six decades of clinical use, with current pediatric series reporting its safe application even in neonates as young as seven days old, demonstrating its enduring value in neuro-oncology and beyond [21].

2.2 Device Architecture and Materials Science

The architecture of a passive reservoir is elegantly simple. It consists of three primary components: a pliable, dome-shaped reservoir body that houses the self-sealing septum; a rigid base, often with holes for suture fixation; and an outlet stem that connects to a flexible silicone or polyurethane catheter. No check-valves, batteries, or electronics govern flow. Instead, the operator directly controls instillation or aspiration by manually puncturing the septum with a noncoring, 25-gauge Huber or butterfly needle. Commercial examples include the Medtronic® Rickham Reservoir, the Integra® CR-Shunt Tap, and various low-profile reservoirs designed for endoscopic placement.

The materials used in modern reservoirs have evolved to enhance safety and durability. Domes are typically constructed from high-grade implantable silicone or, more recently, polyether-ether-ketone (PEEK), which is bonded to a titanium base for structural integrity. The septum remains the critical component, engineered from reinforced silicone to maximize its self-sealing capacity. To improve visualization and safety, the septum or base is often impregnated with barium, rendering it radiopaque for clear identification on imaging studies. Recognizing that infection is the principal complication, manufacturers have also incorporated antimicrobial agents directly into the device materials. Catheters and domes impregnated with antibiotics like rifampicin and minocycline, or antiseptics like silver and chlorhexidine, are now available and have been shown to significantly reduce the risk of bacterial colonization and subsequent device-related infection [10].

2.3 Surgical Implantation Techniques

The surgical placement of a passive reservoir depends on the intended target compartment within the CSF space.

For intraventricular placement, the Ommaya reservoir is most commonly inserted through a small burr hole in the frontal bone. The standard target is the frontal horn of the lateral ventricle on the nondominant side. Using either stereotactic neuronavigation for precision or freehand landmarks (e.g., Kocher's point, located 2–3 cm lateral to the midline and 1 cm anterior to the coronal suture), the surgeon guides the catheter tip through the brain parenchyma into the ventricle, aiming for a trajectory that passes through the foramen of Monro [22]. The dome of the reservoir is then seated firmly on the pericranium in a subgaleal pocket, and a purse-string suture may be used to secure the catheter at the burr hole to prevent migration.

For access to the spinal subarachnoid space, lumbar or thoracic intrathecal ports are utilized. Devices such as the Port-A-Cath® IT or Thecaflex systems consist of a subcutaneous port, similar in concept to those used for intravenous chemotherapy, connected to a spinal catheter. The catheter is threaded into the intrathecal space, typically at the L3/L4 level, using a Tuohy needle under fluoroscopic guidance. The port is then tunneled subcutaneously and anchored to the fascia on the flank or posterior iliac crest. These systems are popular for repeated CSF sampling, for administering test boluses of drugs like clonidine in the management of detrusor hyperreflexia, and for delivering intrathecal chemotherapy in pediatric patients with complex spinal anatomy where repeated lumbar punctures would be difficult or dangerous [23].

2.4 Pharmacokinetics of Pumpless Bolus Administration

Drug delivery via a passive reservoir is fundamentally a bolus administration technique, resulting in a pharmacokinetic profile starkly different from that of a continuous infusion pump. The distribution of the injected drug relies entirely on the volume and speed of the manual injection and the subsequent mixing driven by native CSF pulsatility, which generates flow velocities of approximately 2–5 cm/s in the cervical spine [24]. Positron emission tomography (PET) studies have elegantly demonstrated this principle: a large 10 mL bolus of methotrexate injected through an Ommaya reservoir can reach the lumbar cisterns within 30 min, whereas a smaller 1 mL aliquot remains largely confined to the ventricular system [25].

This bolus delivery creates a "saw-tooth" concentration profile in the CSF over time. Following injection, there is a rapid rise to a peak concentration (Cmax), followed by an exponential decline or "washout" as the drug is distributed, absorbed, and cleared. The rate of this decline is governed primarily by the bulk flow and turnover rate of CSF, which is approximately 0.3–0.4 mL/min in adults. This pharmacokinetic pattern may be suboptimal for time-dependent agents that require sustained concentrations above a therapeutic threshold, but it is well-suited for concentration-dependent drugs or for applications like CSF sampling where a steady-state is not required.

2.5 Clinical Applications and Indications

The versatility and reliability of passive reservoirs have established their role in several key areas of clinical practice. In neuro-oncology, they remain indispensable for the intraventricular delivery of chemotherapeutic agents such as methotrexate, cytarabine, and topotecan for the treatment of leptomeningeal carcinomatosis. They also serve as a platform for delivering experimental agents; for example, Phase I trials of trastuzumab emtansine delivered via an Ommaya reservoir reported CSF drug exposures over 1000-fold higher than those achievable with intravenous administration [26]. In the realm of infectious disease, these devices are life-saving for treating recalcitrant ventriculitis, often related to infected CSF shunts, by allowing the direct intraventricular instillation of high-dose antibiotics like vancomycin or colistin when systemic therapy fails [13].

In hydrocephalus management, particularly in the neonatal population, the Rickham reservoir provides a temporizing measure. It allows for serial ventricular taps to control intracranial pressure in premature infants with posthemorrhagic hydrocephalus, delaying the need for a permanent shunt until the infant is larger and the CSF protein level has normalized, thereby reducing the risk of shunt failure [14]. Finally, in the context of clinical research, reservoirs provide an invaluable tool for pharmacokinetic sampling. In gene therapy and ASO trials, an implanted lumbar or ventricular port allows for repeated CSF collection to measure vector biodistribution and biomarker response without subjecting the patient to the discomfort and risk of serial lumbar punctures [15].

2.6 Complications and Management

Despite their simple design, passive reservoirs are associated with a significant complication profile, which is summarized in Table 1. The most frequent and serious complication is infection. Because the system is repeatedly breached by percutaneous puncture, the lifetime risk of meningitis or cellulitis can be as high as 5%–15%. This risk is elevated in patients requiring prolonged access or frequent punctures [27]. Catheter obstruction due to blood, fibrin, or tumor debris occurs in 3%–10% of cases and may require flushing or surgical revision. More acute surgical complications, such as hemorrhage or seizures upon insertion, are less common (<2%) and are minimized with the use of image guidance. CSF leakage or the formation of a pseudomeningocele can also occur, particularly in patients with thin scalps or a history of radiation therapy [28]. While most infections necessitate device removal, attempts to salvage the reservoir with intensive intrathecal antibiotic therapy are only successful in a minority of cases (<40%) [29].

2.7 Emerging Innovations and Future Directions

While the basic design of the Ommaya reservoir has remained unchanged for decades, new innovations are enhancing its functionality and safety. 3D printing is being used to create custom-contoured domes that match the specific cranial anatomy of infant skulls,

Table 1
Complications associated with passive CSF reservoir systems

Complication	Typical incidence	Primary risk factors	Mitigation and management strategies
Infection (meningitis, ventriculitis, cellulitis)	5%–15% lifetime risk	Prolonged access duration (>12 months), frequent punctures, poor aseptic technique, skin erosion over dome	Strict sterile technique for every access; use of antimicrobial-impregnated devices; prompt device removal for established infection.
Catheter obstruction	3%–10%	Kinking; occlusion by blood, fibrin, tumor cells, or proteinaceous debris from infection	Regular low-dose heparin flushes (controversial); endoscopic repositioning or replacement of catheter.
Hemorrhage or seizure	<2%	Deep or inaccurate catheter trajectory, underlying coagulopathy	Use of stereotactic neuronavigation; correction of coagulopathy (INR < 1.5) prior to surgery.
CSF leak/pseudomeningocele	1%–3%	Thin scalp (e.g., in infants), prior radiation therapy, poor wound healing	Meticulous two-layer galeal closure; creation of a robust subgaleal pocket for the reservoir dome.
Drug extravasation	Rare	Faulty or degraded septum, use of incorrect (coring) needle, needle dislodgement during injection	Exclusive use of non-coring Huber needles; confirmation of needle position with CSF return before injection; ultrasound verification

reducing skin stress and the risk of erosion. The development of silicone impregnated with antibiotics like rifampicin and minocycline has shown a greater than 60% reduction in infection rates in preclinical porcine models [5]. Looking further ahead, research is underway on needle-free access systems that use transcutaneous magnetic connectors to deliver a bolus, eliminating the need for skin puncture entirely. Furthermore, the integration of fiber-optic pressure sensors into the reservoir dome could allow for simultaneous drug delivery and ICP monitoring without the need for additional catheters.

The future of passive reservoirs may lie in their combination with other advanced technologies. They could serve as the access port for convection-enhanced delivery (CED) catheters to perfuse therapeutics directly into brain tumors like brainstem gliomas. They

Table 2
Advantages and limitations of the most relevant approaches and devices for intrathecal drug delivery

Device/approach	Advantages	Limitations
Lumbar puncture	Low cost	Patient discomfort and/or anxiety
	Minimally invasive	Postpuncture headache
	No surgical implantation	
Intrathecal catheters	Controlled drug delivery and dosing	Risk of infection, catheter dislodgement, or occlusion
	Possibility of combination therapies	Routine monitoring
Implantable intrathecal infusion pumps	Continuous, steady-state drug delivery	Surgical implantation
		Pump malfunction, infections, and catheter-related complications

could also be used to inject bioresponsive hydrogels that are liquid at room temperature but solidify into a slow-release drug depot at body temperature, thereby achieving quasicontinuous exposure without a pump. The use of an intraventricular port to activate ultrasound-triggered microbubbles for focal BBB opening is another exciting possibility. These hybrid approaches will ensure that passive access devices remain a relevant and valuable tool in the neurotherapeutic armamentarium. Table 2 summarizes the most relevant first-generation devices and common approaches for IT drug delivery.

3 Part II: Active Programmable Infusion Pumps

Active implantable infusion pumps represent a significant technological leap from passive reservoirs, transforming IT and ICV therapy from a static bolus technique into a dynamic, titratable, and continuous neuromodulation platform.

3.1 Historical Evolution from Constant-Rate to Programmable Infusion

The era of continuous intrathecal infusion began with the development of the Infusaid® pump, first implanted in humans in the 1970s. This ingenious device used a two-chamber system where a volatile fluorocarbon liquid (freon) in an outer chamber vaporized at body temperature, creating a constant pressure that compressed an inner bellows reservoir, expelling the drug at a fixed rate [7]. While revolutionary, its constant flow rate was not titratable without surgical intervention.

The true paradigm shift occurred with the advent of the first programmable, telemetry-enabled pump, the Medtronic SynchroMed™, which received FDA approval in 1991. This device incorporated a lithium-ion battery, an electronic controller, and a peristaltic pump mechanism that could be noninvasively programmed and interrogated using a handheld telemetry wand. This allowed clinicians to precisely titrate the dose, alter flow rates, and administer complex schedules (e.g., variable infusions mimicking circadian rhythms), heralding the modern era of intrathecal neuromodulation [30]. Subsequent innovations, such as the magnetically gated Prometra® II pump, introduced alternative propulsion mechanisms. The Prometra utilizes a high-resistance microvalve that opens and closes in response to a magnetic actuator, a design intended to improve flow accuracy. The PROS-01 registry for this device reported an impressive explant-free survival rate of 96% at 3 years, underscoring the high reliability of modern systems [31]. The clinical indications for these pumps have expanded dramatically from their initial use for intrathecal morphine in refractory cancer pain [32] to include baclofen for severe spasticity [33], the peptide analgesic ziconotide for neuropathic pain, and, more recently, enzyme-replacement therapies and ASOs. While real-world prescription data show that baclofen and opioids still account for the majority of prescriptions, the use of pumps to deliver biologics is the fastest-growing segment [34].

3.2 Device Architecture and Propulsion Mechanisms

A modern programmable pump consists of several core components housed within a hermetically sealed titanium casing. A central reservoir, typically with a volume of 20 or 40 mL, holds the drug. A self-sealing silicone septum allows for percutaneous refills using a Huber needle. The system is powered by a long-life lithium-ion battery (either primary or, in newer designs, inductively rechargeable) and controlled by an on-chip microprocessor and telemetry coil for external communication. The key differentiator between pump models is the propulsion mechanism, as detailed in Table 3. The SynchroMed family uses a motor-driven peristaltic gear pump, while the Prometra family uses a magnetically actuated valve. These sophisticated mechanisms allow for highly precise and programmable flow rates, typically ranging from 0.016 to 2 mL/day.

The pump is connected to a catheter, typically made of radiopaque silicone. Catheters are available in various lengths and configurations, including single-piece and two-piece designs. Specialized catheters may feature multiport distal segments to disperse drug over a wider area, integrated pressure sensors for diagnostic purposes, or high-opening-pressure valves to prevent siphoning.

3.3 Applied Pharmacology and Patient Selection

The choice of drug for intrathecal pump therapy is dictated by its physicochemical properties, potency, and stability in solution. As shown in Table 4, a drug's lipophilicity, measured by its partition coefficient (Log P), is a critical determinant of its distribution. Highly hydrophilic (low Log P) drugs like morphine and baclofen remain in the CSF for longer periods, allowing for broad distribution along the neuraxis. This makes them suitable for treating diffuse nociceptive

Table 3
Comparison of propulsion mechanisms in active infusion pumps

Pump family/type	Flow generation mechanism	Typical flow range/ accuracy	Key characteristics
SynchroMed II (Medtronic)	Battery-powered stepper motor drives a peristaltic gear pump.	0.016–2 mL/day (± 7% accuracy).	Most widely used system; allows for complex, variable-rate programming.
Prometra II (Flowonix)	Magnetically actuated high-resistance valve pulses every 5 min.	Programmable; claims <2% in-vitro flow error [31].	Valve-gated mechanism designed for high accuracy and resistance to external pressure changes.
Infusaid (Historic)/**IsoMed** (Medtronic)	Constant pressure from a vaporizing osmotic gas (e.g., SF_6).	Fixed rate (e.g., 0.5 mL/day); not programmable.	Simple, no battery. Primarily of historical interest or for fixed-dose applications.
Nanofluidic prototypes	Electroosmotic flow through nanochannels (<300 nm).	~0.1 μL/h; <10 μW power consumption [44].	Experimental; offers ultralow flow for potent biologics and minimal power draw.

pain or widespread spasticity. In contrast, highly lipophilic (high Log P) drugs like fentanyl and sufentanil rapidly penetrate the spinal cord tissue. Their action is, therefore, more localized to the dermatomes near the catheter tip, making them suitable for treating focal, segmental pain [2]. Large-molecule biologics like the ASO nusinersen have very long CSF half-lives, allowing for intermittent bolus dosing rather than continuous infusion.

Patient selection is critical to ensure therapeutic success and safety. For chronic pain, candidates typically have severe pain (e.g., Visual Analog Scale ≥7/10) that is refractory to multiple lines of systemic therapy, and a life expectancy of at least 6 months to justify the surgical intervention. For spasticity, patients usually have severe tone (e.g., Modified Ashworth Scale score ≥3) that has failed to respond to oral agents. Absolute contraindications include active infection, uncorrected coagulopathy, and uncontrolled psychiatric disorders. Before implantation, a positive response to a screening trial is mandatory. This usually involves a single intrathecal test bolus (e.g., 50 μg of baclofen) or a short-term trial with an externalized catheter, with a ≥50% improvement in pain or spasticity considered a positive result.

3.4 Surgical Implantation and Postoperative Management

The implantation of a programmable pump is a meticulous surgical procedure. Preoperative planning involves choosing the appropriate

Table 4
Pharmacological properties of common intrathecal pump medications

Drug	Lipophilicity (Log P)	Approx. CSF half-life (t½)	Primary clinical use	Typical starting dose (continuous)
Morphine	−0.1 (Hydrophilic)	90–120 min	Diffuse nociceptive pain (cancer, postsurgical)	200–400 μg/day
Fentanyl	4.1 (Lipophilic)	12–20 min	Localized, segmental neuropathic pain	25–75 μg/day
Ziconotide (Prialt®)	Peptide (Hydrophilic)	4–6 h	Refractory neuropathic pain (nonopioid)	1.2–2.4 μg/day
Baclofen (Lioresal®)	−1.1 (Hydrophilic)	75–90 min	Severe spasticity (spinal or cerebral origin)	50–100 μg/day
Nusinersen (Spinraza®)	Biologic (ASO)	24–48 h	Spinal muscular atrophy (SMA)	12 mg bolus every 4 months (via LP/reservoir)

pump size based on the required drug concentration and desired refill interval, and determining the optimal catheter-tip position based on the drug's properties and the patient's symptoms. The procedure is typically performed with the patient in the lateral decubitus position. Under fluoroscopic guidance, a Tuohy needle is used to access the lumbar intrathecal space, usually via a paramedian approach at L2–L3 or L3–L4. The catheter is advanced to the predetermined spinal level and anchored securely to the lumbodorsal fascia to prevent migration. The pump is then implanted in a subcutaneous pocket created in the anterior abdominal wall, and the catheter is tunneled subcutaneously from the back to the front to be connected. The pump is then filled with the drug and programmed for its initial infusion. Post-operatively, patients are monitored for 24 h for signs of CSF leak, infection, or neurological changes. A crucial aspect of long-term care involves managing MRI scans. Patients must be educated that before any MRI, the pump must be switched to a specific MRI-safe mode to prevent the strong magnetic fields from stalling the motor, and the pump's function must be interrogated and confirmed immediately after the scan [35].

3.5 Complications and Risk Mitigation

While generally safe and effective, programmable pumps carry a risk of specific complications, with infection and catheter-related issues being the most common. Pocket or catheter infections occur in 2%–9% of patients and often require aggressive antibiotic therapy and complete system explantation [36]. The most feared complication specific to opioid infusion is the formation of a catheter-tip granuloma, an inflammatory mass that can cause spinal cord compression. The risk is strongly correlated with high opioid concentration (> 10–20 mg/mL), necessitating the use of the lowest effective concentration and regular neurological monitoring [37]. CSF leaks, pump motor failures, and drug-specific side effects can also occur. For baclofen, abrupt cessation of infusion due to catheter failure or reservoir depletion can precipitate a life-threatening baclofen withdrawal syndrome, characterized by high fever, seizures, and extreme spasticity, requiring emergency medical intervention to restore IT delivery [6].

3.6 Health-Economic Considerations

The high initial cost of an implantable pump system (>$10,000 USD plus surgical fees) necessitates careful consideration of its cost-effectiveness. Health-economic analyses consistently show that the therapy provides value in well-selected patients with a sufficient lifespan. A US-based Markov model for noncancer pain calculated an incremental cost-effectiveness ratio (ICER) of approximately $35,000 per quality-adjusted life year (QALY) gained, a figure well within the accepted range for medical interventions, provided the patient's survival exceeds 18 months. For intrathecal baclofen therapy, the analysis is even more favorable; the high upfront cost becomes cost-saving after approximately 5 years due to a significant reduction in hospitalizations, caregiver burden, and management of spasticity-related crises [38].

4 Comparative Analysis and Conclusion

Passive reservoirs and active pumps offer distinct solutions for CNS drug delivery, each with a unique profile of advantages and limitations. Their key features are contrasted in Table 5. Passive reservoirs offer simplicity, low cost, MRI compatibility, and bidirectional access for both dosing and sampling. Their primary drawbacks are the "saw-tooth" pharmacokinetic profile and a higher risk of infection from repeated skin punctures. Active pumps provide the unparalleled advantage of precise, titratable, continuous infusion, enabling stable CSF drug levels, and complex dosing paradigms. This comes at the cost of higher initial expense, mechanical complexity, battery limitations, and a different set of potential complications, such as motor failure and granuloma formation.

Table 5
Comparative analysis of passive reservoirs versus active infusion pumps

Feature	Passive reservoir system (e.g., Ommaya)	Active programmable pump (e.g., SynchroMed)
Mechanism of action	Manual percutaneous bolus injection/ aspiration	Continuous or variable-rate infusion via an internal, powered pump
Pharmacokinetic profile	"Saw-tooth": High peak (Cmax) followed by exponential decay	Stable, steady-state CSF concentration
Primary advantage	Simplicity, low cost, no battery, bidirectional access (dosing and sampling), MRI safe	Precise, titratable, continuous delivery; complex programming possible
Primary limitation	Pharmacokinetic troughs, pain/ discomfort of repeated access, user-dependent	High initial cost, mechanical complexity, finite battery life, MRI protocols required
Key indications	Intraventricular chemotherapy, CSF sampling, neonatal hydrocephalus taps	Chronic pain, severe spasticity, continuous delivery of biologics
Signature complication	Infection from repeated percutaneous access (5%–15%)	Catheter-tip inflammatory granuloma (with opioids, ~3%)
Approximate device cost	~$800 USD	>$10,000 USD

The choice between a passive and an active system is therefore driven by the specific clinical goal. For applications requiring intermittent bolus chemotherapy, CSF sampling for research, or temporary management of neonatal hydrocephalus, a passive reservoir is the ideal choice. For conditions like chronic pain or spasticity that benefit from a continuous, stable therapeutic effect, a programmable pump is the superior technology.

5 Future Directions and Synthesis

The future of implantable CNS access devices is not a binary choice between passive and active systems, but rather a story of parallel and converging evolution. Passive reservoirs are becoming smarter through material science and hybrid applications. Active pumps are evolving into intelligent, sensor-driven platforms. The next generation of active devices will likely feature closed-loop delivery, integrating biosensors (e.g., electromyography, pressure, or neurotransmitter sensors) to automatically adjust drug delivery in response to a patient's real-time physiological state [39]. Inductively rechargeable solid-state batteries will extend pump life beyond a decade, and nanofluidic metering will enable ultralow, ultra-precise flow for highly potent biologics and gene vectors. Ultimately, these devices are not just drug delivery systems but platforms. The potential to use them to deliver payloads of gene vectors like AAV9 or even therapeutic cells, perhaps in combination with CED, promises to open new frontiers in the treatment of diffuse neurodegenerative and genetic diseases [16, 40].

In conclusion, both passive reservoirs and active pumps are indispensable tools in the neurotherapeutic armamentarium. Their continued refinement and the innovative strategies for their use will ensure that direct access to the central nervous system remains a powerful and expanding avenue for treating its most challenging diseases [41, 42, 43].

Conflict of Interest

The authors have no conflicts of interest to declare that are relevant to the content of this chapter.

References

1. Rickham PP (1949) Further experience with ventriculo-subcutaneous drainage of cerebrospinal fluid in hydrocephalic infants. Acta Paediatr 38:212–219
2. Deer TR, Pope JE, Hayek SM et al. (2017) The Polyanalgesic consensus conference (PACC) for intrathecal therapy: an expert panel's recommendations for the proper use of intrathecal medications. Neuromodulation 20(2):96–132
3. Møllgård K, Beinlich FR, Kusk P et al. (2023) A mesothelium divides the subarachnoid space into functional compartments. Science 379(6627):84–88
4. Stokes C, White EF, Toddes S et al. (2023) 3-D cryo-fluorescence tomography shows CSF clearance along nasal lymphatics. J Imaging 9(2):45
5. Von der Brelie C, Simon K, Gröger M et al. (2017) The effect of rifampin and minocycline impregnated silicone on the risk of Ommaya reservoir associated infections in a porcine model. PLoS One 12(3):e0174154
6. Coffey RJ, Edgar TS, Francisco GE et al. (2002) Abrupt withdrawal from intrathecal baclofen: recognition and management of a potentially life-threatening syndrome. Arch Phys Med Rehabil 83(6):735–741
7. Buchwald H, Rohde TD (2000) Implantable pumps. A historical review. ASAIO J 46(4):479–488
8. Rico RC, Hobika GH, Avellanosa AM, Trudnowski RJ, Rempel J, West CR (1982) Use of intrathecal and epidural morphine for pain relief in malignant disease. J Med 13(3-4):223–231
9. Farid R, Binz K, Emerson JA, Murdock F (2019) Accuracy and Precision of the SynchroMed II Pump. Neuromodulation 22(7):805-810. Epub 2019 Mar 19. PMID: 30889307. https://doi.org/10.1111/ner.12934
10. Zabramski JM, Pittman M, Olson S et al. (2008) A comparison of two antibiotic-impregnated catheters for the reduction of CSF infection in a rabbit model. J Neurosurg Pediatr 2(1):52–56
11. Passini MA, Bu J, Richards AM et al. (2014) Intrathecal delivery of AAV9-SMN1 for spinal muscular atrophy. Hum Gene Ther 25(8):619–630
12. Wermeling DP, Berger JR (2006) Ziconotide infusion for severe chronic pain: case series and review of the literature. Pharmacotherapy 26(3):395–402
13. Bayston R, Hart CA, Barnicoat M (1987) Intraventricular vancomycin in the treatment of ventriculitis associated with cerebrospinal fluid shunting and drainage. J Neurol Neurosurg Psychiatry 50(11):1419-23. PMID: 3694201; PMCID: PMC1032551. https://doi.org/10.1136/jnnp.50.11.1419

14. Jeston S (2020) Ventricular reservoir management in neonates. Royal Children's Hospital Guideline. https://www.rch.org.au/rchcpg/hospital_clinical_guideline_index/Ventricular_reservoir_management_in_Neonates/
15. Oberrauch S, Lu J, Cornthwaite-Duncan L, Hussein M, Li J, Rao G, Velkov T (2022) Intraventricular drug delivery and sampling for pharmacokinetics and pharmacodynamics study. J Vis Exp (181):e63540. PMID: 35435913; PMCID: PMC10400305. https://doi.org/10.3791/63540
16. Madadi AK, Sohn M (2024) Advances in intrathecal nanoparticle delivery. Pharmaceuticals 17(3):1070
17. Bennett CF, Kordasiewicz HB, Cleveland DW (2021) Making sense of antisense drugs for neurological diseases. Annu Rev Pharmacol Toxicol 61:831–852
18. Magill E et al. (2023) Solid implantable devices for sustained drug delivery. Adv Drug Deliv Rev 199:114950
19. Snutch TP (2005) Targeting chronic and neuropathic pain: N-type calcium channels. NeuroRx 2(4):662–670
20. Ommaya AK (1963) Subcutaneous reservoir and pump for sterile access to ventricular cerebrospinal fluid. Lancet 2(7315): 983–984
21. Gaietto A, Panetta JC, Pauley JL, Relling MV, Ribeiro R, Ehrhardt MJ, Pui CH, Inaba H, Swanson HD (2024) Ommaya reservoir use in pediatric ALL and NHL: a review at St. Jude Children's Research Hospital. Cancer Chemother Pharmacol 93(6):617-625
22. StatPearls., 2023. Ommaya reservoir. StatPearls Publishing, Treasure Island (FL)
23. Chartier-Kastler E, Saussine C, Bitker MO (2000) Intrathecal catheter with subcutaneous port for clonidine bolus in spinal-cord injury. Eur Urol 37(1):14–17
24. Bissenas A, Fleeting C, Patel D et al. (2022) CSF dynamics: implications for hydrocephalus and glymphatic clearance. Curr Res Med Sci 1:24–42
25. Gropper, MA, Miller, RD, Eriksson, LI et al., 2020. Miller's anesthesia, 9th edn.. Elsevier, Philadelphia, PA
26. Bendel AE, Ranking A, Cefalo M et al. (2023) A phase 1 study of trastuzumab emtansine (T-DM1) in children with HER2-positive refractory solid tumors: a TACL study. Cancer Med 12(11):12573–12583
27. Szvalb AD, Raad II, Weinberg JS, Suki D, Mayer R, Viola GM (2014) Ommaya reservoir-related infections: clinical manifestations and treatment outcomes. J Infect. 68(3):216-24. Epub 2013 Dec 17. PMID: 24360921. https://doi.org/10.1016/j.jinf.2013.12.002
28. Mead PA, Safdieh JE, Nizza P, Tuma S, Sepkowitz KA (2014) Ommaya reservoir infections: a 16-year retrospective analysis. J Infect 68(3):225-30. Epub 2013 Dec 8. PMID: 24321561. https://doi.org/10.1016/j.jinf.2013.11.014
29. Pfaender M, Häsler R, Pizevica I et al. (2022) Management of Ommaya reservoir infections in cancer patients. J Clin Med 11(19):5909
30. Brill S, Gurman GM, Fisher A (2003) A history of neuraxial administration of local analgesics and opioids. Eur J Anaesthesiol 20(9):682–689
31. Goel V, Yang Y, Kanwar S, Banik RK, Patwardhan AM, Ibrahim M, Sivanesan E, Shankar H (2021) Adverse events and complications associated with intrathecal drug delivery systems: insights from the Manufacturer and User Facility Device Experience (MAUDE) Database. Neuromodulation 24(7):1181-1189. Epub 2020 Dec 11. PMID: 33306248; PMCID: PMC9034464. https://doi.org/10.1111/ner.13325
32. Yaksh TL, Rudy TA (1976) Analgesia mediated by a direct spinal action of narcotics. Science 192(4246):1357–1358
33. Penn RD, Kroin JS (1984) Intrathecal baclofen alleviates spinal cord spasticity. Lancet 1(8386):1078
34. Hao D, Di Capua CA, Dumas C et al. (2024) Real-world prescription data for intrathecal drug delivery. Neuromodulation 27(3):455–467
35. Lee SH, Hofstede RP, Noriega de la Colina A, Gunton JH, Bernstock JD, Traverso G (2025) Implantable systems for neurological chronotherapy. Adv Drug Deliv Rev 221:115574. Epub 2025 Apr 3. PMID: 40187646. https://doi.org/10.1016/j.addr.2025.115574
36. StatPearls., 2024. Implantable intrathecal drug delivery system. StatPearls Publishing, Treasure Island (FL)
37. Yaksh TL, Horais KA, Tozier NA et al. (2006) The effect of concentration on the production of intrathecal inflammatory masses from morphine. Anesthesiology 105(3):581–589
38. Krames ES (1997) The cost-effectiveness of intrathecal therapy for chronic nonmalignant pain. J Pain Symptom Manage 14(1 Suppl):S36–S48
39. (2022) NIH Brain research through advancing innovative neurotechnologies (BRAIN) initiative. https://www.braininitiative.nih.gov/. Accessed 7 Jul 2025
40. Salegio EA, Hancock K, Korszen S (2023) Preclinical delivery of gene therapy products to the CSF. Front Mol Neurosci 16:1248271

41. Moreau C et al. (2024) Intracerebroventricular infusion of anaerobic dopamine in Parkinson's disease. J Neurol Res Ther 8:e2024001
42. Paice JA, Penn RD, Shott S (1996) Intraspinal morphine for chronic pain: a retrospective, multicenter study. J Pain Symptom Manage 11(2):71–80
43. Bejjani GK, Karim NO, Tzortzidis F (1997) Intrathecal granuloma after morphine pump implantation. Surg Neurol 48(3):288–291
44. Lewis TR et al. (2002) Electro-osmotic nanofluidic pump for implantable drug delivery. IEEE Trans Biomed Eng 49(3):211–220

Check for updates

Chapter 4

Implantation and Explantation of Intrathecal Drug Delivery Systems: Surgical Technique and Complications

Manuel Menéndez-González

Abstract

Intrathecal drug delivery systems (IDDS) provide durable access to the spinal subarachnoid space through an implantable pump–catheter construct and have evolved along three principal design lines: programmable electronic pumps with telemetry-controlled dosing, fixed-flow mechanical pumps that deliver a constant rate without electronics, and newer programmable motorless platforms that meter flow via pressurized bellows and valve architecture. This chapter delineates the engineering features, MRI considerations, and maintenance implications that distinguish these devices, including battery longevity and elective replacement for electronic models and scan-specific precautions to prevent inadvertent drug release in motorless systems. It then presents a practical, stepwise surgical workflow for implantation—preoperative trialing and site selection, antibiotic prophylaxis, anesthesia and positioning, fluoroscopy-guided lumbar dural access, cranial catheter advancement and anchoring, subcutaneous pocket creation, tunneling, pump connection, and securement—followed by immediate postoperative surveillance for infection, CSF leak, and neurological change. Explantation and revision strategies are detailed for infections, catheter malfunctions, pump failures, granuloma-related complications, and scheduled battery exchanges, with emphasis on bridging measures to avoid life-threatening withdrawal in patients receiving baclofen or high-dose opioids. The chapter also catalogs surgical, device-related, and pharmacologic adverse events—from hematoma and dural puncture headache to catheter obstruction, motor stalls, overdose, and withdrawal—and links them to prevention and rescue protocols. By integrating device taxonomy with operative technique and complication management, the chapter provides a concise surgical playbook to select, implant, revise, and safely steward IDDS across their life cycle.

Keywords Intrathecal catheterization, Surgical implantation, Explantation, Pump placement, Neuromodulation surgery, Perioperative complications

1 Types of Intrathecal Drug Delivery Devices

Intrathecal drug delivery systems (IDDS) consist of an implantable subcutaneous pump connected to an intrathecal catheter that releases drugs directly into the spinal subarachnoid space [1]. Contemporary platforms fall into two principal families—programmable electronic pumps and fixed-flow mechanical pumps—with a third, hybrid category of programmable motorless devices emerging in the

Manuel Menéndez-González (ed.), *Intrathecal Drug Delivery*, Neuromethods, vol. 234,
https://doi.org/10.1007/978-1-0716-5241-1_4,
© The Author(s), under exclusive license to Springer Science+Business Media, LLC, part of Springer Nature 2026

past decade. Programmable pumps, exemplified by the Medtronic SynchroMed II in 20–40 mL configurations, permit dose changes and complex schedules via external telemetry, enabling precise titration, circadian programming, and rapid response to clinical needs in both analgesia and antispastic therapy [2, 3]. This flexibility is balanced by a finite internal battery life on the order of 5–7 years, after which surgical exchange is required even if all other components remain functional [4, 5]. Fixed-flow pumps are non-electronic, relying on constant-pressure gas reservoirs or bellows to generate a near-constant infusion without user-programmable variability; devices such as the historic Codman 3000 are now largely obsolete as practice has shifted toward titratable systems with finer control and data logging [6, 7].

Hybrid motorless programmable systems, typified by the Flowonix Prometra platform, employ pressurized bellows and flow-activated microvalves to meter drug delivery while still accepting external programming commands [8]. The absence of a rotating motor eliminates stall-related failure modes that have been observed in some electronic devices and reduces power management concerns [9]. At the same time, these valve-regulated, pressurized reservoirs introduce unique operational constraints under strong magnetic fields. Manufacturers and consensus guidance emphasize device-specific MRI protocols, including pre-scan reservoir management to avoid inadvertent bolus release when the pump is exposed to the MRI environment [10]. Table 1 summarizes the principal device classes and characteristic features that influence intraoperative handling, long-term maintenance, and complication profiles.

Table 1
Main types of implantable intrathecal pumps and their characteristics

Pump type	Examples/ manufacturers	Characteristics
Programmable (Electronic)	Medtronic SynchroMed II; Tricumed	Battery-powered; infusion rate adjustable via telemetry; requires replacement when the battery is depleted [2, 3]
Fixed-flow (Nonelectronic)	Codman 3000; Infusaid	Provides continuous infusion at a fixed rate using gas pressure or bellows; no electronic controls [6, 7]
Programmable motorless	Flowonix Prometra	Uses flow-activated valves; no electric motor; [8–10]
External temporary systems	Percutaneous intrathecal catheter + external pump	Used for trialing therapy before permanent implantation; higher infection risk due to percutaneous exit [11]

This chapter focuses exclusively on the surgical aspects of implantation and explantation, as well as complications, and does not address disease-specific indications, drug selection, or dosing algorithms, which are treated elsewhere.

1.1 Preoperative Preparation and General Considerations

A structured preoperative pathway reduces avoidable complications and streamlines the intraoperative course. Most centers perform a therapeutic trial before permanent implantation to demonstrate efficacy, tolerability, and a favorable risk–benefit ratio in the specific patient phenotype [11, 12]. Trials range from single-shot intrathecal dosing to short percutaneous catheter infusions with an external pump, the latter offering a closer approximation to continuous therapy but at the price of higher infection risk and patient encumbrance [11]. Once an adequate response has been documented, definitive implantation is scheduled with attention to anatomic constraints, patient habitus, future refill access, and lifestyle considerations. The abdominal subcutaneous pocket in the lower quadrant is the most common pump location because it balances access for refills, secure fixation to deep fascia, comfort in seated positions, and cosmesis in a broad range of body types [13]. Alternative sites, including supragluteal or lumbar flank pockets, are selected in patients with prior abdominal surgery, very thin soft-tissue envelopes, ostomies, or when a single-position prone operation is planned from start to finish [14].

Perioperative infection prevention begins before incision. An antibiotic with skin flora coverage is administered within an hour of skin incision, with cefazolin 1 g intravenously as a common choice and clindamycin 600 mg for patients with beta-lactam allergy [15]. Skin preparation with an alcohol-based chlorhexidine solution, hair removal with clippers rather than razors, temperature maintenance, and glycemic control are incorporated into a standardized bundle to reduce surgical site infection risk in this prosthesis-bearing operation. Anesthesia selection reflects patient comorbidity and procedural complexity. General anesthesia is favored in patients with severe chronic pain or spasticity to maintain comfort, prevent sudden movement during dural access, and facilitate prolonged fluoroscopy, although implantation under monitored anesthesia care is feasible in selected cases [16]. Positioning decisions are pragmatic. Lateral decubitus accommodates simultaneous abdominal and lumbar access with straightforward ventilation, while the prone position provides optimal fluoroscopic visualization during catheter advancement and dural anchoring; the surgeon's preferred workflow and the patient's cardiopulmonary profile typically determine the choice [14, 17]. Preoperative fluoroscopic localization of the intended interspace reduces needle passes and dural trauma, and a complete set of long and short Tuohy needles, stylets, and soft-tipped guidewires should be immediately available to adjust to anatomic variability [17].

2 Surgical Technique for Intrathecal Pump Implantation

2.1 Intrathecal Catheter Insertion

After positioning and sterile preparation of both the abdominal and lumbar fields, attention turns to the spine. A paramedian approach at L3–L4 or L4–L5 below the conus medullaris is standard, exploiting the wider interlaminar space and minimizing the risk of cord contact [17, 18]. Fluoroscopy in anteroposterior and lateral projections confirms the target level and trajectory. A 2- to 3-cm skin incision exposes the lumbodorsal fascia; a Tuohy needle is introduced toward the interlaminar window using the loss-of-resistance technique, with frequent lateral fluoroscopic checks to avoid epidural redirection. Entry into the intrathecal space is confirmed by free cerebrospinal fluid (CSF) return and by the absence of resistance to gentle saline injection [19]. The intrathecal catheter is advanced cephalad through the needle to the desired vertebral level, which is determined by the clinical target: thoracic levels often provide optimal baclofen dispersion for spasticity of spinal origin, whereas lumbar placement may suffice for lower-extremity analgesia [20]. In long constructs, incremental advancement with fluoroscopic correlation avoids excessive looping and ventral migration.

Once the catheter tip is positioned, the needle is withdrawn over the catheter, and a fascial purse-string suture is placed snugly around the entry point to reduce CSF egress. A dedicated anchor is sutured to the paraspinal fascia several centimeters from the dural entry to decouple skin motion from the intrathecal segment and to prevent migration or kinking during flexion and extension [21]. Patency is verified by observing CSF pulsatility and aspirating a small column of clear fluid from the distal catheter, which also purges air. Careful hemostasis in the lumbar pocket reduces the risk of epidural or subfascial hematoma that could compress neural elements in the immediate postoperative period.

2.2 Subcutaneous Pocket Creation and Pump Placement

Attention then shifts to the abdomen. A 6- to 8-cm transverse or oblique incision in the lower quadrant is deepened to the anterior abdominal fascia. A pocket is developed in the relatively avascular subcutaneous plane sized precisely to the pump dimensions; overly tight pockets predispose to pain, skin ischemia, and device erosion, whereas oversized pockets permit rotation, flipping, and refill inaccuracy [13]. Meticulous hemostasis with bipolar cautery and topical agents is essential to avert hematoma formation, which can impair wound healing and harbor bacteria. A tunneling device is passed from the abdominal pocket to the lumbar incision, keeping to a deep, straight subcutaneous path to minimize catheter kinking across flexion creases and belt lines [22]. The distal catheter is brought into the pocket and trimmed to the manufacturer-specified length after accounting for spinal motion and body habitus.

The pump is prepared on a sterile side table. Air is evacuated from the reservoir and pump mechanism according to the device-specific protocol, and the reservoir is filled with the planned solution or isotonic preservative-free saline if delayed drug initiation is intended. The catheter is connected to the pump port with attention to the connector depth and locking ring engagement. Before seating the pump, the system is tested by aspirating and gently injecting to confirm free flow without subcutaneous swelling. The pump face is oriented to optimize subsequent refill needle access, and the can is secured to the deep fascia with nonabsorbable sutures through manufacturer-provided eyelets when needed to prevent rotation and flipping, which are common causes of refill errors and aborted clinic encounters [23]. Both wounds are irrigated and closed in layers with attention to dead-space obliteration. Dressings are applied to allow visual inspection while maintaining occlusion.

2.3 Intraoperative Programming and Priming

For electronic systems, initial programming is performed before emergence from anesthesia. The pump is primed at a low rate until the calculated catheter and pump tubing volume have been traversed, at which point the therapeutic rate is set, or a bridging strategy is enacted if postoperative monitoring is desired before initiating drug delivery [2–4, 18, 19]. The priming calculation accounts for internal tubing volume and catheter length to avoid under- or overdelivery in the early hours after implantation. Device logs are reviewed to confirm command receipt and reservoir volume reconciliation.

2.4 Postoperative Care

Patients recover with close attention to neurologic examination, wound inspection, and pain control. Early mobilization reduces venous thromboembolism risk and helps identify positional exacerbation that might suggest catheter kinking. Dressings remain occlusive for 48 h, after which the incisions are left open to air unless drainage persists. Over the first week, clinicians remain vigilant for fever, erythema, progressive pain, or fluctuance that could herald infection, as well as orthostatic headache and wound drainage consistent with CSF leak [24]. Initial dose adjustments are conservative. For baclofen, increments are small and spaced in time to avoid hypotonia and somnolence; for opioid or ziconotide therapy, cardiorespiratory monitoring is tailored to comorbid risk and baseline opioid exposure [3, 4, 18, 25]. A structured refill calendar is established at discharge, and patients are educated on pocket care, red flags of overdose and withdrawal, and the need to report planned MRI or surgical procedures that may interact with the device [9, 10, 23].

3 Explantation and Replacement of Intrathecal Drug Delivery Systems

Indications for hardware removal range from infection and device malfunction to catheter-tip granuloma and anticipated battery depletion [26]. Infection often presents with localized pocket erythema and tenderness, fluctuance, fever, or meningitic symptoms if the infection has tracked along the catheter to the intrathecal space. Because biofilm formation on titanium and polymeric surfaces limits the efficacy of systemic antibiotics, complete system explantation is the rule rather than the exception in confirmed deep infections [27]. Cultures are obtained from pocket fluid, catheter tips, and the pump reservoir to tailor antimicrobial therapy. In baclofen-dependent patients, abrupt cessation risks life-threatening withdrawal with hyperthermia, rhabdomyolysis, and multiorgan failure; bridging is therefore planned before explantation using a temporary externalized intrathecal catheter and sterile external pump, or high-dose oral and enteral agents if intrathecal access cannot be safely maintained [28]. Intensive care monitoring is appropriate during transitions, and gradual retitration is undertaken once a new device is implanted after infection eradication.

Elective pump replacement for battery end-of-life is technically straightforward and commonly performed under local anesthesia with sedation. The prior abdominal incision is reopened, the pump is disconnected from a still-functional catheter, and a new pump is connected after standard purge and fill steps. Reservoir volume records and device logs are reconciled to avoid dosing errors at the time of exchange [29]. When catheter-related complications arise—such as disconnection at the anchor, fracture at a fascial bend, obstruction from fibrin or debris, or kinking in a superficial tunnel—revision focuses on the shortest segment that resolves the problem. Fluoroscopic dye studies through the pump side port or via a temporary connector can delineate the failure point and guide limited replacement rather than wholesale system exchange [30]. Catheter-tip granulomas, typically associated with high local opioid concentrations at low flow rates, present with progressive segmental neurologic deficits and imaging findings at the catheter tip. Management includes reducing or discontinuing the offending medication, repositioning the catheter tip, and in selected cases surgical decompression if there is cord or root compression [25].

4 Complications of Intrathecal Therapy

Complications can be conceptualized as surgical, device-related, or pharmacologic, recognizing that individual events often cross categories. Surgical complications cluster around the access sites. Neurologic injury during dural puncture or catheter advancement is rare when using fluoroscopy and strict level selection below the conus, yet practitioners must be alert to new focal deficits in the immediate postoperative period and investigate promptly [31]. Hemorrhagic complications range from subcutaneous pocket hematoma, which impairs wound healing, to epidural or subarachnoid hematoma with back pain, weakness, or cauda equina symptoms; concerning neurologic changes warrant emergent MRI and, when indicated, decompression [32]. CSF leaks present with postural headache and clear drainage from the lumbar incision, sometimes accompanied by pseudomeningocele formation. First-line measures include bed rest, hydration, caffeine, and pressure dressings, with epidural blood patch reserved for persistent cases that fail conservative therapy [33]. Infectious complications encompass superficial cellulitis, deep pocket infection, meningitis, and ventriculitis. Early superficial infections can occasionally be salvaged with antibiotics and close observation, but deep infections almost always mandate complete explantation to eradicate biofilm and prevent central nervous system seeding [27].

Device-related events accumulate over years of therapy. Catheter problems—obstruction by precipitate or tissue ingrowth, migration with loss of efficacy, disconnection at connectors, and frank fracture—are the most frequent reasons for late revision and should be suspected in any patient with abrupt loss of effect, segmental return of pain or spasticity, or unusual alarm logs on the programmer [34]. Pump failures include motor stalls and electronic faults in battery-powered systems, and valve malfunctions in motorless designs. MRI interactions are highly device-specific; electronic programmable pumps may temporarily stall or change state during scanning, while Prometra-type pressurized systems require strict adherence to empty-reservoir or lockout protocols before MRI to avoid unintended drug egress under magnetic influence [9, 10]. Teams should maintain an updated device inventory with MRI guidelines and ensure communication between radiology and the implanting service before scans.

Pharmacologic complications derive from the narrow therapeutic windows of intrathecally delivered agents and from the high potency achieved in the CSF compartment. Overdose may occur through programming error, inadvertent reservoir refill of the wrong concentration, pocket fill due to a missed septum during refill, or device malfunction. Clinical manifestations range from respiratory depression and miosis with opioid overdose to profound hypotonia and coma with baclofen, and psychosis or ataxia with ziconotide [35]. Management is agent-specific and often supportive in an intensive care setting; naloxone reverses opioid toxicity yet must be titrated carefully to avoid acute withdrawal, while baclofen overdose has no direct antidote and is managed with airway protection, ventilatory support, and time. Withdrawal syndromes are particularly dangerous with baclofen and high-dose intrathecal opioids. Catheter occlusion, reservoir depletion, pump stoppage, or abrupt device explantation can precipitate agitation, hyperreflexia, spastic crisis, fever, and autonomic instability in hours. Protocolized rescue includes prompt restoration of intrathecal delivery when feasible, temporary externalization and infusion, aggressive oral or enteral substitution, benzodiazepines for rigidity and seizures, and ICU monitoring until stability returns [36].

4.1 Quality Assurance, Refill Technique, and Long-Term Stewardship

Safe long-term stewardship of IDDS depends on disciplined clinic processes. Refill visits are conducted under sterile conditions with skin antisepsis and precise localization of the pump septum. Standard practice is to aspirate the reservoir before refill, quantify the retrieved volume, and compare it to the programmer-reported expected volume to detect occult pocket fills, flow anomalies, or silent stalls. Only preservative-free solutions appropriate for intrathecal administration are used, and concentration checks by two qualified clinicians reduce compounding errors. Dose changes are documented with the clinical rationale and planned follow-up interval, recognizing that pharmacodynamic steady state may take days to manifest depending on the drug and flow rate. Patients receive written emergency instructions for signs of overdose and withdrawal and are encouraged to carry device identification cards to facilitate emergency care coordination across institutions [3, 4, 18, 25]. For centers with mixed device inventories, regular team training on device-specific MRI protocols and emergency programming maneuvers reduces adverse events in radiology suites and perioperative settings [9, 10, 23].

4.2 Special Populations and Technical Nuances

Pediatric implantation requires careful pocket planning to accommodate growth and to minimize skin tension over relatively thin soft tissue envelopes. Catheter slack is left to avoid traction during growth spurts, and families are counseled about more frequent surveillance for skin breakdown and device prominence [13]. In patients with severe scoliosis or prior posterior instrumentation, fluoroscopic windows may be limited; in these scenarios, ultrasound-assisted level identification and angulated trajectories help achieve safe intrathecal access without traversing hardware [17, 18]. Obesity affects pocket depth and refill accuracy, reinforcing the value of deep fascial fixation to prevent rotation and of intra-clinic image-guided confirmation when septal palpation is challenging. Anticoagulation management follows neuraxial principles from pain and anesthesia guidelines, with individualized timing of interruption and resumption around dural puncture to mitigate hematoma risk while avoiding thrombotic complications [31–33]. Finally, centers should maintain contingency kits containing connectors, extension segments, and occlusion-resistant filters to manage unforeseen intraoperative findings without aborting the case.

Conflict of Interest

The authors have no conflicts of interest to declare that are relevant to the content of this chapter.

References

1. Sparlin JA, de Leon-Casasola OA (2011) Implantable intrathecal pump implantation techniques. Tech Reg Anesth Pain Manag 15(4):158–161
2. Shah N, Di Napoli R, Padalia D (2024) Implantable intrathecal drug delivery system. StatPearls [Internet]. StatPearls Publishing, Treasure Island, FL
3. Prager J, Deer T, Levy R et al (2014) Best practices for intrathecal drug delivery for pain. Neuromodulation 17(4):354–372
4. Deer TR, Pope JE, Hayek SM et al (2017) The polyanalgesic consensus conference (PACC): best practices and guidelines for intrathecal drug infusion systems. Neuromodulation 20(2):96–132
5. Coffey RJ, Owens ML, Broste SK et al (2009) Mortality associated with implantation and management of intrathecal opioid infusion systems for noncancer pain. Anesthesiology 111(4):881–891
6. Flückiger B, Knecht H, Grossmann S, Felleiter P (2008) Device-related complications of long-term intrathecal drug therapy via implanted pumps. Spinal Cord 46(9):639–643
7. Kratzsch T, Stienen MN et al (2015) Catheter-tip granulomas associated with intrathecal drug delivery – a two-center experience. Pain Physician 18(5):E831–E840.
8. Buddeberg BS, Bandschapp O, Girard T (2019) Post-dural puncture headache: risk factors, prevention and treatment. Minerva Anestesiol 85(5):543–553
9. Delhaas EM, Huygen FJ (2020) Complications associated with intrathecal drug delivery systems. BJA Educ 20(2):51–57
10. Frizon LA, Sabharwal NC, Maiti T et al (2018) Removal of intrathecal catheters used in drug delivery systems. Neuromodulation 21(7):665–668
11. Thompson SM et al (2014) Recommendations for trialing of intrathecal drug delivery infusion therapy in chronic pain. Pain Med 15(1):149–160
12. Hassenbusch SJ et al (1997) Clinical realities and economic considerations: economics of intrathecal therapy. J Pain Symptom Manage 14(3 Suppl):S36–S48.
13. Penn RD et al (1992) Intrathecal baclofen for severe spasticity of spinal origin: seven years of experience. J Neurosurg 77(2):236–240
14. Deer TR, Smith HS, Cousins M et al (2010) Consensus guidelines for the selection and implantation of patients with noncancer pain for intrathecal drug delivery. Pain Physician 13(3):E175–E213.
15. Sindt JE, Brogan SE (2016) Interventional treatments of cancer pain. Anesthesiol Clin 34(2):317–339
16. Takagi Y, Yamada H, Ebara H et al (2025) Worsening spasticity due to catheter breakage during intrathecal baclofen therapy: a case report. J Med Case Rep 19(1):22
17. Goel V, Yang Y, Kanwar S et al (2021) Adverse events and complications associated with intrathecal drug delivery systems: Insights from the MAUDE database. Neuromodulation 24(7):1181–1189
18. Prager J et al (2016) Clinical uses of intrathecal therapy and placement in pain care algorithms. Pain Pract 16(8):1092–1106
19. Deer TR, Pope JE, Hayek SM et al (2012) Polyanalgesic consensus conference 2012: recommendations for management of pain by intrathecal drug delivery. Neuromodulation 15(5):436–464
20. Hassenbusch SJ et al (1997) Clinical realities and economic considerations: economics of intrathecal therapy. J Pain Symptom Manage 14(3 Suppl):S36–S48.
21. Buddeberg BS, Bandschapp O, Girard T (2019) Post-dural puncture headache: risk factors, prevention and treatment. Minerva Anestesiol 85(5):543–553
22. Frizon LA, Sabharwal NC, Maiti T et al (2018) Removal of intrathecal catheters used in drug delivery systems. Neuromodulation 21(7):665–668
23. Goel V, Yang Y, Kanwar S et al (2021) Adverse events and complications associated with intrathecal drug delivery systems: insights from the MAUDE database. Neuromodulation 24(7):1181–1189
24. Kratzsch T, Stienen MN et al (2015) Catheter-tip granulomas associated with intrathecal drug delivery – a two-center experience. Pain Physician 18(5):E831–E840
25. Prager J, Deer T, Levy R et al (2014) Best practices for intrathecal drug delivery for pain. Neuromodulation 17(4):354–372
26. Shah N, Di Napoli R, Padalia D (2024) Implantable intrathecal drug delivery system. StatPearls [Internet]. StatPearls Publishing, Treasure Island, FL
27. Kratzsch T, Stienen MN et al (2015) Catheter-tip granulomas associated with intrathecal drug delivery – a two-center experience. Pain Physician 18(5):E831–E840.
28. Coffey RJ, Owens ML, Broste SK et al (2009) Mortality associated with implantation and management of intrathecal opioid infusion systems for noncancer pain. Anesthesiology 111(4):881–891

29. Flückiger B, Knecht H, Grossmann S, Felleiter P (2008) Device-related complications of long-term intrathecal drug therapy via implanted pumps. Spinal Cord 46(9):639–643
30. Deer TR, Pope JE, Hayek SM et al (2017) The polyanalgesic consensus conference (PACC): best practices and guidelines for intrathecal drug infusion systems. Neuromodulation 20(2):96–132
31. Delhaas EM, Huygen FJ (2020) Complications associated with intrathecal drug delivery systems. BJA Educ 20(2):51–57
32. Deer TR, Smith HS, Cousins M et al (2010) Consensus guidelines for the selection and implantation of patients with noncancer pain for intrathecal drug delivery. Pain Physician 13(3):E175–E213.
33. Buddeberg BS, Bandschapp O, Girard T (2019) Post-dural puncture headache: risk factors, prevention and treatment. Minerva Anestesiol 85(5):543–553
34. Goel V, Yang Y, Kanwar S et al (2021) Adverse events and complications associated with intrathecal drug delivery systems: insights from the MAUDE database. Neuromodulation 24(7):1181–1189
35. Thompson SM et al (2014) Recommendations for trialing of intrathecal drug delivery infusion therapy in chronic pain. Pain Med 15(1):149–160
36. Frizon LA, Sabharwal NC, Maiti T et al (2018) Removal of intrathecal catheters used in drug delivery systems. Neuromodulation 21(7):665–668

Chapter 5

Intrathecal Drug Delivery for Chronic Pain

Álvaro José Vázquez Marín, Laura Álvarez Rodríguez and Manuel Alejandro Sánchez-García

Abstract

Chronic pain is highly prevalent and disabling, demanding therapeutic strategies that deliver meaningful analgesia while minimizing systemic toxicity. This chapter reviews intrathecal drug delivery (IDD) as a mature form of chemical neuromodulation for refractory cancer and non-cancer pain, tracing its historical development from early spinal anesthesia and single-shot neuraxial opioids to contemporary fully implantable infusion systems. We outline the conceptual framework of neuromodulation, distinguish surgical from non-surgical approaches, and situate IDD within multidisciplinary pain pathways. Technical elements of modern systems are detailed, including intrathecal catheters, fixed-flow versus programmable pumps, and programming paradigms, with practical comparisons among commonly used devices. Indications, contraindications, and rigorous patient selection are synthesized into a structured pre-implant evaluation that integrates psychological assessment, trialing strategies, and shared decision-making, with separate pathways for cancer-related and chronic non-cancer pain. Pharmacology focuses on agents with the strongest intrathecal evidence—morphine and ziconotide—as well as adjuvants such as bupivacaine and clonidine, emphasizing pharmacokinetics, dosing ranges, stability, and combination rationales. Complications are reviewed across surgical, device-related, and pharmacologic domains, including CSF leakage, infection, catheter malfunction, granuloma formation, overdose, and withdrawal; diagnostic cues and stepwise management algorithms are provided. Finally, we consider future directions that converge on greater precision and safety: smarter pumps and remote monitoring, MRI-aware designs, human-factor improvements to refill and programming workflows, novel non-opioid intrathecal agents, and biomarker-guided personalization alongside synergistic combinations with electrical neuromodulation and rehabilitation. Taken together, intrathecal analgesia emerges as a durable, titratable, and cost-aware option for carefully selected patients, provided it is delivered within a disciplined perioperative framework and longitudinal stewardship that anticipates complications and adapts therapy to evolving clinical goals.

Keywords Intrathecal analgesia, Chronic pain, Neuromodulation, Opioids, Ziconotide, Implantable drug delivery systems

1 Introduction

Chronic pain is a severe, frequent, and current socio-health problem that leads to significant deterioration in the quality of life of patients [1]. It is among the leading causes of suffering and disability, with

Manuel Menéndez-González (ed.), *Intrathecal Drug Delivery*, Neuromethods, vol. 234,
https://doi.org/10.1007/978-1-0716-5241-1_5,
© The Author(s), under exclusive license to Springer Science+Business Media, LLC, part of Springer Nature 2026

an estimated prevalence of about 20% in Europe and 25%–40% in Spain. A major socio-political, medical, and technological effort is necessary to address this problem [2–6].

The International Association for the Study of Pain (IASP), in 2020, defines pain as "an unpleasant sensory and emotional experience associated with or similar to that associated with actual or potential tissue injury" [7].

Furthermore, six key considerations were added:

1. "Pain is always a personal experience that is influenced to varying degrees by biological, psychological, and social factors."
2. "Pain and nociception are different phenomena. Pain cannot be inferred solely from the activity in sensory neurons."
3. "Through their life experiences, individuals learn the concept of pain."
4. "A person's report of and experience as pain should be respected."
5. "Although pain usually serves an adaptive role, it may have adverse effects on function and social and psychological well-being."
6. "Verbal description is only one of several behaviors to express pain; inability to communicate does not negate the possibility that a human or a nonhuman animal experiences pain."

The World Health Organization (WHO) defines health-related quality of life as an individual's perception of their position in life, in the context of their culture and value system, and concerning their goals, expectations, standards, and concerns. This broad concept is influenced by the patient's physical and mental state, social relationships, and environment [8].

Intrathecal drug delivery (IDD) is part of the fourth analgesic step for managing these patients' pain and has been practiced in humans since 1979 [9]. In 1991, the US Food and Drug Administration (FDA) approved the administration of drugs, including morphine, baclofen, and ziconotide, via this route. This technique involves administering drugs into the subarachnoid space to modulate nociceptive information at this level, providing more selective and effective analgesia with lower doses and fewer side effects [10–14]. It is reserved for patients with refractory chronic cancer pain and chronic non-cancer pain (CNCP) who do not respond to other treatments or experience intolerable adverse effects with systemic opioids. Common indications for CNCP include persistent low back pain syndrome and complex regional pain syndrome [13–21].

Throughout the chapter, we will review the history of intrathecal analgesia, the concept of neuromodulation and infusion systems, the indications and selection of patients, and the drugs used, as well as the complications that may arise and their appropriate management. At the end of the chapter, possible future directions will be addressed.

2 History of Intrathecal Drug Delivery for Pain

The analgesic role of the intrathecal route has been developing for over 100 years, providing symptomatic relief in patients with refractory pain and spasticity. Considering its historical evolution, we can distinguish different stages [22, 23]:

- **First Stage: Intrathecal Route as a Therapeutic Target**

The spinal route is considered an alternative route for pain treatment, bypassing the blood–brain barrier and first-pass metabolism, acting directly on the central nervous system. In this phase, various drugs and different techniques will be tested. Initially, its purpose was anesthetic, but it eventually gained popularity for the treatment of chronic refractory pain.

Although the first reference to spinal anesthesia dates back to 1898 (when August Bier self-administered cocaine intrathecally), it wasn't until 1901 that we find the beginnings of the intrathecal route for pain treatment, when the Japanese Otojiro Kitagawa used morphine intrathecally for non-cancer pain control. It was also in 1901 that Sicord and Cathellin described epidural injection of local anesthetics through the sacral hiatus (although it wasn't until 1921 that Fidel Pagés, a Spaniard, published the use of the epidural route for surgical anesthesia). In 1931, the technique, described as the "pendant drop technique" by Digliotti, became widespread.

This single-dose approach was insufficient for certain anesthetic procedures and pain control, so systems began to be designed that would allow for intrathecal administration. This continuous access was first proposed in 1935 by neurosurgeon Grafton Love (Mayo Clinic), who had extensive experience using devices for the treatment of hydrocephalus.

As early as 1940, Dr. William Leonard described the technique of continuous spinal anesthesia, using a catheter that allowed procaine to be infused into the intrathecal space in boluses. Five years later (1945), gynecologist Samuel Manalan used continuous spinal anesthesia via the caudal route [24].

- **Second Stage: Beginnings of Intrathecal Delivery**

Once opioid receptors were discovered (Goldstein, 1970), the possibility of obtaining an analgesic response by administering low doses of morphine into the intrathecal space opened up. Contemporaneously with the latter, Fields and Basbaum described the descending inhibition system, allowing for a greater understanding of this pathophysiology of pain.

In 1979, Dr. Wang [9, 25] described the benefit obtained after administering a single dose of intrathecal morphine for the treatment of pain in a series of patients with pain refractory to systemic opioid treatment. This method began to gain popularity, and months later, Dr. Behar reported good results using epidural morphine.

- **Third Stage: Consolidation of Intrathecal Therapy**

Gradually, all kinds of improvements were incorporated, allowing for adjustments in dose and administration rate.

The moment that marked a true turning point in this case was the appearance of permanent intrathecal and/or epidural catheters, first connected to a subcutaneous reservoir and later to an implantable and programmable device, allowing the administration of a wide variety of drugs with different dosage possibilities. The first implantable delivery device appeared in 1969 (University of Minnesota). The first infusion device was the Infusaid device, and the first human implant was performed in a cancer patient in 1981 (Onofrio). It received FDA approval for cancer pain in 1988 and for chronic non-cancer pain in 1991.

3 Neuromodulation and Intrathecal Drug Delivery Systems

3.1 Concept

When we talk about neuromodulation, we refer to the treatment aimed at modifying the transmission of nerve signals in order to inhibit, modulate, or excite the activity of a neuronal group and thus achieve a therapeutic effect. Intrathecal infusion is a type of chemical surgical neuromodulation, which consists of administering drugs into the subarachnoid space using an implantable device to modulate nociceptive information at this level, providing more selective and effective analgesia with lower doses and fewer side effects.

3.2 Types of Neuromodulation [22, 26, 27]

There are different types of neuromodulation, which can be divided into two main groups: surgical and non-surgical neuromodulation.

Surgical (or Invasive) Neuromodulation

Techniques that use implants that achieve a neuromodulator effect through the use of chemicals and/or electricity. This group includes peripheral nerve stimulators, spinal cord stimulators, radiofrequency, and, of course, spinal drug infusion.

Non-Surgical (or Non-Invasive) Neuromodulation

This includes all techniques that achieve a neuromodulator effect without requiring intervention. Electrical and/or magnetic energy can be used to perform these techniques (TENS, transcranial stimulation, transcutaneous stimulation,...).

We can also differentiate the different types of neuromodulation according to where they exert their effect. Therefore, we would differentiate between:

- Transcutaneous nerve neuromodulation.
- Peripheral nerve neuromodulation.
- Spinal nerve neuromodulation.
- Spinal neuromodulation.

- Deep cerebral and cortical neuromodulation.
- Spinal drug infusion.

3.3 Implantable Devices [22, 27]

Thanks to advances in biotechnology, we have implantable devices capable of infusing pharmacological substances into the intrathecal space. In the early 1980s, the first implantable device appeared (Infusaid®, Infusaid, Inc., Nordwood, Mass., USA). A decade later, the first programmable systems appeared (Synchromed®, Medtronic, Minneapolis, MN).

Components of the Implantable System

These are complex devices made up of the following parts:

- **Intrathecal catheter:** polyurethane catheters (more durable and less kinking) or silicone.
- **Infusion system:** There are two types of infusion systems: external and fully implantable.
  a) **External:** There are two subgroups. Totally external, where the intrathecal catheter is inserted directly into the skin and the device is located externally. Partially external systems, where the intrathecal catheter is tunneled to an area where it ends in a subcutaneous reservoir, where the external device is connected.
  b) **Fully implantable:** The entire device is located inside the patient, with the infusion system usually located in the subcutaneous abdominal plane.
- **Other system elements:** programmers and patient controls.

Types of Programming

Regarding the type of programming, we differentiate between two main groups of systems (fixed flow and programmable).

- **Fixed flow:** This uses uninterrupted, fixed drug delivery. Refilling is based on the reservoir volume and the infusion rate. As this is a battery-free system, it does not require replacement due to battery depletion. In the case of fixed-flow devices with gas (Isomed®, Codman® 3000, Tricumed®), the gas is heated by body temperature, emitting a constant pressure that promotes drug release. Those without gas (AccuRx®) are more precise and less influenced by external factors.
- **Programmable:** Systems with a lithium battery that allow for modification of drug delivery. There are several models on the market: Synchromed II and III® (Medtronic), Prometra I and II® (Cardiva), and Medstream® (Codman).

Table 1 shows the differences between the different models of implantable devices.

Table 1
Intrathecal implantable system comparator

Model	Synchromed I®	Synchromed II®	Isomed®	Codman 3000®	AccuRx®
Material	Titanium Silicone	Titanium Silicone	Titanium Silicone	Titanium Silicone	Silicone elastomer
Weight (g)	20 mL = 205 g	20 mL= 165 g	20 mL= 113 g 35 mL = 116 g 60 mL = 120 g	16 mL = 98 g 30 mL= 137 g 50 mL = 173 g	95 g
Reservoir volume (mL)	10, 18	20, 40	20, 35, 60	16, 30, 50	28
Height (mm)	10 mL=21.6 mm	20 mL= 19.5 mm 18 mL=27.5 mm	20 mL = 24 mm 40 mL = 26 mm	16 mL= 32 mm 30 mL = 29 mm 50 mL = 37 mm	30 mL = 34 mm 50 mL= 37.4 mm
Flow rate (mL/day)	Programmable	Programmable	0.5 1 1.5	0.5; 0.85; 0.86; 1.45; 1.46; 2.0	0.4; 0.5; 0.6; 0.7; 0.8; 0.9; 1.0; 1.2; 1.3; 1.5; 2.0; 2.6; 3.2
Propulsion	Lithium battery	Lithium battery	Gas	Gas	Elastomer
Diameter (mm)	85.2	87.5	87	16 mL = 61 30 mL = 78 50 mL = 87	77
Infusion options	• Single bolus • Multiple boluses • Continuous flow • Continuous flow + Boluses	• Single bolus • Multiple boluses • Continuous flow • Continuous flow + Boluses	Fixed flow	Fixed flow	Fixed flow

Adapted and modified from Cerdá-Olmedo et al. [27]

4 Indications, Contraindications, and Patient Selection [22]

4.1 Indications

We will discuss the main indications for the use of intrathecal therapy in patients with chronic pain. In this type of patient, before deciding on intrathecal therapy, the type of pain, location, life expectancy, age, and so on will be important. We must also assess previous illnesses and medications before deciding whether or not to administer intrathecal therapy. What does seem to be an essential requirement is the presence of pain refractory to pharmacological and/or interventional treatment, with moderate to severe intensity, leading to a worsening of the patient's quality of life.

The 2017 PACC (Polyanalgesic Consensus Conference) guidelines [28] established a list of pathologies that may be susceptible to treatment with intrathecal therapy, which are shown in Table 2. In the new revision of these guidelines in 2024 [14], the indications are not mentioned again, assuming that it is useful in the same scenarios.

There are two different algorithms when assessing these patients and that can guide us towards choosing or not intrathecal analgesic therapy to assess these patients [28]: Non-oncological or non-terminal pain (Fig. 1) and Oncological pain or terminal patient (Fig. 2).

Table 2
Indications for intrathecal analgesic therapy

Axial neck or back pain. Not a candidate for surgery
• **Multiple fractures** • **Discogenic pain** • **Spinal stenosis** • **Multilevel spondylosis**
Persistent low back pain syndrome
Abdominal/pelvic pain
• Visceral pain • Somatic pain
Extremity pain
• Joint pain • Radicular pain
Complex regional pain syndrome
Trunk pain
• Postherpetic neuralgia • Post-thoracotomy pain
Cancer pain, due to direct invasion or chemotherapy
Analgesic efficacy with systemic opioids is hampered by intolerable side effects

Adapted from Deer et al. [14, 28]

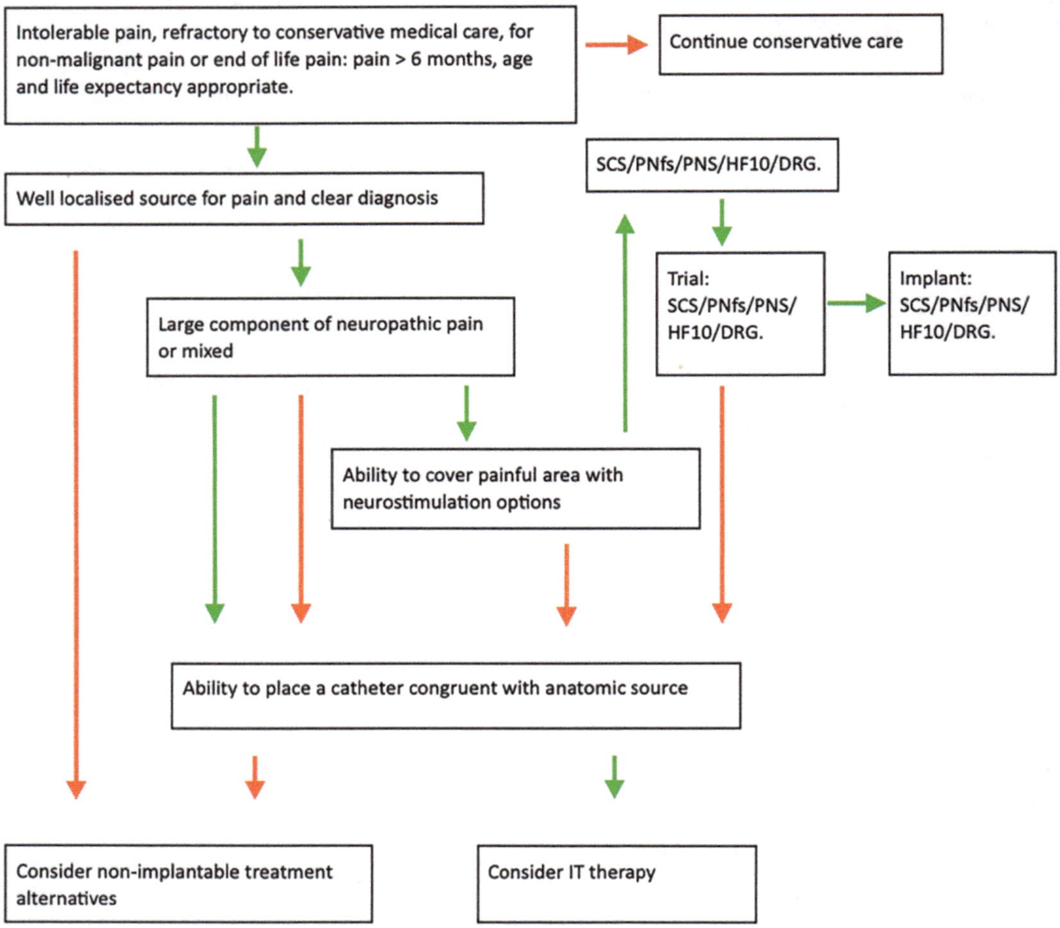

Fig. 1 Algorithm for placement within the pain care algorithm for noncancer or non-end-of-life pain. DRG dorsal root ganglion, HF10 high frequency stimulation, PNfS peripheral nerve field stimulation, PNS peripheral nerve stimulation, SCS spinal cord stimulation. Green arrows indicate affirmation or positive response; red arrows signify negative response. Adapted from Deer et al. [28]

4.2 Contraindications

General contraindications include the following:

- Severe coagulation disorders.
- Allergy to the materials and drugs used.
- Pain with a high psychogenic component.
- Patient refusal and/or unwillingness to rely on a device.
- Psychological pathology that prevents patient supervision and/or refilling.
- Active infection near the implant site or spinal cord.
- Anatomical difficulty that prevents spinal cord access.
- Extremely thin.

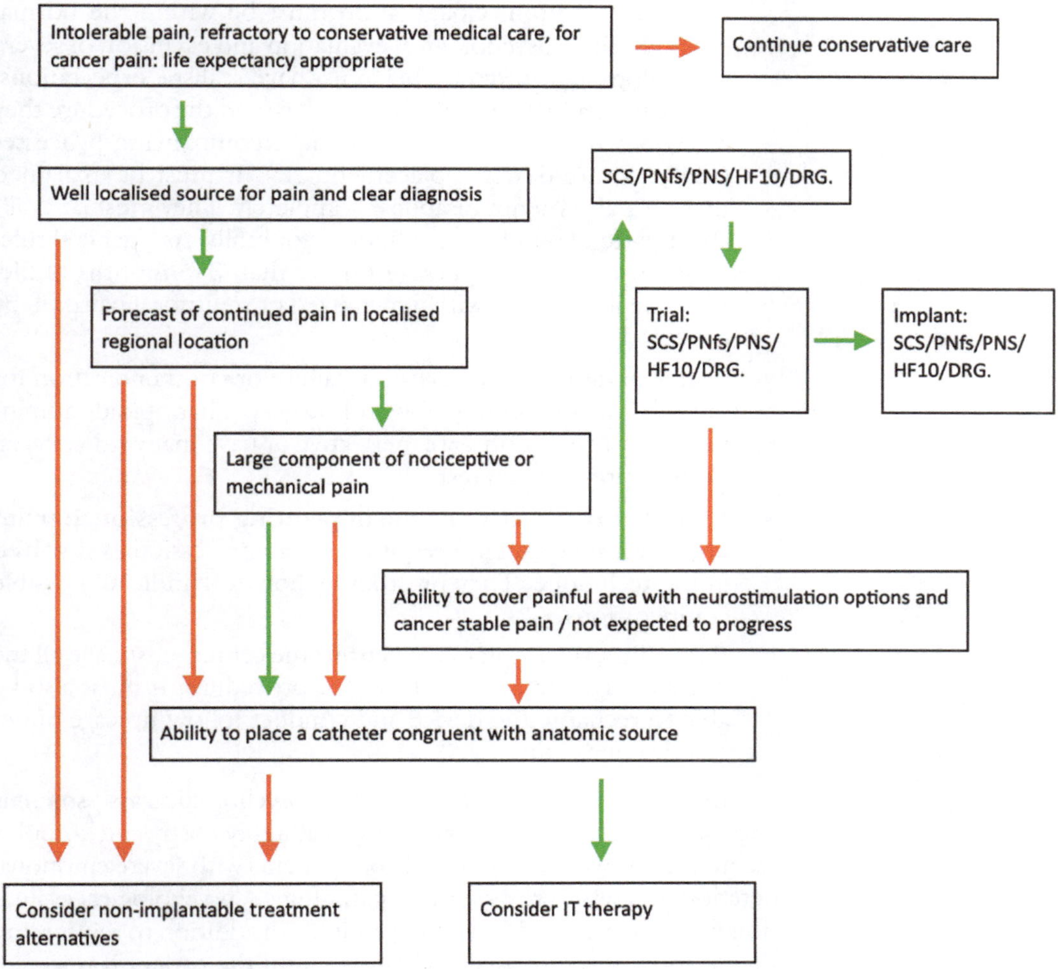

Fig. 2 Pain care algorithm for cancer-related pain. DRG dorsal root ganglion, HF10 high frequency stimulation, PNfS peripheral nerve field stimulation, PNS peripheral nerve stimulation, SCS spinal cord stimulation. Green arrows indicate affirmation or positive response; red arrows signify negative response. Adapted from Deer et al. [28]

4.3 Patient Selection [28]

It is essential to standardize the selection criteria for patients undergoing spinal cord infusion. They should meet a series of criteria to promote the success of the therapy. The involvement of a multidisciplinary team in patient selection and selection is important. Any condition requiring specific treatment must be identified in advance. It is also necessary to confirm this refractoriness to the treatment that has been established (both pharmacological and interventional). We can establish the following general criteria:

- **From the patient's perspective:** they must accept the implant, with prior information and the corresponding signing of the informed consent form. They must maintain their physical condition. Coagulation must be within the normal range. Positive psychological evaluation and exclusion of severe psychological disorders. They must have realistic expectations. The patient and caregivers must understand the procedure they are about to undergo, as well as all accompanying processes (refills, future device replacement, ...). It must be explained that the pain will not disappear completely. They must be available to attend regular consultations for refills. As a general rule, they must have a prognosis of more than 3–6 months of life. And, finally, patients with generalized or multifocal pain will be preferred.

- **Other patient-related criteria:** failure or contraindication for both spinal cord stimulation and surgery. That opioids administered by other routes are ineffective or have many side effects. Positive pre-infusion test.
- **From the perspective of the implanting professional:** training and instruction are required for all professionals involved in the technique. They must know how to handle all possible complications.
- **Regarding the implanting center:** the center must have all the necessary resources to perform the procedure. It must also be able to recharge the device and conduct follow-up care under aseptic conditions.

In addition to the previous criteria, a psychological assessment is necessary, since, let's not forget, this is a highly costly and complex technology. It would serve to rule out patients with severe emotional deterioration or severe psychiatric pathology. A psychological evaluation from a specialized service is required. In addition to reinforcing the decision, it serves as a legal document if the patient is rejected. At this point, the collaboration of another patient in the same situation who has gone through this process will be very useful. And of course, after all this, the final decision must be accepted by the implant committee of each center.

5 Drugs Used

To date, the only FDA-approved drugs are morphine and ziconotide for pain treatment (and baclofen for spasticity). Despite these, the use of other drugs is accepted and promoted even though they are off label, as is the case in the PACC guidelines [14, 28].

- **Morphine:** Like morphine, other opioids act by the same mechanism at the CNS level when administered intrathecally. Although morphine is the only one approved by the FDA, many of them could be used, given their pharmacokinetic characteristics (Table 3).

Table 3
PACC recommendations on the most commonly used intrathecal drugs

Drug	Maximum concentration	Maximum daily dose	Initial dose
Hydromorphone	15 mg/mL (10 mg)	20 mg/mL	Moderate
Morphine	20 mg/mL (15 mg)	2 mg/mL	Moderate
Fentanyl	10 mg/mL	1000 mcg	100 mcg/mL
Sufentanilo	5 mg/mL	500 mcg	2 mg/mL
Bupivacaine	30 mg/mL	20 mg	40 mg/mL
Clonidine	1000 mcg/mL	600 mcg	
Ziconotide	100 mcg/mL	19.2 mcg	

Adapted from González Mesa et al. [22]

Table 4
Differences between Morphine and Ziconotide in intrathecal administration

Morphine	Ziconotide
• Derived from the *Papaver somniferum* plant	• Sea snail conotoxin (*Conus magus*)
• Phenanthrene alkaloid.	• Peptide.
• Hydrophilic substance	• Hydrophilic substance
• Plasma protein binding: 40%–50%	• Plasma protein binding: 53%
• CSF half-life: 2.9–6.5 h	• CSF half-life: 6–12 h.
• Metabolism: hepatic 90%.	• Metabolism: 99% by blood proteases
• Elimination: renal 90%	• Elimination: <1% by kidney and 99% metabolized
• Available in 10, 20, and 40 mg/mL doses for spinal infusion	• Single dosage form: 100 mcg/mL
• Stability in DIT for 90 days as monotherapy	• Stability in IDD for 60 days as monotherapy

CSF cerebrospinal fluid, *IDD* intrathecal device. Adapted from González Mesa et al. [22]

- **Bupivacaine:** An amide-type local anesthetic that acts by blocking voltage-gated sodium channels, preventing neuronal transmission. Its differential blockade makes it a particularly useful local anesthetic. It is useful when combined with morphine for intrathecal administration.
- **Clonidine:** an alpha-2 adrenergic agonist that inhibits nociceptive nerve impulses in the spinal cord. Although not recommended as a single agent, it is considered in combination with morphine.
- **Ziconotide:** a potent analgesic conopeptide isolated from the venom of the Conus magus snail. It only has an analgesic effect when administered intrathecally. Its analgesic effect is achieved

by blocking N-type sensitive calcium channels, which are located in the A delta and C fibers of the dorsal horn of the spinal cord, inhibiting the release of genes related to calcitonin, glutamate, and substance P.

The differences between morphine and ziconotide are shown in Table 4.

6 Complications

6.1 Most Common Complications [14, 22, 27–32]

Intrathecal analgesia devices are not exempt from complications, both due to the surgical procedure and the drugs administered. The complication rate is estimated at 10%–20% each year. We will differentiate these complications into different sections:

6.2 Surgical Complications

- **Hemorrhagic:** These complications usually occur in the immediate postoperative period, especially in the pocket area. We must not forget that manipulation of the intrathecal space can produce a hematoma at this level (epidural hematoma). If symptoms suggest doubt about this diagnosis, an imaging test (CT) will be necessary. If the diagnosis is confirmed, evacuation should be performed.
- **Neurological:** Suspected if pain and unilateral functional impairment are present in the immediate postoperative period. These complications are due to injuries caused by the needle and/or catheter to the spinal cord or a nerve root. If there is suspicion, imaging should be performed to confirm or rule out injury.
- **Seroma**: accumulation of serous fluid under the wound, which prevents healing. It is advisable to drain these under sterile conditions. They are not accompanied by fever or infectious symptoms.
- **Hygroma:** accumulation of CSF (especially in the dorsal area) after leakage from the subarachnoid space through dural puncture, which usually resolves within one to two weeks. This CSF leakage can even reach the pocket area (CSF leak). CSF leakage may or may not be accompanied by post-dural puncture headache, which is initially treated conservatively.
- **Decubitus:** more common in patients with little abdominal fat. Occasionally, the system becomes exposed (requiring its removal).
- **Infection:** This is the most common serious complication. It is more common in the pocket area. If signs of redness and swelling appear within the first week, an inflammatory cause is most likely. However, if they appear between 3 and 5 weeks, it is usually due to an infectious cause due to suture dehiscence. These

must be treated vigorously since contiguity can lead to meningitis. Although the role of intraoperative antibiotics is clear, there is no consensus regarding their continued use postoperatively. Infectious complications secondary to device refilling may also occur.

6.3 Device-Related Complications

- **Catheter problems:**
  - **Catheter migration:** the most common complication. It is confirmed with imaging. If loss of efficacy or withdrawal syndrome occurs, consider this possibility. If it migrates toward the root, it can trigger radicular pain. Treatment consists of the removal and replacement of a new catheter.
  - **Catheter rupture (perforation, breakage, or shear):** occurs either during catheter placement (secondary to manipulation over the needle) or during refills.
  - **Granuloma:** an inflammatory mass, without infection. It appears with long-term morphine infusions (months or years), at the catheter tip.
- **Infusion device problems:**
  - **Rotated device:** occurs mainly due to changes in the patient's weight. If it rotates easily and does not hinder refilling, it can be left as is. If it hinders refilling, replacement should be considered.
  - **Mechanical failures:** primarily due to the device's battery running out. There may also be a failure in one of the components, leading to a malfunction.
- **Human error during refills and programming.**
  - **Error in entering initial data:** catheter length, volume, drugs…
  - **Refilling outside the system:** a very serious situation, causing an overdose in the patient. The use of ultrasound has minimized this risk, but what really reduces the incidence is training and instruction in refilling. If this happens, monitor and supervise the patient with support measures based on the clinical presentation.
  - **Error in medication preparation.**
  - **Error in reprogramming after refilling.**

6.4 Discrepancies with Reservoir Volume

- **Higher volume than expected:** secondary to not infusing the indicated amount. Possible causes: system failure, system shutdown, battery exhaustion, occlusion, programming error, volume remaining after previous refill…
- **Lower volume than expected:** the main causes of this would be: overinfusion secondary to device failure, filling outside the device (at least partially), programming error, …

6.5 Drugs Complications

- **Overdose:** Symptoms and consequences will vary depending on the drug.
    - **Opioids:** Symptoms of opioid overdose may occur (respiratory depression, itching, nausea, seizures, drowsiness…). An antagonist drug (naloxone, naltrexone…) should be added, along with monitoring, supplemental oxygen, and any necessary support measures.
    - **Baclofen:** Hypotension, nausea, respiratory depression, hypotonia, hypothermia, seizures, drowsiness… may occur. As these are severe symptoms, close monitoring and necessary support measures are required. There is no specific reversal agent.
    - **Clonidine:** confusion, sweating, respiratory depression, loss of consciousness, severe high blood pressure…
    - **Ziconotide:** These are mild symptoms that do not require monitoring. Symptoms include dizziness, confusion, ataxia, asthenia, diarrhea, vomiting…
- **Abrupt discontinuation:** As with overdose, this will depend on the drug used.
    - **Opioid:** Classic withdrawal symptoms appear (sweating, nervousness, tachycardia, diarrhea…). Consider reintroducing oral or systemic opioids if necessary or treat with anxiolytics.
    - **Baclofen:** Neuronal overexcitation occurs, which in the most severe cases can lead to malignant hyperthermia (treated with dantrolene). Typical symptoms include changes in mental status, hallucinations, confusion, fever… If withdrawal symptoms occur, administer baclofen and anxiolytics.
    - **Clonidine:** Increased catecholaminergic activity can precipitate acute coronary syndrome or refractory hypertension.

7 Future Lines in the Treatment of Chronic Pain with Intrathecal Drug Infusion

Looking to the future, research and development in this field appear focused on improving efficacy, reducing side effects, and further personalizing treatments for each patient.

One of the most relevant trends is innovation in the drugs used. Currently, opioids and other analgesics, such as ziconotide, are primarily used, but research is exploring new molecules and combinations that may offer longer-lasting relief and lower risks of dependence. For example, ion channel modulators, specific receptor antagonists, and biologic drugs that can more precisely target pain pathways, minimizing adverse effects, are being studied [33].

Another important area of focus is delivery technology and devices. Advances in smaller infusion devices, with greater precision and personalized programming capabilities, will allow doses to be adjusted in real time according to patient needs. Furthermore, the integration of remote monitoring systems will facilitate closer monitoring and early intervention in the event of complications, improving the safety and convenience of treatment [34].

The use of combination therapies is also emerging as a future strategy. Intrathecal infusion could be complemented with other modalities, such as physical therapy, nerve stimulation, or even systemic pharmacological therapies, to address pain from different angles and enhance relief.

Furthermore, treatment personalization will be key. Identifying biomarkers that predict response to certain drugs will allow for the design of more effective and safer therapeutic plans. Genetics, clinical history, and individual patient characteristics will be key in determining which drug, dose, and infusion modality are most appropriate.

Finally, neuroscience research is opening new perspectives on how the brain and nervous system process pain. A better understanding of these mechanisms could lead to intrathecal therapies that not only relieve pain but also modulate the neural pathways involved, achieving more lasting and profound control.

In summary, future directions for chronic pain treatment with intrathecal infusion appear to be moving toward greater precision, pharmacological innovation, advanced technology, and treatment personalization. All this with the aim of improving the quality of life of those who suffer from chronic pain, offering them safer, more effective solutions adapted to their individual needs.

Conflict of Interest

The authors have no conflicts of interest to declare that are relevant to the content of this chapter.

References

1. Sánchez-García MA, Alcázar-Navarrete B, Cortiñas-Saenz M, Cordero Tous N, Gálvez Mateos R (2024) Multidimensional analysis of quality of life in patients with chronic non-cancer pain and short- and long-term intrathecal analgesic therapy. Healthcare 12(18):1870
2. García C VF, Gomez S, Gonzalez M, Gallo FJ, Hermosa JC, Vargas F et al (2015) La atención al paciente con dolor crónico no oncoló gico (DCNO) en atención primaria (AP). Documento de consenso. [Internet]. [cited 2024 May 8]. Available from: https://www.semfyc.es/wp-content/uploads/2016/06/DOCUMENTO-CONSENSO-DOLOR-17-04-A.pdf
3. Torralba A, Miquel A, Darba J (2014) Situación actual del dolor crónico en España: iniciativa "Pain Proposal.". Rev Soc Esp Dolor 21(1):16–22
4. Hayek SM, Deer TR, Pope JE, Panchal SJ, Patel V (2011) Systematic review intrathecal therapy for cancer and non-cancer pain. Pain Physician [Internet] 14:219–248. Available from: www.painphysicianjournal.com

5. Dueñas M, Salazar A, Ojeda B, Fernández-Palacín F, Micó JA, Torres LM et al (2015) A nationwide study of chronic pain prevalence in the general spanish population: identifying clinical subgroups through cluster analysis. Pain Med [Internet] 16:811–822. Available from: https://academic.oup.com/painmedicine/article/16/4/811/2460755
6. Sánchez-García M, Sánchez Guerrero J, Jiménez Rodríguez I (2016) Manual Práctico del manejo del dolor en atención primaria. GAMI editorial, editor. pp11–13
7. Vidal Fuentes J (2020) Versión actualizada de la definición de dolor de la IASP: un paso adelante o un paso atrás. Rev Soc Esp Dolor [Internet] 27(4):232–233. Available from: http://links.lww.com/PAIN/B101
8. Lemus N, Parrado R, Quintana G (2014) Calidad de vida en el sistema de salud. Rev Colomb Reumatol [Internet] 21(1):1–3. Available from: www.elsevier.es/rcreuma
9. Wang JK, Nauss LA, Thomas JE (1979) Pain relief by intratecally applied morphine in man. Anesthesiology 50:149–151
10. Bolash R, Mekhail N (2014) Intrathecal pain pumps: indications, patient selection, techniques, and outcomes. Neurosurg Clin N Am 25(4):735–742
11. Yaksh TL, Santos GG, dos Borges Paes Lemes J, Malange K (2023) Neuraxial drug delivery in pain management: an overview of past, present, and future. Best Pract Res Clin Anaesthesiol 37(2):243–265
12. Rubiu E, Restelli F, Nazzi V, Mazzapicchi E, Bonomo G, Veiceschi P et al (2024) A benefit/risk assessment of intrathecal ziconotide in chronic pain: a narrative review. J Clin Med. 13(6):1644. https://doi.org/10.3390/jcm13061644
13. De Andrés J, Rubio-Haro R, De Andres-Serrano C, Asensio-Samper JM, Fabregat-Cid G. Intrathecal Drug Delivery. Methods Mol Biol. 2020;2059:75–108. https://doi.org/10.1007/978-1-4939-9798-5_3. PMID: 31435916
14. Deer TR, Hayek SM, Grider JS, Hagedorn JM, McDowell GC, Kim P et al (2024) The polyanalgesic consensus conference (PACC)®: intrathecal drug delivery guidance on safety and therapy optimization when treating chronic noncancer pain. Neuromodulation [Internet] 27:1107–1139. https://doi.org/10.1016/j.neurom.2024.03.003. Available from: https://linkinghub.elsevier.com/retrieve/pii/S1094715924000667
15. Abd-Elsayed A, Karri J, Michael A, Bryce D, Sun J, Lee M et al (2020) Intrathecal drug delivery for chronic pain syndromes: a review of considerations in practice management. Pain Physician [Internet] 23:591–617. Available from: www.painphysicianjournal.com
16. Jain S, Malinowski M, Chopra P, Varshney V, Deer TR. Intrathecal drug delivery for pain management: recent advances and future developments. Expert Opin Drug Deliv. 2019 Aug;16(8):815–822. https://doi.org/10.1080/17425247.2019.1642870. Epub 2019 Jul 19. PMID: 31305165
17. Ontario Health (2024) intrathecal drug delivery systems for cancer pain: a health technology assessment. Ont Health Technol Assess Ser 24(2):1–162.
18. Sukul VV. Intrathecal Pain Therapy for the Management of Chronic Noncancer Pain. Neurosurg Clin N Am. 2019 Apr;30(2):195-201. doi: 10.1016/j.nec.2018.12.010. PMID: 30898270 (Update this ref..)
19. Capozza MA, Triarico S, Mastrangelo S, Attinà G, Maurizi P, Ruggiero A (2021) Narrative review of intrathecal drug delivery (IDD): indications, devices and potential complications. Ann Transl Med 9(2):186–186
20. Van Zundert J, Rauck R (2023) Intrathecal drug delivery in the management of chronic pain. Best Pract Res Clin Anaesthesiol [Internet] 37(2):157–169. Available from: https://pubmed.ncbi.nlm.nih.gov/37321764/
21. Staudt MD, Patel S, Hellman A, Platanitis K, DiMarzio M, Khazen O et al (2020) Efficacy of simultaneous usage of spinal cord stimulation and intrathecal therapy for nonmalignant chronic neuropathic pain. World Neurosurg 143:e442–e449
22. González Mesa JM, Gálvez Mateos R, Iglesias Rozas P, Lasuen Aguirre D, López Alarcón MD, Lechuga Carrasco B et al (2012) Guía práctica sobre el manejo de la infusión espinal. Sociedad Española Multidisciplinar del Dolor (SEMDOR), Madrid/Barcelona. www.cegeglobal.com
23. Rizvi S, Kumar K (2015) History and present state of targeted intrathecal drug delivery. Curr Pain Headache Rep 19(2):1
24. Downing GC, Miller M, Durfee RB (1945) The catheter method for continuous caudal anesthesia. Am J Obstet Gynecol 49(3):391–395
25. Onofrio BM, Yaksh TL, Arnold PG (1981) Continuous low-dose intrathecal morphine administration in the treatment of chronic pain of malignant origin. Mayo Clin Proc 56(8):516–520
26. Padrón FJR (2008) Neuromodulación quirúrgica. Nuevos horizontes en Neurocirugía. Neurocirugia 19:143–155
27. Cerdá Olmedo G, Franco Gay ML, Insausti J, et al. Spanish guideline on spinal infusion [Internet]. Madrid: Spanish Group of Neuromodulation (ESRA Spain); n.d. [cited 2026 Apr 26]. Available from: https://www.

esra-spain.org/web/fuentes/guiainfusionespinal.pdf
28. Deer TR, Pope JE, Hayek SM, Bux A, Buchser E, Eldabe S et al (2017) The polyanalgesic consensus conference (PACC): recommendations on intrathecal drug infusion systems best practices and guidelines. Neuromodulation 20(2):96–132
29. Deer TR, Levy R, Prager J, Buchser E, Burton A, Caraway D et al (2012) Polyanalgesic consensus conference-2012: Recommendations to reduce morbidity and mortality in intrathecal drug delivery in the treatment of chronic pain. Neuromodulation 15(5):467–482
30. Deer TR, Prager J, Levy R, Rathmell J, Buchser E, Burton A et al (2012) Polyanalgesic consensus conference-2012: Consensus on diagnosis, detection, and treatment of catheter-tip granulomas (inflammatory masses). Neuromodulation 15(5):483–496
31. Deer TR, Pope JE, Hayek SM, Lamer TJ, Veizi IE, Erdek M et al (2017) The polyanalgesic consensus conference (PACC): recommendations for intrathecal drug delivery: guidance for improving safety and mitigating risks. Neuromodulation 20(2):155–176
32. Shah N, Di Napoli R, Esposito G, Padalia D (2024) Implantable intrathecal drug delivery system. StatPearls Publishing, Treasure Island, FL. Available from: https://www.ncbi.nlm.nih.gov/books/NBK538237/?report=printable
33. Jain S, Malinowski M, Chopra P, Varshney V, Deer TR (2019) Intrathecal drug delivery for pain management: recent advances and future developments. Expert Opin Drug Deliv 16(8):815–822
34. De Andres J, Hayek S, Perruchoud C, Lawrence MM, Reina MA, De Andres-Serrano C et al (2022) intrathecal drug delivery: advances and applications in the management of chronic pain patient. Front Pain Res 3:900566. https://doi.org/10.3389/fpain.2022.900566

Chapter 6

Spasticity and Intrathecal Baclofen Therapy

Manuel Menéndez-González

Abstract

Spasticity is a cardinal manifestation of upper motor neuron (UMN) syndromes and arises from an imbalance between excitatory and inhibitory influences on spinal segmental circuitry. When generalized spasticity remains function-limiting despite optimized conservative care and oral antispasmodics, continuous intrathecal baclofen (ITB) infusion via an implantable pump offers potent, titratable control of tonic hyperactivity with a comparatively favorable systemic side-effect profile. This chapter reviews the neurophysiology of spasticity, baclofen pharmacology, historical development and evidence base of ITB, patient selection and trialing, surgical techniques, dosing paradigms and long-term outcomes, complications and their management, comparisons with other interventions, and future directions in device engineering and neuropharmacology.

Keywords Spasticity, Intrathecal baclofen, Baclofen pump, Spinal cord disorders, Tone management, ITB therapy

1 Introduction

1.1 Definitions and Clinical Phenomenology

Spasticity is best understood as a component of the UMN syndrome characterized by a velocity-dependent increase in tonic stretch reflexes and hyperactive tendon jerks, with coexisting phenomena such as clonus and flexor/extensor spasms. Contemporary physiological syntheses emphasize disinhibition of spinal reflex loops, altered afferent gain, and abnormal muscle mechanics contributing to the clinical picture, rather than a unitary increase in "tone" [1, 2].

1.2 Epidemiology and Burden

Spasticity is common after stroke, spinal cord injury (SCI), traumatic brain injury, multiple sclerosis (MS), and cerebral palsy (CP). Population-based data indicate substantial prevalence across etiologies (e.g., ~40% after severe stroke or traumatic brain injury; high chronic prevalence after SCI) and consistent associations with pain, sleep disturbance, and activity limitations [3, 4]. These sequelae increase

Manuel Menéndez-González (ed.), *Intrathecal Drug Delivery*, Neuromethods, vol. 234,
https://doi.org/10.1007/978-1-0716-5241-1_6,
© The Author(s), under exclusive license to Springer Science+Business Media, LLC, part of Springer Nature 2026

caregiver burden, complicate hygiene and positioning, and accelerate secondary harms such as contractures and skin breakdown [3, 4].

1.3 Rationale for Intrathecal Therapy

Oral agents such as baclofen, tizanidine, benzodiazepines, and dantrolene are foundational but limited by sedation, hypotension, hepatotoxicity, or insufficient efficacy at tolerable doses. Delivering baclofen into cerebrospinal fluid (CSF) via an implanted pump allows microgram-scale dosing at the drug's spinal sites of action with plasma levels orders of magnitude below those reached by oral administration, thereby maximizing efficacy while minimizing central sedation and systemic effects [5, 6].

2 Baclofen: Mechanism of Action

2.1 Receptor Pharmacology and Network Effects

Baclofen is a selective GABA_B receptor agonist. GABA_B receptors are metabotropic and widely distributed presynaptically and postsynaptically in spinal and supraspinal circuits. Their activation hyperpolarizes neurons via inwardly rectifying potassium conductances and suppresses transmitter release by inhibiting voltage-gated Ca^{2+} entry [7, 8]. At the network level, activating GABA_B receptors restores inhibitory tone within segmental interneuronal networks and α-motoneurons, thereby reducing reflex hyperexcitability and spasm generation [7–9].

2.2 Spinal Substrates of Antispastic Action

In intrathecal therapy, the pharmacodynamic emphasis is spinal: baclofen attenuates afferent-mediated polysynaptic excitation in dorsal horn circuits and raises motoneuronal firing thresholds in the ventral horn. Persistent inward currents in motoneurons, a substrate for plateau potentials and self-sustaining firing, are dampened indirectly by reducing synaptic drive and hyperpolarizing the membrane, which helps explain reductions in clonus and spasm frequency without complete abolition of useful tone [9].

3 Historical Development of Intrathecal Baclofen

3.1 From Concept to First-in-Human

After preclinical work on spinal receptor pharmacology and neuraxial drug delivery, Penn and Kroin reported the first continuous IT baclofen infusions in humans in 1985, demonstrating dramatic reductions in severe spasticity with microgram daily doses [10]. A randomized, double-blind, placebo-controlled crossover study in 1989 confirmed robust efficacy in severe spinal-origin spasticity [11].

3.2 Clinical Diffusion and Pediatric Adoption

Early 1990s multicenter series established safety and sustained benefit, while experience broadened to MS, brain injury, and pediatric CP [12, 13]. Pediatric programs reported improvements in ease of care and comfort, laying the groundwork for modern multidisciplinary pathways for trialing and implantation [12, 13].

4 Patient Selection and Trial Algorithm

4.1 Indications, Goals, and Prerequisites

Candidates typically have severe, function-limiting, diffuse spasticity refractory to optimized conservative care and maximally tolerated oral medications, with clearly articulated goals such as pain relief, facilitation of care, or enabling rehabilitation. Randomized and prospective pediatric data support ITB for intractable spastic CP, complementing clinical criteria used in adults with SCI, MS, or mixed etiologies [14, 15].

4.2 Screening Trial

A single-dose intrathecal bolus (commonly 50 μg; repeated at 75–100 μg if equivocal) is administered under monitoring. Clinically meaningful response thresholds include ≥1-grade reduction on the Modified Ashworth Scale and significant decreases in spasm frequency, without dose-limiting adverse effects. Positive trials predict benefit from pump implantation; lack of response prompts reassessment for fixed contracture or non-neural contributors.

4.3 Multidisciplinary Evaluation and Expectations

Evaluation spans neurological stability, CSF pathway patency, anesthesia risk, and psychosocial capacity for long-term pump management. Realistic expectations are essential: ITB reduces hypertonia and spasms; it does not restore corticospinal strength. Nonetheless, by suppressing cocontraction and spasm-triggered pain, ITB can reveal latent voluntary control and increase therapy participation [3, 4, 14, 15].

5 Surgical Technique

5.1 Catheter Placement and Targeting

Under fluoroscopic guidance, a flexible catheter is inserted into the lumbar subarachnoid space and advanced rostrally. Thoracic tips suffice for predominantly lower-extremity spasticity; cervical positioning may be selected for quadriplegic patterns to enrich cervical exposure. Distribution models and in vitro/ex vivo experiments demonstrate that CSF pulsatility, catheter geometry, and injection/infusion parameters materially shape segmental drug concentrations, supporting individualized targeting [16–20].

5.2 Pump Pocket, Connection, and Programming

A subcutaneous abdominal pocket houses a programmable pump coupled to the catheter. Reservoir concentration is selected to balance daily dose and refill interval. Initial continuous rates are commonly derived from the effective bolus dose scaled over 24 h, then iteratively titrated. Flex dosing with daytime up-titration and nocturnal down-titration is used when clinically helpful.

5.3 Immediate Postoperative Management

Flat bed rest reduces low-pressure headache risk, and early surveillance focuses on wound integrity and signs of overdelivery or underdelivery. Education covers alarm responses, refill schedules, and emergency procedures for suspected withdrawal.

6 Dosing, Titration, and Outcomes

6.1 Dose Ranges and Pharmacokinetics

Adult maintenance requirements commonly fall in the low hundreds of micrograms per day, with higher averages in cerebral-origin spasticity than in spinal etiologies. A population PK/PD model integrating CSF and plasma data confirms the steep CSF: plasma exposure gradient with intrathecal delivery and supports conservative titration to effect with 10%–20% adjustments at intervals dictated by clinical response [13]. Longitudinal cohorts show modest annual dose escalations related to growth, disease evolution, or tolerance [21].

6.2 Objective Antispastic Efficacy

Across etiologies, ITB reduces Ashworth grades by approximately one to two points and markedly decreases spasm frequency. In stroke, reductions in spastic hypertonia translate into more efficient gait and decreased energy cost in ambulant subsets [22]. In pediatric CP, controlled and prospective series report durable improvements in tone and dystonia, with knock-on benefits for positioning and therapy tolerance [14, 23, 24].

6.3 Function, Comfort, and Quality of Life

Quality-of-life metrics such as CPCHILD and KINDL improve after pump implantation in nonambulant children, with sustained benefits at multiyear follow-up in domains of comfort, ease of care, and caregiver-reported burden [23, 24]. These gains are paralleled by reductions in pain and nocturnal spasms and by facilitation of daily caregiving tasks such as dressing and hygiene.

7 Complications

7.1 Pharmacological Adverse Effects

Sedation, dizziness, hypotonia, and urinary retention are dose-related and typically reversible with small rate reductions. Overdose may present with profound somnolence, hypotonia, respiratory depression, and hemodynamic instability; management centers on immediate pump deactivation/reprogramming, airway protection, and supportive care, with CSF exchange if necessary [25].

7.2 Withdrawal as a Medical Emergency

Abrupt cessation of intrathecal delivery due to catheter disruption, empty reservoir, or pump failure can precipitate a hyperexcitability syndrome with escalating spasms, hyperthermia, rhabdomyolysis, and multi-organ compromise. Early recognition, restoration of intrathecal baclofen delivery, and interim GABAergic sedation are lifesaving. The canonical clinical description emphasizes the importance of rigorous refill logistics and 24/7 access to pump expertise [26].

7.3 Hardware-Related Events

Catheter kinking, fracture, obstruction, or disconnection necessitates imaging and often revision. Pump malfunctions are less frequent than catheter problems, with battery end-of-life addressed by elective replacement. Surgical-site infection and meningitis—more common in early months and in pediatrics—often mandate explantation and antibiotics, followed by staged reimplantation after eradication [27].

7.4 Orthopedic Considerations

Reports describe accelerated scoliosis progression after ITB in some children with CP, though the literature remains mixed and confounded by the high baseline risk of deformity in nonambulant CP. Clinicians should monitor spinal alignment longitudinally and counsel families regarding this possibility and its management in collaboration with orthopedic teams [28, 29].

8 Comparative and Adjunctive Therapies

8.1 ITB vs Other Neuraxial and Ablative Options

Intrathecal opioids or clonidine provide inferior global spasticity control and are limited by sedation, hypotension, or analgesia-centric mechanisms. Phenol neurolysis can target focal problems but lacks reversibility and is impractical for diffuse involvement. Selective dorsal rhizotomy may be preferred in carefully selected ambulant spastic diplegia but is irreversible and largely limited to lower limbs; in nonambulant populations, both SDR and ITB reduce tone, with ITB favored when adjustability and trunk/upper-limb involvement are priorities [14]. Deep brain stimulation of the internal globus pallidus is superior to ITB for primary generalized dystonia, reflecting distinct network pathophysiology [30].

8.2 Combining ITB with Focal Chemodenervation

In diffuse spasticity with residual focal drivers of disability, botulinum toxin injections can be layered onto ITB to treat recalcitrant muscle groups or task-specific patterns. Evidence from pediatric cohorts and case-based reports supports improved comfort and function when intrathecal tone reduction is complemented by targeted chemodenervation [31, 32].

8.3 Rehabilitation and Technology Adjuncts

ITB often enables more effective stretching, strengthening, and task-specific training by removing the "noise" of cocontraction and spasms. Programs may synchronize pump flex dosing with therapy schedules. Functional electrical stimulation, standing programs, and orthotic strategies are frequently more tolerable and efficacious once hypertonia is controlled.

9 Future Directions

9.1 Delivery Science: Distribution, Devices, and Dosing Intelligence

Computational fluid dynamics, benchtop spinal replicas, and in vivo/ex vivo work are refining how infusion rate, catheter design, and CSF pulsatility shape segmental exposure, opening paths to multisegment catheters and schedules that more precisely match individual topography and circadian needs [16–20, 33]. Closed-loop concepts coupling EMG or posture sensors to adaptive dosing remain investigational but conceptually align ITB delivery with the temporal dynamics of spasticity.

9.2 Pharmacology: Beyond Baclofen and Alongside It

Formulation science is extending concentration stability to lengthen refill intervals. Parallel pipelines explore long-acting GABA-mimetics, positive allosteric modulators to widen the therapeutic window, and agents acting outside GABAergic pathways that may complement or substitute for baclofen in selected phenotypes [6, 20, 15].

9.3 Alternative Compartments and Regenerative Strategies

In selected anatomies with impaired rostral spread, intraventricular baclofen has been piloted, although its use is restricted to expert centers. At the same time, vectors that enhance inhibitory neurotransmission within spinal circuits and cell-based approaches aiming to rebuild inhibitory interneuronal networks are progressing preclinically. Imaging advances in neurofluid dynamics may guide patient-specific delivery strategies and signal impending distribution failures before clinical deterioration [34, 35].

9.4 Real-World Practice Patterns

Large-scale prescription analyses in intrathecal drug delivery underscore how off-label concentrations and combinations are commonplace in pain indications and highlight the need for similarly rigorous real-world surveillance in antispastic ITB—particularly as next-generation pumps and higher-concentration formulations diffuse into practice [36].

10 Conclusion

Intrathecal baclofen provides robust, sustained suppression of pathologic tone and spasms when generalized spasticity remains disabling despite conservative care. Its success rests on sound neurophysiology, predictable pharmacology, and programmable delivery that brings drug precisely to its spinal targets. Outcomes are maximized by rigorous selection, an evidence-based trial, meticulous implantation, and longitudinal stewardship that anticipates dose evolution and rare emergencies such as withdrawal. Within multidisciplinary rehabilitation, ITB enables function, reduces pain, eases caregiving, and improves quality of life. Ongoing advances in delivery science, sensors, and pharmacology promise further personalization while preserving the safety and reliability that have made ITB a mainstay of modern spasticity management.

Conflict of Interest

The authors have no conflicts of interest to declare that are relevant to the content of this chapter.

References

1. Dietz V, Sinkjaer T (2007) Spastic movement disorder: impaired reflex function and altered muscle mechanics. Lancet Neurol 6(8):725–733. https://doi.org/10.1016/S1474-4422(07)70193-X
2. Nielsen JB, Crone C, Hultborn H (2007) The spinal pathophysiology of spasticity—from a basic science point of view. Acta Physiol (Oxf) 189(2):171–180. https://doi.org/10.1111/j.1748-1716.2006.01652.x
3. Rizzo MA, Hadjimichael OC, Preiningerova J, Vollmer TL (2004) Prevalence and treatment of spasticity reported by multiple sclerosis patients. Mult Scler 10(5):589–595. https://doi.org/10.1191/1352458504ms1085oa
4. Zahavi A, Geertzen JHB, Middel B, Staal M, Rietman JS (2004) Long-term effect of ITB on impairment, disability, and quality of life in patients with severe spasticity. J Neurol Neurosurg Psychiatry 75:1557–1561. https://doi.org/10.1136/jnnp.2003.014282
5. Calias P, Banks WA, Begley D, Scarpa M, Dickson P (2014) Intrathecal delivery of protein therapeutics to the brain: a critical reassessment. Pharmacol Ther 144(2):114–122. https://doi.org/10.1016/j.pharmthera.2014.05.009
6. Magill E et al (2023) Solid implantable devices for sustained drug delivery. Adv Drug Deliv Rev 199:114950. https://doi.org/10.1016/j.addr.2023.114950
7. Bowery NG (1993) GABA_B receptor pharmacology. Annu Rev Pharmacol Toxicol 33:109–147. https://doi.org/10.1146/annurev.pa.33.040193.000545
8. Peet MJ, McLennan H (1986) Pre- and postsynaptic actions of baclofen in CA1 hippocampal neurones. Exp Brain Res 61(3):567–574. https://doi.org/10.1007/BF00237582
9. Heckman CJ, Gorassini MA, Bennett DJ (2008) Persistent inward currents in motoneuron dendrites: implications for motor output. Neuroscientist 14(3):232–251. https://doi.org/10.1177/1073858408314986
10. Penn RD, Kroin JS (1985) Intrathecal baclofen alleviates spinal cord spasticity. Lancet 2(8452):1092–1094. https://doi.org/10.1016/S0140-6736(85)90163-1
11. Penn RD et al (1989) Intrathecal baclofen for severe spinal spasticity. N Engl J Med 320(23):1517–1521. https://doi.org/10.1056/NEJM198906083202303
12. Meythaler JM et al (2001) Continuous intrathecal baclofen in cerebral-origin spasticity: a randomized, controlled, cross-over trial in stroke. Stroke 32(3):591–597. https://doi.org/10.1161/01.STR.32.3.591
13. Heetla HW et al (2016) Pharmacokinetic–pharmacodynamic modeling of intrathecal baclofen. Br J Clin Pharmacol 82(5):1418–1431. https://doi.org/10.1111/bcp.12781
14. Hoving MA, van Raak EPM, Spincemaille GHJJ et al (2009) Safety and one-year efficacy of ITB in intractable spastic CP: randomized controlled trial. Eur J Paediatr Neurol 13:240–246. https://doi.org/10.1016/j.ejpn.2008.05.002
15. Delgado MR, Tilton A, Russman B et al (2010) Practice parameter: pharmacologic treatment of spasticity in children with CP. Neurology 74(4):336–343. https://doi.org/10.1212/WNL.0b013e3181cbcd2f

16. Tangen KM, Leval R, Mehta AI, Linninger AA (2017) Intrathecal drug distribution: effects of injection volume, CSF pulsatility, and uptake. Anesth Analg 124(5):1686–1696. https://doi.org/10.1213/ANE.0000000000002011
17. Tangen KM, Hsu Y, Zhu DC, Linninger AA (2015) CNS-wide simulation of flow resistance and drug transport due to spinal microanatomy. J Biomech 48(10):2144–2154. https://doi.org/10.1016/j.jbiomech.2015.02.018
18. Lueshen E, Tangen K, Mehta AI, Linninger AA (2017) Backflow-free catheters for efficient and safe convection-enhanced delivery. Med Eng Phys 45:15–24. https://doi.org/10.1016/j.medengphy.2017.02.018
19. Burla GKR, Shrestha D, Bowen M et al (2024) Effect of injection protocols on intrathecal solute dispersion: in vitro cynomolgus CSF system. Fluids Barriers CNS 21:61. https://doi.org/10.1186/s12987-024-00556-2
20. Madadi AK, Sohn M (2024) Advances in intrathecal nanoparticle delivery. Pharmaceuticals (Basel) 17(8):1070. https://doi.org/10.3390/ph17081070
21. Coffey RJ, Cahill D, Steers W et al (1993) Intrathecal baclofen for intractable spasticity: multicenter experience. J Neurosurg 78(2):226–232. https://doi.org/10.3171/jns.1993.78.2.0226
22. Meythaler JM, McCary A, Hadley MN (2001) Open-label randomized study of ITB after stroke. Stroke 32(3):591–597. https://doi.org/10.1161/01.STR.32.3.591
23. Kraus T, Gegenleitner K, Svehlik M, Novak M, Steinwender G, Singer G (2017) Long-term ITB improves QoL in severe spastic CP. Eur J Paediatr Neurol 21(3):565–569. https://doi.org/10.1016/j.ejpn.2017.01.016
24. Buizer AI, Martens BHM, van de Pol LA et al (2019) Continuous ITB in children: long-term effects and determinants. Dev Med Child Neurol 61(3):363–369. https://doi.org/10.1111/dmcn.14005
25. Watve SV, Sivan M, Raza WA, Phadke CP (2012) Intrathecal baclofen overdose: recognition and management. Spinal Cord 50(1):19–22. https://doi.org/10.1038/sc.2011.112
26. Coffey RJ, Edgar TS, Francisco GE et al (2002) Abrupt withdrawal from intrathecal baclofen. Arch Phys Med Rehabil 83(6):735–741. https://doi.org/10.1053/apmr.2002.31822
27. Motta F, Buonaguro V, Stignani C (2007) Complications of ITB in children. J Neurosurg 107(3 Suppl):32–35. https://doi.org/10.3171/PED-07/09/032
28. Hart AR, Tripathi C, Rowland DJ, Broadley P, Mordekar SR (2007) Rapid progression of scoliosis complicating ITB pump insertion. Dev Med Child Neurol 49(9):717. https://doi.org/10.1111/j.1469-8749.2007.00717.x
29. Lee S, Hyun C, Kim K et al (2023) Effect of ITB pump on scoliosis in children with CP: meta-analysis. Ann Rehabil Med 47:11–18. https://doi.org/10.5535/arm.22108
30. Vidailhet M, Vercueil L, Houeto J-L et al (2005) Bilateral pallidal stimulation for primary generalized dystonia. N Engl J Med 352(5):459–467. https://doi.org/10.1056/NEJMoa042187
31. Delgado MR, Hirtz D, Aisen M et al (2010) Combined intrathecal baclofen and botulinum toxin therapy. J Child Neurol 25(12):1510–1516. https://doi.org/10.1177/0883073810371007
32. Santamato A, Panza F, Ranieri M et al (2010) Combined ITB and botulinum toxin type A in paraplegia: case report. J Rehabil Med 42(7):708–710. https://doi.org/10.2340/16501977-0596
33. Salegio EA, Hancock K, Korszen S (2023) Preclinical delivery of gene therapy to CSF: challenges for translation. Front Mol Neurosci 16:1248271. https://doi.org/10.3389/fnmol.2023.1248271
34. Muccio M, Sun Z, Chu D, Damadian BE, Minkoff L, Bonanni L, Ge Y (2024) Body position and neurofluid dynamics: imaging advances. Front Aging Neurosci 16:1454282. https://doi.org/10.3389/fnagi.2024.1454282
35. Draulans N, Vermeersch K, Degraeuwe B et al (2013) Intrathecal baclofen in severe spasticity: evolution of dosage and complications. Clin Rehabil 27(12):1137–1148. https://doi.org/10.1177/0269215513497146
36. Hao D, Di Capua CA, Dumas C, Rosenblum SB, Thorne R, Narang S (2025) Real-world prescription data for intrathecal drug delivery: analysis in 32,784 patients. Neuromodulation 28(7):1054–1058. https://doi.org/10.1016/j.neurom.2024.07.007

Check for updates

Chapter 7

Intrathecal Therapy for CNS Infections

Manuel Menéndez-González

Abstract

The central nervous system is protected by the blood–brain barrier (BBB) and the blood–CSF barrier (BCSFB), structures that maintain neuronal homeostasis but impede therapeutic access. Invasive infections such as health care-associated ventriculitis and meningitis, and those due to multidrug-resistant organisms, frequently defy systemic therapy because most antimicrobials achieve subtherapeutic cerebrospinal fluid (CSF) concentrations even at toxic serum levels. This chapter reviews intrathecal and intraventricular administration as a strategy that circumvents the BBB/BCSFB to deliver reliably high local drug exposures while limiting systemic toxicity. We outline core pharmacokinetic and pharmacodynamic principles governing drug behavior in the CSF, emphasizing that the CSF is a flowing, nonwell-mixed compartment in which the site of administration, CSF production and drainage, device configuration, and drug physicochemistry dictate exposure at the site of infection. Practical technique is detailed for intermittent lumbar puncture and for ventricular access via Ommaya or external ventricular drains, with a comparative analysis of distribution, patient burden, and risk. Clinical applications focus on post-neurosurgical and nosocomial infections and MDR Gram-negative disease, summarizing experience with intrathecal/intraventricular colistin, tigecycline, aminoglycosides, and vancomycin, including typical dosing ranges, clearance considerations with drains, and strategies such as temporary clamping to enhance distribution. Fungal CNS infections receive special attention: intrathecal amphotericin B remains the only curative therapy for coccidioidal meningitis and serves as salvage in refractory cryptococcosis and selected Candida or rare molds. The chapter also delineates the limited and largely rescue role of intrathecal approaches in viral and parasitic encephalitides, where pathophysiology and available agents lessen the pharmacologic rationale. Complications are systematically reviewed, from procedure-related infection and hemorrhage to drug-induced neurotoxicity (chemical arachnoiditis, encephalopathy, seizures, myelopathy), with guidance on recognition and management. Looking forward, we describe the transition from barrier-bypassing to intrathecal pharmacokinetic control, highlighting sustained-release formulations (liposomes, nanoparticles, hydrogels), programmable pumps, and next-generation catheters designed to improve residence time and spatial distribution. Emerging gene-targeted and molecular platforms suggest future anti-infective strategies within the CSF space. Throughout, the chapter underscores the necessity of multidisciplinary care, PK-guided dosing, and rigorous technique to maximize efficacy and safety as intrathecal therapy evolves from rescue intervention to precision treatment.

Keywords Intrathecal antibiotics, Central nervous system infections, Meningitis, Ventriculitis, Targeted antimicrobial delivery

Manuel Menéndez-González (ed.), *Intrathecal Drug Delivery*, Neuromethods, vol. 234,
https://doi.org/10.1007/978-1-0716-5241-1_7,
© The Author(s), under exclusive license to Springer Science+Business Media, LLC, part of Springer Nature 2026

1 Introduction

1.1 The Blood–Brain Barrier (BBB) and Blood–CSF Barrier (BCSFB) as Brain Defense

The central nervous system (CNS) is often described as a "pharmacological sanctuary," an immunologically privileged environment shielded from the systemic circulation by a series of highly selective barriers [1]. The most important of these are the blood–brain barrier (BBB) and the blood-cerebrospinal fluid barrier (BCSFB). These barriers, composed of tight junctions between endothelial cells and a complex network of lipid layers, are designed to protect the CNS from toxins, pathogens, and fluctuations in systemic compounds [2]. While this protection is fundamental for neuronal homeostasis, it represents a formidable obstacle for the treatment of CNS infections. Most anti-infective agents, especially large or hydrophilic molecules, cannot cross these barriers in therapeutic concentrations [2]. Consequently, treating infections like meningitis or ventriculitis with intravenous (IV) therapy often requires extremely high systemic doses, which increases the risk of toxicity to other organs and may still fail to achieve effective pathogen eradication within the CNS [2].

The dynamics of these barriers are not static. Inflammation, a hallmark of meningitis, can alter the integrity of the BBB, opening the tight junctions and paradoxically increasing permeability to certain drugs, particularly hydrophilic antibiotics [2]. However, this effect is unpredictable and varies considerably depending on the pathogen, the severity of the infection, and the host's immune response [2]. This variability introduces an unacceptable element of uncertainty into the treatment of life-threatening infections. Relying on an unpredictable increase in barrier permeability is a high-risk strategy, especially when facing multidrug-resistant (MDR) pathogens, where time is critical and the margin for error is minimal. This situation creates a clinical paradox: the very inflammation that could theoretically aid systemic drug penetration also underscores the need for a strategy that ensures reliable and rapid therapeutic concentrations. Intrathecal therapy eliminates this uncertainty, ensuring high drug levels in the cerebrospinal fluid (CSF) regardless of the inflammatory state of the meninges. Therefore, the severity of inflammation, rather than being a reason to avoid intrathecal therapy, becomes one of the strongest justifications for its use, providing indispensable therapeutic certainty in critical clinical situations.

1.2 Rationale for Intrathecal Administration: Achieving Therapeutic Concentrations

Intrathecal (IT) administration, a generic term encompassing injection into the lumbar subarachnoid space (IT-L) or directly into the cerebral ventricles (intraventricular or IVT), is a therapeutic strategy designed to directly bypass the BBB and BCSFB [1–9]. By delivering the anti-infective agent directly into the CSF, extremely high local concentrations can be achieved at the site of infection, often orders of magnitude greater than what could be safely achieved with systemic administration [1].

This targeted delivery offers several fundamental advantages. First, it allows for the use of significantly lower total drug doses, which drastically reduces the risk of systemic toxicity, such as the nephrotoxicity or ototoxicity associated with aminoglycosides, or myelosuppression with certain antivirals [1]. Second, and crucially, IT therapy has become a rescue and sometimes first-line option for CNS infections caused by MDR pathogens, where systemic options are limited or completely ineffective due to poor CNS penetration [4].

2 General Indications, Advantages, and Disadvantages

The primary indications for IT therapy include healthcare-associated ventriculitis and meningitis (HAVM), which are severe complications of neurosurgical procedures, head trauma, or the placement of CNS devices like shunts or drains [4]. Another paramount indication is the treatment of infections caused by MDR organisms that are refractory to IV therapy [4]. Furthermore, for certain fungal infections, such as coccidioidal meningitis, IT therapy with amphotericin B is the only known curative treatment [8].

The advantages of IT therapy are clear: bypassing the BBB, achieving high local drug concentrations, and reducing systemic toxicity [1–4]. However, the therapy is not without significant disadvantages and risks. It is an invasive procedure that carries the risk of complications such as secondary infections, hemorrhage, and direct drug-induced neurotoxicity, which can manifest as chemical meningitis, arachnoiditis, seizures, or permanent neurological deficits [3, 10–12]. Additionally, the evidence base for many agents and pathologies remains limited, often relying on case series, case reports, and retrospective studies, which underscores the need for expert clinical judgment and a multidisciplinary approach to its implementation [2].

3 Pharmacokinetic and Pharmacodynamic (PK/PD) Principles of Intrathecal Drug Administration

3.1 The Cerebrospinal Fluid (CSF) Compartment: A Dynamic Environment

The success of intrathecal therapy hinges on a deep understanding of the pharmacokinetics (PK) and pharmacodynamics (PD) within the CSF compartment, a system that is far more complex than a simple fluid reservoir.

3.2 Volume and Composition

The volume of the CSF compartment (VCSF) is a critical determinant for the volume of distribution (Vd) of intrathecally administered drugs. This volume is not constant; it varies significantly with age, the presence of pathologies like hydrocephalus (which can dramatically increase VCSF), and the presence of blood clots or inflammatory exudates, which can reduce the effective volume of distribution and create isolated compartments [2].

3.3 Production and Flow Dynamics

CSF is produced at a rate of approximately 20 mL/h in adults, primarily by the choroid plexuses in the ventricles, with a total volume turnover several times a day [10]. The flow of CSF follows a directional path: from the lateral ventricles to the third ventricle, through the aqueduct of Sylvius to the fourth ventricle, and then into the cisterna magna and the subarachnoid space surrounding the brain and spinal cord [2]. This net, directional flow is of paramount pharmacokinetic importance. It implies that the CSF is not a single, well-mixed compartment—a conceptual error that can lead to therapeutic failures. Treating the CSF space as a single, homogenous compartment is a critical pharmacokinetic fallacy. Evidence conclusively demonstrates that the site of administration and the site of infection must be closely aligned. A negative lumbar CSF culture after IT-L administration can give a false sense of security if the primary infection is ventricular, as therapeutic concentrations may never have been reached at the site of ventriculitis. This reality underscores the necessity of IVT administration via an Ommaya reservoir or an external ventricular drain (EVD) to treat ventriculitis, not just for convenience, but for fundamental pharmacokinetic and pharmacodynamic reasons.

4 Factors Influencing Drug Distribution and Clearance in CSF

4.1 Site of Administration

As derived from CSF flow dynamics, the site of injection is perhaps the most important factor determining drug distribution. Intraventricular (IVT) administration is pharmacokinetically superior for the treatment of ventriculitis and infections affecting the cerebral surfaces [1]. Pharmacokinetic studies have shown that drugs injected directly into the ventricles are distributed more uniformly throughout the neuroaxis, achieving concentrations up to 10 times higher than the same dose administered via lumbar puncture (IT-L) [1]. Conversely, IT-L dosing often fails to achieve adequate therapeutic concentrations in the ventricles, making it a suboptimal strategy for ventriculitis [2]. Therefore, the choice of administration route (IT-L vs. IVT) is not a mere procedural preference but a fundamental determinant of therapeutic success or failure, dictated by the location of the infection.

4.2 CSF Drainage and Clearance

The clearance of drugs from the CSF is governed by three main mechanisms: bulk CSF outflow into the venous circulation, retrograde diffusion across the BBB/BCSFB, and active transport mediated by efflux transporters [2]. The presence of drainage devices, such as an

EVD, can dramatically increase drug clearance, shortening its half-life in the CSF. This is a crucial factor that must be considered when determining the dosing frequency [2]. A common clinical strategy to mitigate this rapid clearance and facilitate drug distribution is to clamp the drain for a period of time (e.g., 15–120 min) after drug administration [2].

4.3 Impact of Drug Properties on PK/PD

The physicochemical characteristics of the anti-infective agent profoundly influence its behavior within the CSF.

4.4 Lipophilicity and Molecular Weight

Large, hydrophilic molecules, such as aminoglycosides (gentamicin, amikacin) and glycopeptides (vancomycin), have very poor penetration from the blood into the CSF. Once in the CSF, their clearance is primarily by bulk CSF outflow [2]. These properties make them ideal candidates for intrathecal administration, as systemic concentrations contribute minimally to CSF levels, and their CSF concentrations can be maintained for longer periods. Conversely, smaller, more lipophilic drugs can diffuse more readily across cell membranes into adjacent tissues, resulting in a larger volume of distribution and potentially faster clearance by retrograde diffusion back into the blood [2].

4.5 Protein Binding

In the systemic circulation, only the free, unbound fraction of a drug can cross the BBB. While this factor is less critical when the drug is administered directly into the CSF, which has a very low protein content, it can influence the diffusion of the drug from the CSF into the brain parenchyma [2].

4.6 Techniques and Devices for Intrathecal Administration

The safe and effective administration of intrathecal therapy requires meticulous technique and appropriate selection of the access device that have been described in other chapters of the book. Basically, there are two primary methods: lumbar puncture (LP) and the use of intraventricular catheters and reservoirs.

Lumbar Puncture (LP): Technique and Procedural Considerations

LP is the most common method for intermittent administration of intrathecal drugs.

Patient Preparation and Positioning

Before the procedure, it is imperative to obtain informed consent and perform a thorough evaluation, including a review of potential contraindications such as coagulopathies or severe thrombocytopenia (platelet count $<50 \times 10^9/L$) [12]. The patient is placed in the lateral decubitus position with knees flexed toward the chest (fetal position) or sitting and leaning forward to maximize the opening of the lumbar intervertebral spaces [12]. Strict aseptic technique must be followed, including the use of sterile gloves, cleaning the skin with an antiseptic solution, and using sterile drapes [12].

Needle Insertion and CSF Collection

The L3-L4 or L4-5 intervertebral space is identified, and the skin and subcutaneous tissues are infiltrated with a preservative-free local anesthetic [12]. A spinal needle with a stylet is inserted in the midline, advancing slowly until CSF flow is obtained, confirming entry into the subarachnoid space. At this point, CSF samples can be collected for microbiological and biochemical analysis [12].

Drug Administration

The drug, which must be a **preservative-free** formulation specifically prepared for intrathecal use, is injected slowly over 1–2 min to avoid sudden increases in intracranial pressure [12]. After injection and removal of the needle, a sterile dressing is applied. The patient is usually instructed to remain supine for 30–60 min to minimize the risk of post-dural puncture headache, a common complication [12].

Intraventricular Catheters and Reservoirs (e.g., Ommaya, Rickham)

For patients requiring prolonged intrathecal therapy or for those with ventriculitis, a ventricular access system is the preferred method.

Device Description

An Ommaya reservoir is a surgically implanted device consisting of a small, dome-shaped silicone reservoir placed under the scalp, connected to a catheter that is threaded into one of the lateral ventricles of the brain [13,14]. This system allows for repeated, less invasive access to the CSF for sampling and drug administration, eliminating the need for repeated lumbar punctures [15].

Surgical Placement

The placement of an Ommaya reservoir is a neurosurgical procedure typically performed under general anesthesia. The preferred site for the incision is the non-dominant right frontal region. Modern neuronavigation techniques, guided by computed tomography (CT) or magnetic resonance imaging (MRI), have largely replaced freehand placement, significantly improving accuracy and reducing the risk of complications such as catheter malposition [15]. The catheter tip is ideally aimed toward the foramen of Monro to ensure optimal drug distribution within the ventricular system [15].

Reservoir Aspiration and Injection

Access to the reservoir is performed percutaneously. After aseptic preparation of the scalp, the dome of the reservoir is palpated and punctured with a fine-gauge needle (e.g., 25-gauge) [15]. A volume of CSF (usually equal to the volume of the drug to be injected) can be aspirated before slowly administering the therapeutic agent [15].

Comparative Analysis: Lumbar vs. Intraventricular Routes

The choice between the lumbar and intraventricular route is not trivial and has profound pharmacokinetic and clinical implications.

Efficacy and Distribution

As mentioned, IVT administration via an Ommaya reservoir provides more consistent and uniformly distributed drug concentrations throughout the CNS. This makes it the route of choice for the treatment of ventriculitis, meningitis affecting the cerebral convexities, or any therapy requiring reliable distribution throughout the neuroaxis [1].

Patient Comfort and Complications

An Ommaya reservoir avoids the discomfort and anxiety associated with repeated lumbar punctures, which is a considerable advantage for patients needing long-term treatment [14]. However, reservoir placement is an invasive surgical procedure that carries its own risks, such as device infection (in 5.5%–8% of cases), catheter malfunction (obstruction, disconnection), intracranial hemorrhage, and drug-related leukoencephalopathy [1]. LP, while less invasive per procedure, can be poorly tolerated if repeated frequently and is associated with a higher risk of post-dural puncture headache.

4.7 Intrathecal Therapy for Bacterial Meningitis and Ventriculitis

IT/IVT therapy has found its most critical and well-documented application in the treatment of severe bacterial CNS infections, especially those arising in the nosocomial setting and those caused by MDR pathogens.

Treatment of Post-Neurosurgical and Nosocomial Infections

CNS infections following neurosurgery or head trauma are a primary indication for IT/IVT therapy. These infections, often termed HAVM, occur when bacteria gain direct access to the meninges through structural defects [4]. The risk increases dramatically, up to 20%, in patients with ventricular drainage catheters [6]. A meta-analysis demonstrated that the addition of IT/IVT antibiotics to systemic IV therapy in patients with postoperative meningitis or ventriculitis was associated with a significant decrease in mortality (Odds Ratio 0.27; 95% confidence interval [CI]: 0.15–0.49) and an increased cure rate (OR 3.4; 95% CI: 1.6–7.22) [6]. Also, the combination of local intraventricular/intrathecal polymyxin B sulfate administration and systemic antibiotic therapy offers a more efficient treatment approach for these complex infections in NICU patients after neurosurgery 54.

Management of Multidrug-Resistant (MDR) Gram-Negative Infections

The growing threat of MDR Gram-negative bacteria, such as carbapenem-resistant *Acinetobacter baumannii* (CRAB), *Pseudomonas aeruginosa*, and *Klebsiella pneumoniae*, has made IT/IVT therapy a life-saving last-resort strategy [4]. Many systemic antibiotics are ineffective against these pathogens or cannot achieve adequate concentrations in the CSF.

- **Colistin:** It is one of the most studied IT/IVT agents for MDR *A. baumannii* infections. Systemic colistin has notoriously poor CSF penetration, not exceeding 25% even with inflamed meninges [5]. IT/IVT administration of colistin has proven effective, with cure rates of 74% in pediatric populations and high microbiological cure rates in adults [5]. Dosing is highly variable and often empirical. Adult doses range from 1.6 to 20 mg/day, while pediatric doses of 0.16–0.24 mg/kg/day have been used [5]. A systematic review found the median time to CSF sterilization was 4.1 days [17-21].
- **Tigecycline:** Combined IV and IT/IVT therapy with tigecycline has been shown to be effective against MDR *A. baumannii*. A

retrospective study reported a significantly higher microbiological clearance rate in the combination therapy group (83.3%) compared to the IV-only therapy group (22.2%) [22].

- **Aminoglycosides (Amikacin, Gentamicin):** These are used for various MDR Gram-negative infections. One study reported microbiological eradication in postoperative meningitis (POM) due to *K. pneumoniae* in a mean of 3 days with a treatment of 30 mg/day of IVT amikacin [4]. Typical adult IT doses are 4–8 mg for gentamicin and 30 mg (range 5–50 mg) for amikacin [18].
- **Vancomycin:** This is the mainstay of IT/IVT treatment for Gram-positive coccal infections, such as methicillin-resistant *Staphylococcus aureus* (MRSA) and CNS device infections. IVT administration of vancomycin (daily doses of 2–20 mg) achieves CSF concentrations (2.5–292.9 mg/L) far exceeding those achieved with IV administration (0.06–22.3 mg/L) [9, 23, 24]. Although current evidence suggests it is effective and safe, optimal regimens are not yet clearly defined, and no clear relationship has been found between CSF vancomycin levels and efficacy or toxicity [9].

Table 1 summarizes commonly used IT/IVT antibiotic regimens, serving as a quick clinical reference guide.

Special Considerations for Tuberculous Meningitis (TBM)

TBM is a neurological emergency with a high rate of mortality and morbidity [3]. Treatment is complicated by the poor CNS penetration of crucial first-line drugs like rifampicin (RIF), which only reaches 10%–20% of plasma concentrations in the CSF [3]. Although isoniazid (INH) and pyrazinamide (PZA) penetrate well (80%–100%), sub-exposure to RIF can lead to treatment failure and the emergence of resistance [3]. IT administration of INH and RIF has been documented since the 1950s and 1970s, respectively, and shows promise in refractory cases, allowing for reduced systemic doses and associated side effects [3]. The future of TBM treatment may rely on the development of sustained-release intrathecal delivery systems, such as osmotic pumps or nanoparticle-laden hydrogels, which could maintain constant and effective drug concentrations at the site of infection, overcoming the limitations of current therapy [3].

4.8 Intrathecal Therapy for Fungal CNS Infections

Fungal infections of the CNS are devastating diseases, often associated with high rates of morbidity and mortality, especially in immunocompromised hosts [25] and paediatric population [26]. Intrathecal therapy plays a critical, and in some cases curative, role in the management of these difficult-to-treat infections.

The primary challenge in treating CNS mycoses is the extremely poor penetration of most systemic antifungals into the CSF [25]. Amphotericin B (AmB), a mainstay in the treatment of many severe fungal infections, only achieves CSFCerebrospinal fluid (CSF) concentrations that are 2%–4% of serum levels [8]. This

Table 1
Intrathecal/intraventricular antibiotic regimens for common pathogens in CNS infections

Pathogen	Drug	Typical adult daily dose (IVT/IT)	Typical pediatric daily dose (IVT/IT)	Reported efficacy/ cure rate	Key toxicities	References
A. baumannii (MDR/XDR)	Colistin	10 mg (125,000 IU); range 1.6–20 mg	0.16–0.24 mg/kg; up to 10 mg	89% success rate; 83% clinical cure rate	Chemical meningitis/ventriculitis (11%–16%), seizures (rare)	[5]
P. aeruginosa (MDR/XDR)	Colistin	5–10 mg	0.16–0.24 mg/kg	High cure rates in case series	Chemical meningitis	[5]
K. pneumoniae (MDR/XDR)	Amikacin	30 mg (range 5–50 mg)	Not well-established	Microbiological eradication in ~3 days	Ototoxicity, nephrotoxicity (theoretical)	[4]
S. aureus (MRSA), CoNS	Vancomycin	5–20 mg	5–20 mg	High clinical and microbiological efficacy	No serious adverse effects reported	[9]
Gram-negative bacilli (general)	Gentamicin	4–8 mg	1–2 mg	Effective as adjunctive therapy	Brainstem lesions, ototoxicity	[21]
Gram-negative bacilli (general)	Tigecycline	5–10 mg	Not well-established	Superior microbiological clearance with combined (IV+IVT) therapy	Neurotoxicity not well-defined	[4]

CoNS coagulase-negative staphylococci, *MDR* multidrug-resistant, *MRSA* methicillin-resistant Staphylococcus aureus, *XDR* extensively drug-resistant pharmacokinetic limitation often renders systemic administration alone insufficient, necessitating direct administration into the CNS to achieve effective infectionCentral nervous system infections control.

Coccidioidal Meningitis (CM): Intrathecal Amphotericin B as Curative Therapy

Coccidioidal meningitis is the most feared complication of coccidioidomycosis. Historically, it was a 100% fatal disease within 2 years [8]. The introduction of intrathecal amphotericin B deoxycholate (AmBd) in the 1950s revolutionized its treatment, reducing mortality to approximately 30% [8]. Importantly, while oral azoles (e.g., high-dose fluconazole 800–1200 mg/day) are used as first-line therapy and can suppress the infection, they are not curative and require lifelong treatment [8]. Currently, IT AmBd remains the only known curative option for CM [8-27].

IT AmBd therapy is typically reserved for patients who fail high-dose azole therapy or who cannot tolerate it [8]. The treatment is a long and complex process that requires careful drug preparation and a meticulous dosing schedule.

Preparation and Administration

The preparation of AmBd for intrathecal use is a high-risk procedure that must be performed under strict aseptic conditions in a certified laminar flow hood [8]. It is crucial to use only preservative-free (PF) products. The formulation often includes a corticosteroid, such as methylprednisolone, to mitigate the intense inflammatory reaction (chemical arachnoiditis) that AmBd can induce in the subarachnoid space [8].

Table 2 details a preparation and administration protocol based on the extensive experience of Kern Medical Center, a reference center for the treatment of CM.

Cryptococcal Meningitis: The Role of Intrathecal Amphotericin B in Refractory Disease

Unlike CM, intrathecal therapy is not a first-line treatment for cryptococcal meningitis. The IDSA and WHO practice guidelines recommend a systemic induction treatment, preferably with a combination of liposomal AmB and flucytosine, followed by consolidation and maintenance therapy with fluconazole [32–31]. IT AmB is reserved as salvage therapy for severe or refractory cases where systemic treatment has failed or is not tolerated [32]. One case report described the successful treatment of a refractory cryptococcal encephalitis by continuous intrathecal administration of liposomal AmB via a lumbar drain, after IV treatment failed. The patient received a total dose of 28 mg of IT liposomal AmB over 1 month, leading to clinical recovery and CSF sterilization [32].

Candida *and Other Fungal Pathogens*

For *Candida* meningitis, the IDSA guidelines also focus on systemic therapy with echinocandins, lipid-formulation AmB, or fluconazole, depending on the species and susceptibility [33]. IT therapy with AmB is not routinely recommended but may be considered as salvage therapy, especially in pediatric patients, in cases

Table 2
Intrathecal amphotericin B preparation and administration protocol for coccidioidal meningitis

Part A: Ingredients and compounding preparation	
Ingredients	**Quantity**
Amphotericin B deoxycholate 50 mg vial (preservative-free)	1
Sterile water for injection (preservative-free) 10 mL vial	2
Methylprednisolone (Solu-Medrol) 40 mg vial (preservative-free)	1
Dextrose 5% 50 mL	1
Empty sterile vial 10 mL	1
Preparation instructions	
1. Reconstitute the 50 mg AmBd vial with 10 mL of sterile water for injection to yield a final concentration of 5 mg/mL. Shake well.	
2. Reconstitute the 40 mg methylprednisolone vial with 1.6 mL of sterile water for injection to yield a concentration of 25 mg/mL.	
3. In the empty 10 mL sterile vial, inject: • 0.2 mL (5 mg) of methylprednisolone. • 3.8 mL of 5% dextrose. • The desired dose of AmBd suspension (e.g., 0.1 mg = 0.02 mL).	
Part B: Dose titration and tapering schedule	
Phase	**Frequency**
Phase I (Induction)	Three times per week
Phase II (Consolidation)	Twice per week (3 months), then once per week (3 months)
Phase III (Tapering)	Every 2 weeks (3 months), then every 3, 4, and 6 weeks (3 months each)

Adapted from Florez-Perdomo et al. [8]

of device-associated ventriculitis, or when infected devices cannot be removed [34, 35, 36].

For other rare and often fatal fungal infections, such as cerebral mucormycosis, IT/IVT therapy with AmB has been used anecdotally as an adjunct to aggressive surgical resection and systemic therapy, although data are very limited [28,37]. Emerging experience also suggests the utility of intrathecal caspofungin for multidrug-resistant fungal pathogens; success has been reported in treating shunt infections due to *Candida auris* [2].

4.9 Intrathecal Therapy in Viral and Parasitic CNS Infections

The role of intrathecal therapy in viral and parasitic CNS infections is considerably less defined and is based on a much sparser evidence base compared to bacterial and fungal infections. This reflects fundamental differences in the pathophysiology of these diseases and the pharmacology of the available therapeutic agents. Whereas IT therapy for bacterial and fungal infections is an established strategy to overcome a known pharmacokinetic barrier, its use in viral and parasitic infections is often an act of clinical desperation or a rescue intervention rather than an evidence-based strategy.

Many systemic antivirals, such as acyclovir, have inherently better CNS penetration than most antibiotics and antifungals [29]. Furthermore, viral CNS disease often manifests as a diffuse parenchymal encephalitis rather than a purulent meningitis, a target that is less accessible to drugs administered into the CSF [30]. This diminishes the fundamental pharmacokinetic rationale for intrathecal antiviral administration. For extremely rare and fulminant parasitic infections like primary amebic meningoencephalitis (PAM), the rapid course and high mortality make controlled trials impossible, so any intervention is inherently experimental. Therefore, the role of IT therapy in these infections is fundamentally different: it is not a primary strategy to overcome a known PK barrier, but a rescue attempt in the face of systemic failure or for ultra-rare, devastating diseases.

Viral Encephalitis (Cytomegalovirus—CMV)

CMV disease of the CNS, which includes encephalitis, ventriculitis, and myelitis, is a rare but often fatal complication, primarily affecting severely immunocompromised patients, such as hematopoietic stem cell transplant (HSCT) recipients and patients with AIDS [38, 40, 41, 42, 44–43]. Standard treatment relies on systemic antivirals like ganciclovir and/or foscarnet [31]. However, the CNS penetration of ganciclovir is poor, reaching only 15%–17% of plasma concentrations, which can lead to a dissociation between control of the virus in the blood and its persistence in the CSF [30].

Evidence for the direct intrathecal administration of antivirals is virtually nonexistent. Instead, case reports have focused on the use of intrathecal high-titer anti-CMV immunoglobulin as a rescue therapy. In some cases of CMV meningitis or encephalitis refractory to systemic treatment, the addition of IT immunoglobulin has led to the clearance of CMV DNA from the CSF [44]. However, success is not guaranteed, and failures of IT immunoglobulin have also been reported [45]. This strategy represents an attempt to provide passive immunity directly into the infected compartment when the host's immune response and systemic antiviral chemotherapy are insufficient.

Parasitic Infections (Primary Amebic Meningoencephalitis—PAM)

PAM, caused by the free-living amoeba *Naegleria fowleri*, is an acute, fulminant, and almost uniformly fatal brain infection contracted from exposure to contaminated freshwater [51–48]. Since death usually occurs within 1–18 days of symptom onset, there is no standard treatment based on clinical trials [38].

Current treatment is a multi-drug cocktail administered systemically, typically including amphotericin B, miltefosine, azoles (such as fluconazole or voriconazole), and rifampicin [49]. Intrathecal drug administration is not part of standard recommendations but has been used in a few reported survival cases worldwide. In what has been described as a "heroic dose" or "kitchen sink" approach, clinicians have attempted every possible route to get drugs into the brain [50]. Notably, a 10-year-old girl in Spain who survived PAM was treated with amphotericin B both intravenously and intrathecally [50]. This use represents a desperate measure in an extreme clinical scenario, rather than an established therapeutic strategy.

4.10 Complications of Intrathecal Therapy for CNS infections and Their Management

Although intrathecal therapy can be life-saving, it is an invasive procedure with a significant risk profile. Complications have been extensively described in other chapters of the book. Basically, complications can be related to the administration procedure itself or to the direct toxicity of the drugs injected into the delicate CNS environmProcedure-Related Complicationent.

Procedure-Related Complications

- **Infection:** Paradoxically, the introduction of pathogens during lumbar puncture or reservoir access for treating CNS infections is an inherent risk. Ommaya reservoir-related infections occur in 5.5%–8% of patients, often within 10 days of access, and can manifest as scalp cellulitis, meningitis, or ventriculitis [15]. Management requires device removal and aggressive systemic and/or intrathecal antibiotic therapy.
- **Hemorrhage and Device Malfunction:** Suboptimal placement of a ventricular catheter can cause injury to CNS structures like the basal ganglia or thalamus and lead to hemorrhage. Although clinically relevant hemorrhage is rare (<1% of cases), it is a potentially devastating complication [15]. Catheter malfunction, such as obstruction or disconnection, is another mechanical complication that can lead to therapeutic failure [1].
- **Post-Dural Puncture Headache:** This is a common complication of LP, thought to be caused by CSF leakage through the dural hole. It is characterized by a postural headache that worsens when standing and improves when lying down. Management includes bed rest, hydration, and, in severe cases, an epidural blood patch [14].

Drug-Induced Neurotoxicity

The administration of high concentrations of drugs directly into the CSF carries an inherent risk of neurotoxicity. It is critical for clinicians to distinguish the symptoms of neurotoxicity from those of a worsening infection, as the treatment is diametrically opposed.

- **Chemical Arachnoiditis and Aseptic Meningitis:** This is one of the most common complications and is characterized by an acute syndrome of headache, back pain, vomiting, fever, and meningism, accompanied by CSF pleocytosis [39]. It is crucial to differentiate it from an infectious meningitis; the diagnostic key is obtaining sterile CSF cultures [51]. Colistin and methotrexate are commonly implicated agents [39]. Treatment consists of discontinuing or reducing the dose of the causative agent, and symptoms are usually reversible [51].

- **Direct Parenchymal and Radicular Toxicity:** The most severe complications involve direct damage to neurons and nerve roots. These can include seizures, encephalopathy, myelopathy (spinal cord injury), paraplegia, and cauda equina syndrome [11–12, 52].
    - **Methotrexate:** Is associated with a broad spectrum of neurotoxicities, ranging from mild, transient symptoms (headache, nausea) to severe, permanent sequelae such as seizures, paralysis, and leukoencephalopathy [41]. The risk increases with prolonged exposure to high CSF concentrations [41].
    - **Aminoglycosides:** Intrathecal gentamicin has been linked in case reports to inflammatory lesions in the brainstem [21].
    - **Cephalosporins:** Have been associated with encephalopathy and non-convulsive status epilepticus, especially in patients with renal failure who cannot effectively clear the drug [21].

Table 3 summarizes the most important complications of intrathecal therapy, aiming to provide a clinical safety tool to help clinicians anticipate, recognize, and manage these potentially devastating adverse effects.

5 The Future of Intrathecal Therapy for CNS Infections

The field of intrathecal therapy is at an inflection point, evolving from a "barrier-bypassing" strategy to a "space-mastering" one. Research is no longer solely focused on how to get drugs into the CSF, but on how to precisely control their behavior—distribution, duration, and concentration—once they are there. This paradigm shift is driven by advances in drug delivery systems (DDS), therapeutic platforms, and device technology, which promise to transform intrathecal therapy from a rescue intervention to a highly refined and controlled treatment modality.

The primary challenge is no longer simply overcoming the BBB. The new frontier is managing the pharmacokinetics within the CNS to achieve a sustained therapeutic effect while minimizing toxicity. This involves addressing the problems of rapid CSF clearance and the peak-and-trough dynamics of bolus injections, which can result in both sub-therapeutic troughs and toxic peaks. Future

Table 3
Common complications of intrathecal therapy: agents, manifestations, and management

Complication type	Common implicated agents	Clinical presentation	Diagnostic clues	Management strategy	References
Chemical arachnoiditis/ Aseptic meningitis	Colistin, Methotrexate, Cytarabine, Amphotericin B	Acute onset (hours post-injection) of headache, back pain, vomiting, fever, nuchal rigidity	CSF pleocytosis with sterile cultures	Discontinue or reduce drug dose, symptomatic treatment	[39]
Encephalopathy	Methotrexate, Cephalosporins (esp. Cefepime), Aminoglycosides	Altered mental status, confusion, lethargy, coma	EEG shows diffuse slowing or triphasic waves	Discontinue drug, consider hemodialysis in renal failure	[21]
Seizures/Nonconvulsive status epilepticus	Methotrexate, Cephalosporins, Colistin (rare)	Clinical seizure activity or subtle (e.g., encephalopathy only)	EEG is diagnostic for non-convulsive status	Discontinue drug, administer antiepileptics (e.g., benzodiazepines)	[21]
Myelopathy/Paraplegia	Methotrexate, Cytarabine	Progressive lower extremity weakness, sensory deficits, and sphincter dysfunction	Spinal MRI to rule out other causes	Discontinue drug, rehabilitation; recovery is often poor	[41]
Cauda equina syndrome	Methotrexate, Cytarabine	Lower extremity weakness, saddle anesthesia, areflexic bladder, and bowel dysfunction	Electrodiagnostic studies show lumbosacral polyradiculopathy	Discontinue drug, rehabilitation; prognosis for sphincter function is poor	[41]

technologies are not just delivery systems, but intrathecal pharmacokinetic modulators.

5.1 Novel Drug Delivery Systems (DDS)

The goal of new DDS is to improve drug residence time, reduce the frequency of injections, and minimize the toxicity associated with high peak concentrations [43].

- **Sustained-Release Formulations:**
  - **Liposomes and Nanoparticles:** These nanocarriers can encapsulate drugs, protecting them from degradation, prolonging their half-life in the CSF, and potentially enhancing parenchymal penetration [6, 53, 54, 55, 56] The design of these particles is an active field of research, with key parameters such as size (ideally <100 nm for better distribution), surface charge, and morphology influencing their fate in the CNS [57]. Brain-targeting lipid nanoparticles are being developed to be delivered via intrathecal injection, opening the door to new antimicrobial therapies for CNS infections [58].
  - **Hydrogels and Polymeric Matrices:** Injectable hydrogels that gel in situ can form a depot in the intrathecal space, providing a constant, uninterrupted release of the drug over a prolonged period. This technology is being actively explored for the treatment of TBM, where maintaining stable concentrations of anti-tuberculous drugs is critical [3].
- **Advanced Therapeutic Platforms:**
  - **Gene-Targeted Therapies:** The intrathecal delivery of antisense oligonucleotides (such as nusinersen for spinal muscular atrophy [46]) and of viral vectors that transport gene-editing tools like CRISPR or RNA interference (RNAi) serves as a blueprint for future anti-infective strategies. Theoretically, therapies could be designed to silence viral replication genes or bacterial virulence factors directly within the CNS [47].
  - **Targeted Toxins:** While primarily explored for the treatment of chronic pain, the concept of delivering toxins that selectively target specific receptors on neurons via the IT route could be adapted for anti-infective purposes, for example, to modulate the neuronal inflammatory response to infection [47].

5.2 Innovations in Delivery Technology

- **Programmable Pumps:** State-of-the-art implantable infusion pumps are programmable, allowing for complex, patient-specific infusion protocols. They can deliver drugs at variable flow rates, in scheduled boluses, or even allow for patient-controlled boluses (PCA) to manage breakthrough pain [59, 60, 61].

These more complex dosing regimens are thought to reduce the development of tolerance and inflammatory granuloma formation at the catheter tip compared to fixed-rate pumps [48].

- **Next-Generation Catheters:** Innovation in catheter design is critical for improving drug distribution. Future designs may include multi-lumen catheters, allowing for the simultaneous but separate administration of incompatible drugs (e.g., an opioid and a local anesthetic) [47]. Another area of development is catheters with multiple micropores along their length. This would allow for a more uniform and sustained distribution of the drug over a larger segment of the neuroaxis, diluting the local concentration and reducing the risk of toxicity and granuloma formation [47, 62].

6 Conclusion: A Multidisciplinary Approach to a Complex Intervention

Intrathecal therapy represents a critical and often life-saving intervention in the arsenal against severe CNS infections. Its ability to bypass the brain's protective barriers and deliver high concentrations of anti-infective agents directly to the site of infection makes it an indispensable tool, especially in the current era of increasing antimicrobial resistance. However, its application is not without risks and complexities. Success depends on a nuanced understanding of CSF pharmacokinetics, careful patient selection, meticulous administration technique, and vigilant monitoring for potential neurotoxic complications.

The safe and effective implementation of intrathecal therapy demands a multidisciplinary team approach, integrating the expertise of neurosurgeons for CNS access, infectious disease specialists for drug selection and therapeutic strategy, neurologists for monitoring response and neurological toxicity, and clinical pharmacologists to optimize dosing [5]. As the field advances, intrathecal therapy is evolving from being primarily a rescue therapy to a more refined, evidence-based intervention. With innovations in drug delivery systems, new therapeutic platforms, and device technology on the horizon, the future of intrathecal therapy promises even greater control over the CNS environment, offering hope for better outcomes for patients with these devastating diseases [10, 13, 23, 24, 26, 19, 20, 63, 35, 36, 37, 16, 62, 64].

Conflict of Interest

The authors have no conflicts of interest to declare that are relevant to the content of this chapter.

References

1. Le Rhun E, Guckenberger M, Smits M et al (2021) EANO-ESMO clinical practice guidelines for diagnosis, treatment and follow-up of patients with brain metastasis from solid tumours. Ann Oncol 32(11):1332–1347
2. Nau R, Blei C, Eiffert H (2020) Intrathecal antibacterial and antifungal therapies. Clin Microbiol Rev 33(3):e00190–19.
3. Bhaskar A, Kumar P, Kumar A et al (2024) An optimized framework for the management of tuberculous meningitis: a comprehensive review. J Pers Med 14(4):412
4. Turgut N, Alabaz D, Yilmaz N et al (2021) Intrathecal/intraventricular antibiotic administrations in postoperative meningitis caused by multidrug-resistant gram-negative bacteria. Can J Infect Dis Med Microbiol 2021:8419485
5. Branch of Bacterial Infection and Drug Resistance Prevention of Chinese Medical Association; Infectious Diseases Specialty Committee, China Medicine Education Association. [Guidelines for the diagnosis and treatment of carbapenem-resistant pseudomonas aeruginosa infection(2026 edition)]. Zhonghua Yi Xue Za Zhi. 2026 Feb 24;106(7):601-617. Chinese. https://doi.org/10.3760/cma.j.cn112137-20250721-01803. PMID: 41688167.
6. Mastandrea C, Pavan M, Varna M et al (2021) Nanoparticles for drug delivery to the central nervous system. Molecules 26(12):3567
7. Al-Hussaini M, Al-Harbi A, Al-Otaibi F et al (2022) Intrathecal/intraventricular colistin for antibiotic-resistant bacterial CNS infections in pediatric population: a systematic review. Children (Basel) 9(3):339
8. Florez-Perdomo A, Echeverria-Pimienta F, Garcia-Duque O et al (2023) Antibiotic therapy by intrathecal or intraventricular approach for postsurgical meningitis or ventriculitis: a systematic review and meta-analysis. Egypt J Neurosurg 38(1):20
9. Tunkel AR, Hasbun R, Bhimraj A et al (2017) 2017 Infectious Diseases Society of America's clinical practice guideline for healthcare-associated ventriculitis and meningitis. Clin Infect Dis 64(6):e34–e65
10. Thomas SK, Smith M, Wilson J et al (2011) Aminoglycoside-induced neurotoxicity. Curr Opin Neurol 24(5):519–525
11. Kim DY, Lee KW, Kim YJ et al (2013) Cauda equina syndrome after intrathecal chemotherapy with methotrexate and cytarabine arabinoside for primary central nervous system lymphoma: a case report. Ann Rehabil Med 37(3):428–433
12. Ali R, O'Brien M, Kelleher T et al (2019) Complications of intrathecal chemotherapy in adults: single-institution experience in 109 consecutive patients. J Oncol Pract 15(5):e436–e444
13. Ommaya AK (1963) Subcutaneous reservoir and pump for sterile access to ventricular cerebrospinal fluid. Lancet 2(7316):983–984
14. Maniar T, Laxmidas M (2013) Intrathecal drug delivery systems for chronic pain. J Anaesthesiol Clin Pharmacol 29(1):17–22
15. Bragina NV, Gorbachev VI, Netesin ES, Petrova IL. Intrathecal antibiotic therapy for neurosurgical infectious complications. Russian Journal of Anesthesiology and Reanimatology. 2023;(1):63-74. (In Russ.) https://doi.org/10.17116/anaesthesiology202301163
16. Song J et al (2021) Efficacy analysis of intraventricular/intrathecal polymyxin B sulphate injections in the treatment of Post-neurosurgical CNS infections in neurocritical ill patients: a prospective observational study. Eur J Clin Microbiol Infect Dis. https://doi.org/10.1007/s10096-025-05234-1.
17. Xu X, Chen C, Chen H et al (2024) Efficacy of intravenous combined with intrathecal/intracerebral ventricle injection of tigecycline in the treatment of post-neurosurgical intracranial infections caused by multidrug-resistant Acinetobacter baumannii. Exp Ther Med 27(6):261
18. Kyriakidis I, Vasileiou E, Pana ZD, Tragiannidis A (2021) Acinetobacter baumannii antimicrobial resistance, and treatment options. J Clin Med 10(11):2347
19. Ince E, Taha F, Yilmaz A et al (2022) Intrathecal colistin induced chemical meningitis in a case of multi-drug resistant Acinetobacter baumannii meningitis. Infect Dis Clin Microbiol 2(1):28–31
20. Khawcharoenporn T, Apisarnthanarak A, Mundy LM (2010) Intrathecal colistin for drug-resistant Acinetobacter baumannii central nervous system infection: a case series and systematic review. Clin Microbiol Infect 16(7):888–894
21. Cascio A, Conti A, Sinardi L et al (2010) Post-neurosurgical multidrug-resistant Acinetobacter baumannii meningitis successfully treated with intrathecal colistin. A new case and a systematic review of the literature. Int J Infect Dis 14(Suppl 3):e240–e244.
22. Shire Human Genetic Therapies (2013) Intrathecal delivery of protein therapeutics to treat genetic diseases involving the CNS
23. Wu Y, Yang J, Yang T et al (2022) Efficacy, safety, and pharmacokinetics of vancomycin for the treatment of central nervous system infections: A systematic review. Front Pharmacol 13:1034335

24. Miyamatsu Y, Tanizaki R. Primary pyogenic ventriculitis due to Streptococcus dysgalactiae treated with intralumbar vancomycin administration: A case report. J Infect Chemother. 2025 Dec;31(12):102854. doi: 10.1016/j.jiac.2025.102854. Epub 2025 Nov 8. PMID: 41213427.
25. Tunkel AR, Hartman BJ, Kaplan SL et al (2004) Practice guidelines for the management of bacterial meningitis. Clin Infect Dis 39(9):1267–1284
26. Kardas N, Wyroba N, Sordyl R, Antkowiak L, Mandera M. Fungal Infections of Cerebrospinal Fluid Shunt in Pediatric Patients: A Systematic Literature Review With a Case Report. World Neurosurg. 2025 Dec;204:124567. https://doi.org/10.1016/j.wneu.2025.124567. Epub 2025 Oct 15. PMID: 41106546.
27. Galgiani JN, Ampel NM, Blair JE et al (2016) 2016 Infectious Diseases Society of America (IDSA) clinical practice guideline for the treatment of coccidioidomycosis. Clin Infect Dis 63(6):e112–e146
28. World Health Organization, 2018. Guidelines for the diagnosis, prevention and management of cryptococcal disease in HIV-infected adults, adolescents and children. World Health Organization, Geneva
29. Williamson PR, Jarvis JN, Panackal AA et al (2017) Cryptococcal meningitis: epidemiology, immunology, diagnosis and therapy. Nat Rev Neurol 13(1):13–24
30. Perfect JR, Dismukes WE, Dromer F et al (2010) Clinical practice guidelines for the management of cryptococcal disease: 2010 update by the Infectious Diseases Society of America. Clin Infect Dis 50(3):291–322
31. Saag MS, Graybill RJ, Larsen RA et al (2000) Practice guidelines for the management of cryptococcal disease. Infectious Diseases Society of America. Clin Infect Dis 30(4):710–718
32. Falagas ME, Bliziotis IA, Tam VH (2007) Intraventricular or intrathecal use of polymyxins in patients with gram-negative meningitis: a systematic review of the available evidence. Int J Antimicrob Agents 29(1):9–25
33. Ng J, Gosbell IB, Kelly JA et al (2006) Cure of multiresistant Acinetobacter baumannii central nervous system infections with intraventricular or intrathecal colistin: case series and literature review. J Antimicrob Chemother 58(5):1078–1081
34. Felton T, Troke PF, Hope WW (2014) Tissue penetration of antifungal agents. Clin Microbiol Rev 27(1):68–84.
35. Pappas PG, Kauffman CA, Andes DR et al (2016) Clinical practice guideline for the management of candidiasis: 2016 update by the Infectious Diseases Society of America. Clin Infect Dis 62(4):e1–e50
36. Zhao Y, Chen Y, Wu X et al (2021) Intrathecal amphotericin B in the treatment of Candida meningitis in an infant: a case report and literature review. BMC Infect Dis 21(1):955
37. Patel R, Greenlee JE (2014) Cerebral mucormycosis: a case report and review of the literature. J Neurol Neurosurg Psychiatry 85(10):1178–1179
38. Cinque P, Marenzi R, Ceresa D et al (2014) Cytomegalovirus central nervous system compartmentalization in a patient presenting with AIDS. Antivir Ther 19(6):627–630
39. Bou-Antoun S, Azar L, El-Khoury M et al (2022) Elsberg syndrome caused by cytomegalovirus primary infection in an immunocompetent patient. Neurol Neuroimmunol Neuroinflamm 9(5):e200079
40. Pang I, Singhabahu S, Novitzky-Basso I et al (2022) Intrathecal cytomegalovirus immunoglobulin for CMV encephalitis post allogeneic stem cell transplantation. IDCases 29:e01608
41. Li Y, Guo L, Jiang E (2021) Early diagnosis and treatment of cytomegalovirus ventriculoencephalitis post-haploidentical donor hematopoietic stem cell transplantation: a case report and systematic review. Front Neurol 12:649429
42. Shah NN, Pandya S, Nanjappa S et al (2018) Ganciclovir-resistant cytomegalovirus encephalitis in a patient with AIDS. Consultant 58(1):22–26
43. Chemaly RF, Chou S, Einsele H et al (2019) Definitions of resistant and refractory cytomegalovirus infection and disease in transplant recipients for use in clinical trials. Clin Infect Dis 68(8):1420–1426
44. Toprak D, Öcal Demir S, Kadayifci EK, Türel Ö, Soysal A, Bakir M. Recurrent Candida albicans Ventriculitis Treated with Intraventricular Liposomal Amphotericin B. Case Rep Infect Dis. 2015;2015:340725. https://doi.org/10.1155/2015/340725. Epub 2015 Oct 8. PMID: 26558119; PMCID: PMC4617703.
45. van den Berg R, van der Eijk AA, Pas SD et al (2024) CNS viral infections—what to consider for improving drug treatment: a plea for using mathematical modeling approaches. Pharmaceuticals (Basel) 17(4):450
46. Centers for Disease Control and Prevention. Amebic Meningitis. Accessed 20 Aug 2024
47. Merck Manual Professional Version. Amebic Brain Infection. Accessed 20 Aug 2024
48. Wikipedia. Naegleriasis. Accessed 20 Aug 2024
49. Baghban A, Malinis M. Ganciclovir and foscarnet dual-therapy for cytomegalovirus encephalitis: A case report and review of the literature. J Neurol Sci. 2018 May 15;388:28-36. https://doi.org/10.1016/j.jns.2018.02.029. Epub 2018 Feb 20. PMID: 29627026.
50. Nakajima M, Akahoshi Y, Takanashi M et al (2014) Intrathecal administration of high-titer

cytomegalovirus immunoglobulin for cytomegalovirus meningitis. Case Rep Hematol 2014:259318
51. Canberra Health Services (2022) Intrathecal chemotherapy administration - Adults and Adolescents at Canberra Hospital
52. Jamil K, Asghar A, Khan S et al (2023) Neurological adverse effects of intrathecal methotrexate in pediatric cancer patients: a cross-sectional study. J Univ Coll Med Dent 13(2):1–6
53. Torchilin VP (2005) Recent advances with liposomes as pharmaceutical carriers. Nat Rev Drug Discov 4(2):145–160
54. Tang WL, Tang WH, An J (2021) The application of nanoparticles in the central nervous system drug delivery. Int J Nanomedicine 16:6623–6644.
55. Householder KT, Dharmaraj S, Sandberg DI et al (2019) Fate of nanoparticles in the central nervous system after intrathecal injection in healthy mice. Sci Rep 9(1):12587
56. Liu J, Chen C, Zhang Y et al (2025) Lipid Nanoparticles Enhance mRNA Delivery to the Central Nervous System Upon Intrathecal Injection. Adv Mater 37(19):e2417097
57. Tunkel AR, Glaser CA, Bloch KC et al (2008) The management of encephalitis: clinical practice guidelines by the Infectious Diseases Society of America. Clin Infect Dis 47(3):303–327
58. Pang I, Singhabahu S, Novitzky-Basso I et al (2022) Intrathecal Cytomegalovirus Immunoglobulin for CMV Encephalitis Post Allogeneic Stem Cell Transplantation. IDCases 29:e01608
59. Yaksh TL, Fisher CJ, Hockman T et al (2022) The future of intrathecal drug delivery and the potential role for new devices and drug-device combination products. Front Pain Res (Lausanne) 3:900566
60. Fitz-Henry J (2014) Intrathecal drug delivery. BJA Educ 14(1):27–31
61. Yaksh TL, Farb DH, Leeman SE et al (1991) Intrathecal capsaicin and other sensory neurotoxins. Ann N Y Acad Sci 632:190–202
62. Lagarce F, Benoit JP (2004) Sustained release formulations for spinal drug delivery. Int J Pharm 285(1-2):13–33.
63. Xie H, Luo P, Li Z, Li R, Sun H, Wu D. Continuous intrathecal administration of liposomal amphotericin B for treatment of refractory Cryptococcus neoformans encephalitis: A case report. Exp Ther Med. 2017 Jul;14(1):780-784. doi: https://doi.org/10.3892/etm.2017.4554. Epub 2017 Jun 7. PMID: 28672999; PMCID: PMC5488679.
64. Chapuis AG, Orozco JJ, Milano F. Intrathecal rituximab for the treatment of Epstein-Barr virus-associated encephalitis. Front Immunol. 2025 Dec 11;16:1736017. https://doi.org/10.3389/fimmu.2025.1736017. PMID: 41459525; PMCID: PMC12739377.

Check for updates

Chapter 8

Classic Intrathecal Therapy in Oncology

Manuel Menéndez-González

Abstract

Intrathecal therapy is a cornerstone of modern neuro-oncology because it delivers drugs directly into cerebrospinal fluid (CSF), circumventing the blood–brain barrier and achieving cytotoxic concentrations within the central nervous system that are rarely attainable systemically. This chapter outlines the physiological and pharmacological foundations that justify this route, emphasizing that CSF is a dynamic, non-homogeneous compartment whose volume, production, circulation, and potential flow obstructions critically shape drug distribution and clearance. It compares lumbar puncture with intraventricular administration via an Ommaya reservoir, highlighting the pharmacokinetic and patient-comfort advantages of ventricular access for prolonged regimens or diffuse leptomeningeal disease, versus the lower procedural invasiveness of the lumbar route for short courses or prophylaxis. The therapeutic armamentarium is reviewed with a focus on approved agents—methotrexate, conventional and liposomal cytarabine, and thiotepa—covering rationale for combinations, induction–consolidation–maintenance schedules, and toxicity profiles. Particular attention is given to chemical meningitis, myelopathy, and methotrexate-associated leukoencephalopathy, as well as the arachnoiditis characteristic of liposomal cytarabine that mandates corticosteroid prophylaxis. Indications are discussed across hematologic malignancies, where intrathecal therapy is embedded in curative protocols, and leptomeningeal metastases from solid tumors, where goals are primarily palliative and response assessment integrates neurological examination, CSF cytology, and contrast-enhanced MRI. Ongoing controversies are addressed, notably the relative roles of intrathecal prophylaxis versus high-dose systemic methotrexate in diffuse large B-cell lymphoma in light of predominantly parenchymal relapse patterns. Finally, international guideline recommendations are synthesized for patient selection, drug choice, and safety monitoring, underscoring the importance of multidisciplinary decision-making guided by intrathecal pharmacokinetics and individual clinical context.

Keywords Intrathecal chemotherapy, Oncology, Methotrexate, Cytarabine, CNS malignancies, Neuraxial cancer therapy

1 Introduction: Fundamentals of Intrathecal Therapy in the Central Nervous System Sanctuary

Intrathecal (IT) therapy is a cornerstone therapeutic strategy in neuro-oncology, designed to deliver pharmacological agents directly into the cerebrospinal fluid (CSF). Its development and application are rooted in the need to overcome one of the greatest challenges in

Manuel Menéndez-González (ed.), *Intrathecal Drug Delivery*, Neuromethods, vol. 234,
https://doi.org/10.1007/978-1-0716-5241-1_8,
© The Author(s), under exclusive license to Springer Science+Business Media, LLC, part of Springer Nature 2026

treating central nervous system (CNS) neoplasms: the blood–brain barrier (BBB). This section explores the physiological and pharmacological principles that underpin intrathecal administration as an indispensable modality in the oncological arsenal.

1.1 The Blood–Brain Barrier (BBB): A Fundamental Therapeutic Obstacle

The CNS is an immunologically privileged and anatomically isolated organ, protected by a highly selective structure known as the blood–brain barrier. The BBB is composed of a complex network of capillary endothelial cells, joined by tight junctions, that line the blood vessels of the brain and spinal cord. This barrier prevents the free exchange of ions, molecules, and cells between the bloodstream and the neural parenchyma [1, 2]. Its primary function is to protect the CNS from fluctuations in plasma composition and from the entry of potentially neurotoxic agents [1, 3].

While this protective function is vital for neurological homeostasis, it represents a formidable obstacle for pharmacotherapy. The vast majority of chemotherapeutic agents administered systemically (intravenously or orally) are hydrophilic or large molecules that cannot effectively cross the BBB [4, 5]. As a result, the concentrations they reach in the CSF and brain tissue are sub-therapeutic, often insufficient to exert a cytotoxic effect on tumor cells. This limitation turns the CNS into a "pharmacological sanctuary," a compartment where neoplastic cells can survive and proliferate despite a favorable systemic response to treatment [5, 6]. This is why diseases such as acute leukemias or high-grade lymphomas, although systemically controlled, can present with isolated relapses in the CNS [7, 8]. There are notable exceptions, such as temozolomide, a lipophilic molecule, or methotrexate administered at high intravenous doses, which achieve clinically significant penetration into the CNS [9]. However, for most drugs, the BBB remains an insurmountable barrier, establishing the fundamental need for alternative administration routes like intrathecal therapy.

1.2 Cerebrospinal Fluid (CSF): Physiology, Dynamics, and Therapeutic Vehicle

IT therapy uses the cerebrospinal fluid as a vehicle for drug distribution. The CSF is a clear, colorless fluid that bathes the brain and spinal cord, occupying the subarachnoid space and the cerebral ventricles. In an adult, the total volume of CSF is approximately 125–150 mL [1]. This volume is not static; the CSF is continuously produced and reabsorbed at a rate of about 20 mL/h, resulting in a complete turnover of the total volume approximately three times a day [1]. This circulation dynamic is crucial, as it is the primary mechanism by which a drug injected into the intrathecal space is distributed throughout the neuroaxis [6].

The composition of the CSF is maintained in a strict balance to regulate the chemical environment of the CNS [1]. In pathological conditions, such as neoplastic infiltration of the meninges (leptomeningeal metastases), CSF dynamics can be profoundly altered.

Tumor cells can form deposits that obstruct CSF flow pathways, causing communicating or obstructive hydrocephalus. This leads to an increase in intracranial pressure, manifesting with symptoms such as headache, nausea, vomiting, and altered level of consciousness [10, 11]. Understanding CSF physiology is therefore essential not only for diagnosis but also for therapeutic planning, as a flow obstruction can prevent the proper distribution of the intrathecal drug and is a contraindication for its administration until resolved [10].

1.3 Rationale for Intrathecal Administration: Bypassing the Barrier to Achieve Cytotoxic Concentrations

The fundamental logic of intrathecal therapy is simple and direct: if the drug cannot cross the barrier, it must be administered on the CNS side of that barrier. IT administration involves the direct injection of chemotherapeutic agents into the subarachnoid space, completely bypassing the BBB [1, 6, 12]. This approach allows for drug concentrations in the CSF that are orders of magnitude higher than what could be achieved with systemic administration, even with very high doses [6].

By administering a relatively low dose directly into the CSF, cytotoxic levels are achieved at the site of meningeal disease, while minimizing systemic exposure and, consequently, associated systemic toxicity (myelosuppression, mucositis, etc.) [6]. The goal is to treat existing disease in the CNS or prevent its appearance (prophylaxis) in neoplasms with a high risk of dissemination to the CSF [4, 13, 14].

However, this strategic approach introduces a new set of challenges. The central principle of IT therapy can be understood as a trade-off: the problem of drug *penetration* across the BBB is solved, but a new challenge is created, that of drug *distribution* within the complex CSF environment. Once the drug is injected into the intrathecal space, its efficacy depends entirely on its ability to circulate with the CSF, reach all meningeal surfaces (including the ventricles, basal cisterns, and deep sulci of the cerebral cortex), and remain at therapeutic concentrations for a sufficient time before being cleared by the rapid CSF turnover [1, 6]. This issue of distribution and intrathecal pharmacokinetics is the driving force behind the evolution of different administration techniques (lumbar puncture versus Ommaya reservoir) and the development of advanced pharmaceutical formulations (such as liposomal cytarabine), all designed to optimize the drug's permanence and circulation in the CNS sanctuary [1].

2 Modalities of Intrathecal Administration: Techniques and Clinical Considerations

The choice of method for administering intrathecal chemotherapy is a crucial clinical decision that impacts the drug's pharmacokinetics, treatment efficacy, patient safety, and quality of life. The two main techniques are lumbar puncture (LP) and administration through an intraventricular Ommaya reservoir.

2.1 Lumbar Puncture (LP): Procedure, Technique, and Optimization

Lumbar puncture is the most common and least invasive method to access the subarachnoid space. It is a technique frequently performed for both diagnostic (obtaining CSF for analysis) and therapeutic (drug administration) purposes [1, 4].

The procedure must be performed under strict aseptic conditions to minimize the risk of infection [4, 15]. Patient positioning is key to the success of the technique; in adults, the lateral decubitus position with knees flexed towards the chest (fetal position) is preferred, while in children, a sitting position is often used [1, 4]. A local anesthetic may be used to reduce the pain associated with the puncture [1].

From a therapeutic perspective, a critical step is the withdrawal of a volume of CSF before the instillation of the drug. To avoid a sharp increase in intracranial pressure and its associated complications (such as severe headache or brain herniation in cases of already elevated pressure), it is recommended to withdraw a volume of CSF equivalent to the volume of chemotherapy to be administered [1]. After withdrawal, the drug should be injected slowly, over 3–5 min, to allow for gradual distribution and minimize meningeal irritation [1].

The patient's position after the procedure is also relevant. Although there is variability in practice, a period of rest in the supine position (lying on the back) for at least one hour is recommended to reduce the incidence of post-dural puncture headache, a common complication caused by CSF leakage through the dural puncture site [1, 4]. Some protocols suggest a prone position (lying on the stomach) to theoretically facilitate the diffusion of the drug towards the cerebral ventricles, although the evidence supporting this practice is limited [1].

2.2 Ommaya Reservoir: Surgical Implantation, Use, and Pharmacokinetic Advantages

For patients requiring frequent or prolonged intrathecal treatment, repeated lumbar punctures can be painful, technically difficult, and psychologically distressing. In these cases, the placement of an Ommaya reservoir may be chosen [1]. This device consists of a small silicone dome surgically implanted under the scalp, connected to a catheter that is inserted into one of the lateral ventricles of the brain [1, 16].

Once implanted, the reservoir allows for easy and repeated access to the ventricular system. Chemotherapy administration is performed by puncturing the dome through the skin with a fine needle [13]. This method not only improves patient comfort but also offers significant pharmacokinetic advantages. By introducing the drug directly into the lateral ventricle, it takes advantage of the natural flow of CSF from its production point in the choroid plexuses, achieving a more homogeneous and predictable distribution throughout the CNS [1, 6]. Pharmacokinetic studies have shown that intraventricular administration via an Ommaya reservoir can achieve drug concentrations in the CSF up to ten times higher than those obtained with the same dose administered via the lumbar route [17].

Despite its advantages, the placement of an Ommaya reservoir is a neurosurgical procedure that carries risks, such as intracranial hemorrhage, infection (meningitis or ventriculitis), catheter malfunction (obstruction or displacement), and seizures [6, 13].

2.3 Comparative Analysis: Lumbar Puncture vs. Ommaya Reservoir

The choice between a lumbar puncture and an Ommaya reservoir is not merely a technical decision but a strategic clinical one that must be based on the underlying disease, the treatment goal, the expected duration of therapy, and the patient's overall condition. A comparison between lumbar puncture and Ommaya reservoir can be found in Table 1, while most commonly used IT chemotherapies can be found in Table 2. Finally, thhe classification and clinical manifestations of neurotoxicity from intrathecal chemotherapy is summarized in Table 3.

Lumbar puncture is less invasive and suitable for short-term prophylaxis or treatment regimens, such as those used in many hematologic malignancies. For example, a patient with acute lymphoblastic leukemia may require a defined and limited number of intrathecal doses as part of their curative protocol, for which LP is often the preferred method [4, 7].

On the other hand, patients with established leptomeningeal metastases from solid tumors face a grim prognosis and often require long-term palliative treatment to control neurological symptoms and improve quality of life [10, 18]. In this scenario, the Ommaya reservoir becomes a superior option. It offers reliable and less painful access for repeated administrations and, more importantly, ensures optimal drug distribution, which is critical for maximizing efficacy

Table 1
Comparison of intrathecal administration modalities (lumbar puncture vs. Ommaya reservoir)

Feature	Lumbar puncture (LP)	Ommaya reservoir (intraventricular)
Invasiveness	Minimally invasive; does not require surgery	Invasive; requires a neurosurgical procedure for implantation [17]
Patient comfort	Can be painful and cause anxiety with each procedure [1]	Once implanted, access is less painful and better tolerated [10]
Drug distribution	Variable and less homogeneous; depends on patient position and CSF flow [1]	Superior; more uniform and predictable distribution through CSF circulation [1, 17]
Drug concentration	Lower concentration in the cerebral ventricles [1]	Significantly higher (up to ten times) throughout the CNS for the same dose [17]
Risk of failed injection	Risk of extrathecal injection (approximately 10%) [17]	Virtually nil if the device is functioning correctly
Procedure risks	Post-dural puncture headache, lumbar pain, nerve irritation, infection, bleeding [1, 13, 15]	Surgical risks: intracranial hemorrhage, infection, cerebral edema [13]
Long-term complications	Minimal, related to repeated punctures (e.g., fibrosis)	Device infection (5%–10%), catheter obstruction or migration [10, 13, 17]
Typical use case	Prophylaxis or short-term treatment (e.g., leukemias, lymphomas) [4, 7]	Prolonged or frequent treatment (e.g., leptomeningeal metastases from solid tumors) [1]

in a diffuse disease. Evidence suggests that this pharmacokinetic advantage may translate into better clinical outcomes; a study in patients with leptomeningeal metastases from solid tumors showed a significantly longer overall survival in those treated via an Ommaya reservoir compared to those treated by LP (9.2 vs. 4.0 months) [17]. Furthermore, intraventricular administration avoids the risk of a failed extrathecal injection, which can occur in approximately 10% of lumbar punctures, and is associated with a lower incidence of complications in this patient population [17].

Therefore, the decision to implant an Ommaya reservoir often reflects a commitment to a more intensive and prolonged intrathecal treatment strategy. It implicitly acknowledges a clinical scenario where maximizing drug delivery to the CNS is of utmost importance, even in a palliative context, weighing the initial surgical risks against the long-term benefits of reliable access and superior pharmacokinetics.

3 Standard Chemotherapeutic Agents for Intrathecal Use

The arsenal of chemotherapeutic drugs approved for intrathecal administration is remarkably limited, reflecting the strict safety and formulation requirements for direct injection into the CNS. Historically, three cytotoxic agents have formed the pillar of this therapy, summarized in Table 2.

3.1 Methotrexate (MTX): The Gold Standard Agent

Methotrexate is a folic acid antagonist antimetabolite that inhibits the enzyme dihydrofolate reductase, thereby blocking the synthesis of purines and thymidylate, essential components of DNA [10]. It has been the most widely used and studied drug in intrathecal therapy for decades, becoming the gold standard for the treatment and prophylaxis of CNS infiltration in a wide range of neoplasms [10, 17, 19].

Its indications cover both hematologic malignancies, such as acute lymphoblastic leukemia and high-grade lymphomas [6, 17, 20, 21], and leptomeningeal metastases from solid tumors [19]. Administration protocols typically follow a phased schedule: an *induction* phase with more frequent doses (e.g., 10–15 mg twice a week), followed by a *consolidation* phase (once a week), and finally, a long-term *maintenance* phase (once a month) to prevent recurrence [10]. Its long-term use is well-characterized, but so is its toxicity profile, which ranges from an acute, transient chemical meningitis to a chronic, progressive, and often irreversible leukoencephalopathy, especially when combined with cranial radiotherapy [10, 22, 23].

Table 2
Standard intrathecal chemotherapy drugs

Drug	Mechanism of action	Main indications	Typical dosing regimen	Key/dose-limiting toxicity	Special considerations
Methotrexate (MTX)	Folic acid antagonist; inhibits DNA synthesis [10]	Leukemias, Lymphomas, Leptomeningeal metastases (solid) [10, 19]	Induction: 10–15 mg, 2/week. Maintenance: 1/month [10]	Neurotoxicity (acute, subacute, chronic), chemical meningitis, myelopathy, leukoencephalopathy [10, 22, 23]	Risk of leukoencephalopathy increases with concomitant cranial radiotherapy [22]
Cytarabine (Ara-C) standard	Pyrimidine analog; inhibits DNA polymerase [10]	Leukemias, Lymphomas [19, 45]	25–100 mg, 2/week (induction). Maintenance: 1/month [10]	Chemical meningitis, myelopathy, cerebellar neurotoxicity (with high doses) [10, 22]	Requires frequent administration due to its short half-life in the CSF [10]
Liposomal cytarabine (DepoCyt®)	Sustained release of cytarabine [10, 24]	Lymphomas, Leukemias (meningeal) [24, 25]	50 mg every 2 weeks (induction). Maintenance: every 4 weeks [10]	Severe chemical arachnoiditis (17%), cauda equina syndrome, headache [10, 25]	Mandatory prophylaxis with corticosteroids (dexamethasone) is required to reduce arachnoiditis [24]
Thiotepa	Alkylating agent; forms cross-links in DNA [10]	Leptomeningeal metastases (solid) [10]	10 mg, 2/week (induction). Maintenance: 1/month [10]	Myelosuppression, chemical meningitis, myelopathy [10, 46]	Less widely used than MTX and Ara-C, but it is a valid option [10]

3.2 Cytarabine (Ara-C): Standard and Liposomal Formulations

Cytarabine, also known as Ara-C, is a pyrimidine analog that, once phosphorylated to its active form, is incorporated into DNA and inhibits DNA polymerase, halting cell replication [10]. It is a mainstay in the treatment of acute leukemias and is also commonly used intrathecally, often in combination or alternation with methotrexate, for managing CNS disease [19].

A significant advancement in intrathecal pharmacotherapy was the development of a sustained-release formulation: liposomal cytarabine (DepoCyt®). In this formulation, the drug is encapsulated in microscopic lipid particles that dissolve slowly in the CSF [10, 24]. While standard cytarabine has a short half-life in the CSF and requires frequent administrations (several times a week), the liposomal formulation maintains cytotoxic concentrations for up to 14 days. This allows for a much more convenient dosing schedule, typically one administration every two weeks, significantly improving the patient's quality of life [10, 24].

However, this pharmacokinetic innovation comes at a toxicological cost. The prolonged exposure of the meninges to the drug and its lipid excipients causes a more intense inflammatory reaction. Liposomal cytarabine is associated with a significantly higher incidence of severe chemical meningitis and arachnoiditis, which can manifest as cauda equina syndrome in up to 17% of patients [10, 25, 26]. This toxicity is so predictable that the concomitant use of corticosteroids (such as dexamethasone) is considered a standard and necessary prophylaxis to mitigate the inflammatory reaction [24]. This example illustrates a fundamental principle in the development of intrathecal therapy: optimizing pharmacokinetics to overcome rapid CSF clearance can introduce new and more severe toxicity profiles, creating a delicate balance between efficacy, convenience, and safety.

3.3 Thiotepa and Other Cytotoxic Agents

Thiotepa is an alkylating agent that acts by forming cross-links in DNA, which interferes with cell replication and transcription. It is the third classic drug approved for intrathecal use, although its use is less common than that of methotrexate and cytarabine [10]. Clinical studies suggest that its efficacy in treating neoplastic meningitis is comparable to that of other standard agents, and it has a different toxicity profile, with less acute neurotoxicity but a potential for myelosuppression [10].

In addition to these three drugs, the search for new options has led to the evaluation of other cytotoxic agents. Topotecan, a topoisomerase I inhibitor, has been studied in clinical trials for intrathecal administration [10]. Other drugs such as etoposide have also been explored in preclinical or early clinical research settings, although they have not achieved widespread standard use [20]. The limited variety of available agents underscores the ongoing need to develop new drugs that are safe and effective for direct administration into the CNS. The use of cell therapies such as CAR T-cell immunotherapy [27, 28, 29, 30] or nanoparticles [31, 32, 33] for brain and spinal cord tumors are beyond the scope of this chapter.

4 Clinical Indications for Intrathecal Therapy in Neoplasms

The application of intrathecal chemotherapy is well-defined for a specific set of oncological scenarios where the disease affects or has a high risk of affecting the central nervous system. The indications are primarily divided into two major categories: hematologic malignancies and metastases from solid tumors.

4.1 Hematologic Malignancies: Prophylaxis and Treatment of CNS Infiltration

In the field of hematologic oncology, IT therapy is not just an option but an integral and often indispensable component of treatment with curative intent.

Acute Leukemias

In Acute Lymphoblastic Leukemia (ALL), CNS infiltration was historically one of the main causes of treatment failure. The introduction of CNS prophylaxis and treatment with intrathecal chemotherapy (often combined with high-dose systemic chemotherapy and/or cranial radiotherapy in the past) transformed the prognosis of the disease, making it one of the success stories of modern oncology [4, 8, 15]. IT therapy is, therefore, an absolute standard of care in all ALL treatment regimens to prevent CNS relapse. In Acute Myeloid Leukemia (AML), the risk of CNS involvement is lower, and IT prophylaxis is reserved for high-risk subtypes or in cases of confirmed meningeal disease.

High-Grade Lymphomas

Similar to leukemias, certain aggressive non-Hodgkin lymphomas have a high propensity to infiltrate the CNS. For very high-grade lymphomas, such as Burkitt lymphoma and lymphoblastic lymphoma, the risk of CNS involvement can reach 30%, making IT prophylaxis a standard and mandatory part of treatment [19].

The role of IT prophylaxis in Diffuse Large B-Cell Lymphoma (DLBCL), the most common type of aggressive lymphoma, is more nuanced and controversial. The overall risk of CNS relapse is approximately 5%, which does not justify universal prophylaxis for all patients [7]. Instead, clinical practice is based on risk stratification. Factors that increase the likelihood of CNS involvement have been identified, such as an elevated serum lactate dehydrogenase (LDH) level along with the presence of more than one extranodal site, or infiltration of specific "sanctuaries" like the testes, breast, or epidural space [7, 19]. In patients presenting with these high-risk factors, CNS prophylaxis is recommended. IT therapy is also used to treat established CNS disease, either at diagnosis or at relapse [19, 34].

4.2 Leptomeningeal Metastases (Meningeal Carcinomatosis) from Solid Tumors

Unlike its often curative role in hematologic malignancies, intrathecal therapy in the context of solid tumors has a predominantly palliative goal. Leptomeningeal metastasis (LM), also known as meningeal carcinomatosis, is the dissemination of cancer cells through the CSF, with diffuse seeding in the leptomeninges. It is a devastating and late-stage complication of advanced cancer, associated with severe neurological morbidity and a very poor prognosis, with a median survival often measured in months [10].

Lung and Breast Cancer

LM is most frequently observed in patients with breast and lung cancer, although it can occur with almost any solid tumor, including melanoma [11, 17, 35]. In these patients, who already have stage IV disease, the goal of intrathecal treatment is not cure, but control of neurological symptoms (such as headache, cranial nerve palsies, radiculopathy), improvement of quality of life, and, in some cases, prolongation of survival [10, 18]. IT therapy is one of the main treatment modalities for LM, often in combination with radiotherapy targeted to areas of bulky or symptomatic disease and systemic chemotherapy that has some CNS penetration [18].

Other Solid Tumors and Primary CNS Tumors

IT therapy is also used in the treatment of leptomeningeal dissemination from other solid tumors, such as melanoma. Additionally, it is a crucial component in the treatment of certain primary CNS tumors, especially in the pediatric population. Tumors like medulloblastoma and ependymoma have a high tendency to spread through the CSF, and intrathecal chemotherapy is used both to treat disseminated disease and to prevent it after surgery and radiotherapy [24].

It is essential to recognize the dichotomy in the purpose and efficacy of IT therapy between hematologic malignancies and solid tumors. In the former, IT prophylaxis is a standard component of a curative strategy, with a solid evidence base demonstrating its ability to prevent CNS relapses and improve overall survival [4, 8, 15]. In the context of LM from solid tumors, the disease is already metastatic and incurable. IT therapy becomes a palliative tool. The evidence supporting a significant survival benefit is weaker and more heterogeneous, and the main goals are the control of neurological symptoms and the cytological clearance of the CSF [18, 35, 36]. This distinction is vital for setting realistic expectations with patients and their families and for making informed decisions about the use of an invasive and potentially toxic treatment in a population with a limited life expectancy.

5 Toxicity and Management of Complications Associated with Intrathecal Therapy

Administering drugs directly into the delicate environment of the central nervous system is not without risks. The complications of intrathecal therapy can be classified into two main categories: those related to the procedure of accessing the subarachnoid space and those arising from the inherent neurotoxicity of the drugs themselves.

5.1 Procedure-Related Complications

Both lumbar puncture and the placement and use of an Ommaya reservoir carry specific risks.

For lumbar puncture, the most common complication is post-dural puncture headache, which is believed to be caused by CSF leakage through the dural hole, reducing CSF pressure and volume. Other common adverse effects include back pain at the puncture site and, less frequently, irritation of the lumbar nerve roots (radiculopathy) [15]. Serious complications are rare but can include epidural or subdural hematomas (especially in patients with coagulopathies or thrombocytopenia) and infections such as iatrogenic meningitis [1, 15].

The Ommaya reservoir involves a neurosurgical procedure, with the inherent risks of brain surgery, such as hemorrhage, edema, and infection [13]. In the long term, the device itself can be a source of complications. The infection rate associated with the reservoir can be as high as 5%–10%, a serious complication that often requires removal of the device and prolonged antibiotic treatment [10]. Other mechanical complications include catheter obstruction, disconnection, or migration, which can prevent proper drug administration and may require surgical revision [13, 17].

5.2 Drug-Induced Neurotoxicity: A Spectrum of Clinical Syndromes

Neurotoxicity induced by intrathecal chemotherapy is the main concern and the dose-limiting factor for this treatment modality. The manifestations can be acute, subacute, or chronic, and vary depending on the drug used.

Chemical Meningitis and Arachnoiditis

This is a sterile (non-infectious) inflammatory reaction of the meninges that occurs shortly after drug administration. The symptoms are similar to those of infectious meningitis and include headache, fever, neck stiffness, nausea, and vomiting [10, 22]. It is a relatively common complication with intrathecal methotrexate. However, it is particularly frequent and severe with liposomal cytarabine, due to the prolonged exposure of the meninges to the drug. In this case, the inflammatory reaction can be so intense that it manifests as severe arachnoiditis, with inflammation and scarring of the nerve roots of the cauda equina, requiring the prophylactic use of corticosteroids for its prevention [10, 15, 25].

Myelopathy and Cauda Equina Syndrome

One of the most devastating, though rare, complications is acute transverse myelopathy, which involves direct damage to the spinal cord. It can manifest with the sudden onset of paraplegia or tetraplegia, loss of sensation below the level of the lesion, and sphincter dysfunction [10, 15, 22]. It has been described with all three standard agents (methotrexate, cytarabine, and thiotepa) and is often irreversible [10, 15]. Cauda equina syndrome, a form of severe radiculopathy affecting the lumbosacral nerve roots, is a particular manifestation of this toxicity, especially associated with liposomal cytarabine [10, 25].

Leukoencephalopathy and Acute/Subacute/Chronic Encephalopathy

Methotrexate is the drug most notoriously associated with a spectrum of encephalopathy syndromes.

- **Acute toxicity:** Can occur hours after administration, manifesting with drowsiness, confusion, or seizures [23].
- **Subacute neurotoxicity:** This distinctive syndrome typically appears between 2 and 14 days after methotrexate administration (both intrathecal and high-dose systemic). It is characterized by the onset of transient focal neurological deficits that mimic a stroke, such as hemiparesis, aphasia, or dysarthria [23, 37]. Although the symptoms can be alarming, this syndrome is generally reversible, with complete recovery in most patients within 48–72 h [23].
- **Chronic leukoencephalopathy:** This is the most severe form of methotrexate neurotoxicity. It is a late complication, developing months or years after treatment, and is largely irreversible. It is characterized by progressive cognitive decline leading to dementia, gait ataxia, and sphincter incontinence [10, 22, 23]. The risk of developing leukoencephalopathy increases dramatically in patients who receive concomitant or prior cranial radiotherapy, suggesting a synergistic toxic effect between the two modalities [15, 22].

The management of these complications requires a high clinical suspicion and close neurological monitoring. It is crucial to differentiate the symptoms of neurotoxicity from those of disease progression, as the clinical presentation can be identical. Treatment is largely supportive, and the most important decision is usually to discontinue the causative agent (Table 3).

Table 3
Classification and clinical manifestations of neurotoxicity from intrathecal chemotherapy

Toxic syndrome	Typical clinical presentation	Typical latency	Most common drugs	Reversibility	Clinical management
Chemical meningitis /arachnoiditis	Headache, fever, neck stiffness, nausea, vomiting, radicular pain [10, 25]	Hours to days [22]	Liposomal cytarabine (frequent/severe), Methotrexate, Standard cytarabine [10, 25]	Generally reversible	Corticosteroids, NSAIDs. Prophylaxis with corticosteroids for liposomal cytarabine [10]
Acute myelopathy	Acute onset paraplegia/ tetraplegia, sensory loss, sphincter dysfunction [10, 15]	Days to weeks [22]	Methotrexate, Cytarabine, Thiotepa [10]	Often irreversible [15, 22]	Discontinue drug, supportive care. No specific treatment [22]
Acute encephalopathy	Confusion, drowsiness, lethargy, seizures [22, 23, 38]	Hours [23]	Methotrexate, Ifosfamide (systemic), Cytarabine (high-dose) [22]	Generally reversible in 10–14 days, but can be fatal [22]	Discontinue drug, supportive care, seizure management
Subacute neurotoxicity (stroke-like)	Hemiparesis, aphasia, dysarthria, transient focal neurological deficits [23]	2–14 days [23]	Methotrexate (IT or high-dose systemic) [23]	Generally reversible in 48–72 h [23]	Supportive care. Dextromethorphan use has been reported [23]
Chronic leukoencephalopathy	Progressive dementia, ataxia, incontinence, personality changes [10, 22, 23]	Months to years [23]	Methotrexate [10, 22]	Irreversible [15, 22]	No effective treatment exists. Leucovorin is not preventive [22]

6 Clinical Controversies and Efficacy Evaluation

Despite being an established treatment modality, intrathecal therapy is not without controversy, especially regarding its role in CNS prophylaxis in certain types of lymphoma and the methods for evaluating its efficacy in leptomeningeal metastases.

6.1 The Debate: Intrathecal Prophylaxis vs. High-Dose Systemic Chemotherapy in Lymphomas

One of the most significant controversies in current hematologic neuro-oncology revolves around the optimal strategy for CNS prophylaxis in patients with high-risk DLBCL. Historically, intrathecal chemotherapy was the standard approach, based on the logic that conventional-dose systemic chemotherapy did not cross the BBB [4]. However, accumulating evidence in the rituximab era has questioned the efficacy of this strategy.

The fundamental problem lies in pharmacokinetics and the pattern of relapse. Pharmacokinetic studies have shown that methotrexate administered intrathecally achieves excellent therapeutic concentrations in the CSF but very poor penetration into the brain parenchyma [8, 19]. At the same time, analyses of relapse patterns in DLBCL patients in the modern era have revealed that the majority of CNS relapses (approximately 70%) are parenchymal (tumor masses within the brain tissue), not leptomeningeal (CSF infiltration) [8, 19]. This creates a critical mismatch: intrathecal prophylaxis effectively protects a compartment (the CSF) that is not the main site of relapse, while leaving the highest-risk compartment (the parenchyma) unprotected.

This observation has spurred the development of an alternative strategy: the administration of high-dose systemic chemotherapy, primarily with methotrexate (HD-MTX). At sufficiently high doses (typically $\geq 3\ g/m^2$), methotrexate can overcome the BBB and achieve therapeutic concentrations in both the CSF and the brain parenchyma [8, 38]. Therefore, theoretically, HD-MTX provides more comprehensive prophylaxis, covering both compartments.

Although definitive randomized clinical trials are lacking, multiple retrospective studies and systematic reviews have found that isolated intrathecal prophylaxis does not significantly reduce the risk of CNS relapse in DLBCL patients treated with modern immunochemotherapy [19, 38]. Conversely, several studies suggest that prophylaxis with HD-MTX is more effective. A retrospective analysis in the high-risk subgroup of primary adrenal/renal lymphoma, for example, found that HD-MTX was superior to IT therapy in both preventing CNS relapse and improving overall survival [38].

This debate represents a broader strategic question: Is it better to bypass the BBB locally (IT therapy) or force passage through it systemically (high-dose therapy)? The current evidence in DLBCL prophylaxis increasingly favors the systemic approach with HD-MTX, relegating IT therapy to a secondary role for patients who are not candidates for HD-MTX (due to advanced age, comorbidities, or renal insufficiency) or as a supplement in cases with already confirmed leptomeningeal involvement [8, 19].

6.2 Response Evaluation in Leptomeningeal Metastases

Evaluating the response to treatment in LM is a complex challenge due to the diffuse and non-nodular nature of the disease. Unlike solid tumors that form measurable masses, LM is a "layer" of tumor cells coating the surfaces of the brain and spinal cord. Therefore, standard response criteria based on tumor measurement (such as RECIST criteria) are not applicable.

Instead, response evaluation in LM requires a multimodal approach that integrates three key components [10]:

1. Clinical Evaluation: A detailed neurological examination is performed to assess the improvement, stabilization, or worsening of the patient's neurological signs and symptoms (e.g., headache, cranial nerve function, motor strength, mental status).
2. CSF Analysis: Serial lumbar punctures or Ommaya reservoir taps are performed to monitor markers in the CSF. The primary goal is to achieve "cytological clearance," i.e., the disappearance of malignant cells in the CSF. Other parameters, such as cell count, protein, and glucose levels, are also monitored.
3. Radiological Evaluation: Magnetic resonance imaging (MRI) with gadolinium of the entire neuroaxis (brain and spinal cord) is used to assess changes in leptomeningeal enhancement. A radiological response is defined by a decrease in this enhancement.

A complete response to treatment is defined as the resolution of neurological signs and symptoms, negativization of CSF cytology, and the disappearance of meningeal enhancement on MRI. The disease is considered stable if there are no significant changes, and progressive if there is a deterioration in any of these three domains [10]. This tripartite evaluation is essential for making decisions about continuing, modifying, or discontinuing intrathecal treatment.

7 Clinical Practice Guidelines and Expert Recommendations (ESMO, NCCN, EANO)

The clinical practice of intrathecal therapy in oncology is guided by the recommendations of several international and national professional organizations. These guidelines, based on the best available evidence and expert consensus, provide a framework for patient selection, drug choice, and treatment management.

7.1 Patient Selection Criteria and Treatment Recommendations

Organizations such as the European Society for Medical Oncology (ESMO), the European Association of Neuro-Oncology (EANO), and the U.S. National Comprehensive Cancer Network (NCCN) have published clinical practice guidelines that address intrathecal therapy.

The joint EANO-ESMO guidelines for the management of leptomeningeal metastases from solid tumors recommend a multidisciplinary approach [17, 39]. They state that systemic therapy (chemotherapy or targeted therapy that can cross the BBB) and focal radiotherapy to symptomatic or bulky lesions are pillars of treatment. Intrathecal therapy is considered an appropriate option for patients with a good performance status and positive CSF cytology, especially if there is no bulky disease or CSF flow blockages [17]. The standard recommended drugs for IT therapy in this context are methotrexate, cytarabine (both standard and liposomal), and thiotepa [10, 17]. The choice among them depends on the primary tumor, prior treatments, and the toxicity profile.

The NCCN guidelines for CNS cancers, such as primary CNS lymphoma, also provide detailed recommendations [4, 40, 41]. For primary CNS lymphoma, first-line treatment is based on high-dose systemic chemotherapy regimens with methotrexate, which has excellent CNS penetration [41]. Intrathecal chemotherapy may be considered in cases with confirmed leptomeningeal involvement by CSF cytology [41]. For prophylaxis in high-risk DLBCL, the guidelines reflect the existing controversy, often favoring the use of systemic HD-MTX over isolated IT therapy, although the latter remains an option for patients who are not candidates for HD-MTX.

The Spanish Society of Medical Oncology (SEOM) also publishes clinical guidelines for the management of CNS tumors and treatment complications, such as neurotoxicity, which serve as a reference for clinical practice in Spain [22, 42].

In general, there is a consensus that the decision to initiate intrathecal therapy should be based on a careful evaluation of the tumor type, the patient's functional status, the presence of controlled or uncontrolled systemic disease, and the absence of CSF flow obstructions. IT therapy is not recommended for patients with a very poor functional status or rapidly progressive systemic disease, as they are unlikely to obtain a benefit that outweighs the risks and burden of treatment [43, 44].

Conflict of Interest

The authors have no conflicts of interest to declare that are relevant to the content of this chapter.

References

1. Allué MC, Gascón MP, Lomba MLA (2017) Quimioterapia intratecal: revisión de la práctica farmacéutica [Intrathecal chemotherapy: a review of pharmaceutical practice]. Farm Hosp 41(1):105–123
2. Ross ME, Sal-Man N (2022) Advancements in drug delivery methods for the treatment of brain disease. Front Vet Sci 9:1039745
3. Generalitat Valenciana, Conselleria de Sanitat Universal i Salut Pública (n.d.) Administración de quimioterapia intratecal.https://www.san.gva.es/documents/d/portal-del-paciente/administracion-de-quimioterapia-intratecal?download=true. Accessed 19 Jul 2025
4. Wang Y, Chen D, Chen F, Ouyang W, Tang H (2025) A review on intrathecal administration of medications for leptomeningeal metastases in solid tumors. Front Pharmacol 16:1472945
5. Varios Autores (2024) Terapias combinadas en el tratamiento de tumores del sistema nervioso central: Revisión bibliográfica [Conference paper]. Fórum Científico Estudiantil de Ciencias Médicas de Cienfuegos 2024, Cienfuegos, Cuba
6. Grupo Español de Linfomas y Trasplante Autólogo de Médula Ósea (GELTAMO) (2017) Guía de práctica clínica sobre linfoma B difuso de célula grande. GuíaSalud
7. Frishman-Levy L, Izraeli S (2017) Advances in understanding the pathogenesis of CNS acute lymphoblastic leukaemia and potential for therapy. Br J Haematol 176(2):157–167
8. Guillén Ponce C, Molina Garrido MJ (n.d.) ¿Qué es, cómo funciona y tipos de quimioterapia? Sociedad Española de Oncología Médica (SEOM). https://seom.org/seomcms/index.php?option=com_content&task=view&id=157&. Accessed 19 Jul 2025
9. Sepúlveda-Sánchez JM, Pardo-Moreno P (2021) Meningitis neoplásica en tumores sólidos: revisión actualizada de diagnóstico, pronóstico, manejo terapéutico y direcciones futuras [Neoplastic meningitis in solid tumours: an updated review of diagnosis, prognosis, therapeutic management, and future directions]. Neurologia 36(9):706–716
10. Memorial Sloan Kettering Cancer Center (n.d.) Información sobre la metástasis leptomeníngea. https://www.mskcc.org/es/cancer-care/patient-education/leptomeningeal-metastases.Accessed 19 Jul 2025
11. National Cancer Institute (n.d.) Quimioterapia intratecal. NCI Diccionario de Cáncer. https://www.cancer.gov/espanol/publicaciones/diccionarios/diccionario-cancer/def/quimioterapia-intratecal. Accessed 19 Jul 2025
12. IBM Micromedex (2025, June 1) Quimioterapia intratecal. Drugs.com
13. Martínez de Castro E et al. (2016) Carcinomatosis leptomeníngea como forma de debut de un adenocarcinoma de pulmón con mutación de EGFR. Concurso +mir de Casos Clínicos para Residentes de Oncología Médica 2016. Sociedad Española de Oncología Médica (SEOM), Spain
14. Universidad de Chile, Facultad de Medicina (n.d.) Quimioterapia Intratecal. Síntesis. https://sintesis.med.uchile.cl/condiciones-clinicas/hemato-oncologia/hemato-oncologia-procedimientos-diagnosticos-y-terapeuticos/11460-quimioterapia-intratecal. Accessed 19 Jul 2025
15. Memorial Sloan Kettering Cancer Center (2024, February 2) Información sobre la colocación del depósito Ommaya
16. American Cancer Society (2019, October 24) Quimioterapia por inyección
17. American Cancer Society (2020, June 22) Quimioterapia para el linfoma no Hodgkin
18. Sociedad Española de Oncología Médica (SEOM) (n.d.) Neurotoxicidad por quimioterapia. In Guía SEOM Uso de los Fármacos Antiepilépticos en Oncología
19. Delgado EF, Ruggeri RA, Calvo C (2023) Neurotoxicidad subaguda por metotrexato en pacientes con enfermedad oncohematológica. Reporte de tres casos. Arch Argent Pediatr 121(2):e202202649
20. Marín, M, 2017. Profilaxis y tratamiento de la infiltración del SNC en los linfomas [Presentation]. Grupo Oncológico para el Tratamiento y Estudio de los Linfomas (GOTEL), Spain
21. National Cancer Institute (2023, March 9) Tratamiento del linfoma primario del sistema nervioso central (PDQ®)–Versión para profesionales de salud
22. Navajas A, Pardo C, Diez B, Sarrion I (2016) Citarabina liposomal para el tratamiento de la diseminación leptomeníngea en tumores del sistema nervioso central pediátricos [Liposomal cytarabine for the treatment of leptomeningeal dissemination in paediatric central nervous system tumours]. An Pediatr 85(2):98–104
23. García-Escrivá A (2016) Estudio de la seguridad de la administración intratecal de fármacos antineoplásicos en pacientes adultos. Dialnet
24. Stewart DJ (2006) A critique of the role of the blood-brain barrier in the chemotherapy of human brain tumors. Crit Rev Oncol Hematol 60(3):205–223
25. Agencia Española de Medicamentos y Productos Sanitarios (AEMPS) (2023, October) Ficha técnica: Citarabina Pfizer 100 mg/ml solución inyectable y para perfusión EFG

26. Sociedad Española de Farmacia Hospitalaria (SEFH) (n.d.) Libro de medicamentos
27. Zugasti I, Espinosa-Aroca L, Fidyt K et al. (2025) CAR-T cell therapy for cancer: current challenges and future directions. Sig Transduct Target Ther 10:210. https://doi.org/10.1038/s41392-025-02269-w
28. Seattle Children's (2023, July) CAR T-cell immunotherapy clinical trials for brain and spinal cord tumors
29. Vitanza NA et al. (2024) CAR T cell therapy for pediatric central nervous system tumors: a review of the literature and current North American trials. J Immunother Cancer 12(11):e009845
30. Bagley SJ et al. (2024) Intrathecal bivalent CAR T cells targeting EGFR and IL13Rα2 in recurrent glioblastoma: phase 1 trial interim results. Nat Med 30(3):764–773.
31. Al-Obaidi M, Zaky W (2024) Advances in intrathecal nanoparticle delivery: targeting the blood–cerebrospinal fluid barrier for enhanced CNS drug delivery. Biomedicines 12(8):2021
32. Bobo D et al. (2024) Intrathecal delivery of nanoparticle PARP inhibitor to the cerebrospinal fluid for treatment of metastatic medulloblastoma. Nat Commun 15: 3891
33. Lavan DA, Tuan J (2020) Intrathecal drug delivery in the era of nanomedicine. Adv Drug Deliv Rev 160:138–156
34. Alam U, Rath S, Ahmad O, Hamed BM, Sajjad F (2025) Assessing efficacy of intrathecal chemotherapy in patients with leptomeningeal metastases: A systematic review and meta-analysis. J Clin Oncol 43(16_suppl):e14000
35. Le Rhun E, Weller M, van den Bent M et al. (2023) Leptomeningeal metastasis from solid tumours: EANO–ESMO clinical practice guideline for diagnosis, treatment and follow-up. ESMO Open 8(5):101624
36. Horbinski C, Nabors LB, Portnow J et al. (2023) NCCN guidelines® insights: central nervous system cancers, Version 2.2022. J Natl Compr Canc Netw 21(1):12–20
37. Sociedad Española de Oncología Médica (SEOM) (2021) Guía clínica SEOM-GEINO de manejo de metástasis cerebrales focalizado en mama, pulmón y melanoma
38. Li Y, Wang Y, Zhang Y, Wang J (2025) Intrathecal administration of PD-1 inhibitor combined with pemetrexed for leptomeningeal metastases from breast cancer: a case report. Front Immunol 16:1567324
39. Tomuleasa C et al. (2023) Therapies beyond physiological barriers and drug resistance: a pilot study and review of the literature investigating if intrathecal trastuzumab and new treatment options can improve oncologic outcomes in leptomeningeal metastases from HER2-positive breast cancer. Cancers 15(9):2508
40. Kunte S et al. (2023) A systematic review and meta-analysis of intrathecal versus alternate routes of delivery for HER2-targeted therapies in patients with HER2+ breast cancer leptomeningeal metastases. J Clin Oncol 41(16_suppl):2034
41. Le Rhun E et al (2024) IT-IO: Intrathecal administration of nivolumab and ipilimumab in combination with systemic combination of nivolumab and ipilimumab in patients with non-small cell lung cancer or melanoma and newly diagnosed leptomeningeal metastasis, a multicentric phase I study [Abstract]. ASCO Annual Meeting
42. Ferraro D et al. (2024) CAR-T cells for the treatment of central nervous system tumours: known and emerging neurotoxicities. Cells 14(12):1220
43. Dana-Farber Cancer Institute (2024, June 1) Clinical trials show promise in treating central nervous system lymphoma, breast cancer, and glioblastoma
44. Pan Z, Li L, Zhang W (2023) A review on the efficacy and safety of intrathecal administration of novel medications for leptomeningeal metastases in solid cancers. Curr Cancer Drug Targets 23(6):456–469
45. Sepúlveda Sánchez JM (2019) Neurotoxicidad asociada a los nuevos fármacos oncológicos [Neurotoxicity associated with new oncologic drugs]. Rev Cáncer 33(2):83–94
46. Xie J et al. (2021) Systemic vs. intrathecal central nervous system prophylaxis in primary adrenal/renal diffuse large B-cell lymphoma: a multi-institution retrospective analysis and systematic review. Clin Lymphoma Myeloma Leuk 21(11):e899–e908

Part III

New Therapeutics and Formulations

Check for updates

Chapter 9

Modern Intrathecal Therapy in Oncology

Manuel Menéndez-González

Abstract

Intrathecal (IT) therapy is entering a pivotal phase in neuro-oncology. Building on decades of experience with a small set of cytotoxics, current programs are redefining what can be delivered to the cerebrospinal fluid (CSF), how long active drug persists, and which compartments are meaningfully exposed. This chapter surveys the field's forward edge across four domains: next-generation cytotoxic and targeted agents selected for meningeal disease biology; intrathecal immunotherapy, including checkpoint blockade and chimeric antigen receptor (CAR) T cells; materials-driven delivery innovations that regulate residence time, spread, and cellular engagement; and the operational framework for personalized, multimodal care that coordinates IT therapy with systemic treatment and focal radiotherapy. We emphasize practical lessons from contemporary trials and guidelines, the pharmacokinetic and safety guardrails that govern dosing in a dynamic CSF space, and emerging tools—CSF liquid biopsy, model-informed scheduling, and composite endpoints—that can shorten learning cycles. Together, these advances point to a future in which intrathecal therapy is no longer a uniform add-on but a mechanism-matched, exposure-controlled intervention integrated with modern systemic regimens to improve outcomes in leptomeningeal and mixed compartment disease.

Keywords Modern oncologic therapeutics, Intrathecal immunotherapy, Targeted therapy, Antibody delivery, CSF pharmacology, Neuro-oncology

1 Introduction

The original rationale for IT therapy—bypassing the blood–brain barrier (BBB) to reach cytotoxic levels in the CNS—remains sound, but the strategy now extends beyond simple barrier circumvention. Contemporary analyses of CNS pharmacology and clinical guidelines for leptomeningeal disease (LMD) frame IT delivery as a compartment-specific modality whose value depends on disease biology, CSF hydrodynamics, and the partner systemic regimen [1-4]. Recent reviews and translational work have consolidated practical insights: sustained CSF exposure correlates with symptom control and cytologic clearance; ventricular access often yields more predictable distribution than repeated lumbar administrations; and response must be judged with composite measures that integrate neurology, CSF biomarkers, and MRI rather than cytology alone [3, 5, 6].

Manuel Menéndez-González (ed.), *Intrathecal Drug Delivery*, Neuromethods, vol. 234,
https://doi.org/10.1007/978-1-0716-5241-1_9,
© The Author(s), under exclusive license to Springer Science+Business Media, LLC, part of Springer Nature 2026

2 New Cytotoxic Agents and Targeted Therapies

A new wave of agents is being selected for intrathecal use, not because they are broadly cytotoxic, but because their mechanisms match meningeal disease biology and their formulations can be engineered for CSF kinetics (Table 1).

Early-phase experience with pemetrexed in LMD—most frequently in non-small-cell lung cancer—shows that carefully titrated intraventricular dosing can achieve neurological benefit and cytologic responses when coordinated with systemic targeted therapy. Practical considerations include slow administration via Ommaya reservoir, interval CSF sampling to confirm exposure, and MRI surveillance of ventricular size and flow to avoid underdosing in patients with drains or shunts [3, 5]. Case-level signals suggest that pairing intrathecal exposure with concurrent checkpoint blockade may be feasible in selected settings, although such combinations remain investigational and require vigilant neurotoxicity monitoring [7-9].

Targeted biologics delivered into CSF exemplify the shift toward mechanism-matched therapy. In HER2-positive breast cancer with LMD, intrathecal trastuzumab has produced symptom improvement

Table 1
Representative agents under investigation for intrathecal oncologic use and key evidence anchors

Agent or class	Rationale in the CSF compartment	Current evidence contours
Pemetrexed (antifolate)	Multitarget folate pathway inhibition with schedules adaptable to CSF kinetics	Early-phase benefit in LMD cohorts when delivered intraventricularly with PK monitoring and systemic co-therapy [3, 5]
Trastuzumab (HER2 mAb)	Mechanism-matched for HER2-positive LMD; direct CSF dosing overcomes BBB	Clinical benefit in series; meta-analysis suggests role is selective relative to modern systemic HER2 agents [10, 11]
Intrathecal checkpoint blockade	Local immune activation within CSF sanctuary	Feasibility with nivolumab/ipilimumab; heterogeneous survival signals; requires composite response assessment [5, 12]
CAR-T cells (systemic or intrathecal)	Cellular effectors reach or are placed into CNS; potential for deep compartment control	High activity in CNS lymphoma (systemic CD19 CAR-T); intraventricular delivery under study for glioma with defined neurotoxicity management [4, 13-16]
Nanoparticle carriers	Programmable release, stability, and targeting in low-protein CSF milieu	Preclinical and translational signals for sustained exposure and payload protection; manufacturing and dosing guardrails defined [6, 17, 18]

and durable disease control in subsets, particularly when used with ongoing systemic HER2 therapy. At the same time, meta-analytic work highlights that newer systemic HER2-directed agents with better CNS reach can provide survival comparable to or exceeding intrathecal antibody delivery, arguing for selective, context-dependent use of the intrathecal route rather than routine substitution for optimized systemic therapy [10, 11]. Broader classes of targeted agents are being explored intrathecally, but their adoption hinges on formulation stability in CSF, dosing schedules compatible with bulk CSF outflow, and a clear advantage over contemporary systemic options [5].

3 Intrathecal Immunotherapy: Checkpoint Blockade and CAR-T Cells

Checkpoint inhibitors administered intrathecally are now in phase I/II evaluation for melanoma and lung cancer LMD. Feasibility has been demonstrated with intraventricular nivolumab and ipilimumab alongside systemic therapy, with pharmacodynamic evidence of immune activation within the CSF compartment. Reported survival outcomes are heterogeneous, but these studies validate the broader principle that immune effectors can be engaged safely within the CNS sanctuary when dose, interval, and CSF flow are controlled and when response is assessed with multimodal criteria [5, 12].

CAR-T therapy has altered assumptions about CNS access. Systemically infused CD19-directed products achieve high complete response rates in primary and secondary CNS lymphoma, establishing that effector cells can traverse the BBB and control disease in parenchyma and meninges [13, 14]. Direct intraventricular administration pushes concentration higher at the site of disease and is under active study for glioblastoma and pediatric CNS tumors, including bivalent constructs targeting EGFR and IL13Rα2. Early reports show feasibility and local activity, with neurotoxicity profiles—immune effector cell-associated neurotoxicity syndrome (ICANS) and tumor inflammation-associated neurotoxicity (TIAN)—that are manageable within standardized neurocritical pathways [8, 9, 15, 16]. Program design now routinely incorporates neuromonitoring protocols, judicious corticosteroid rescue, and cytokine-directed mitigation to preserve antitumor function while limiting toxicity [4, 14-16].

4 Advances in Drug Delivery Systems: Nanotechnology

Materials science is redefining intrathecal pharmacology. Polymeric, lipid, and micellar nanoparticles permit programmable release that flattens peak concentrations and prolongs therapeutic plateaus, reducing dosing frequency and potentially lowering the risk of inflammatory neurotoxicity associated with steep boluses. Particle engineering—size, surface charge, and hydrophilic coronas—can stabilize payloads in low-protein CSF and modulate interaction with meningeal surfaces. Ligand functionalization, including antibody fragments or peptides, is being used to enrich drugs at meningeal tumor clusters or circulating tumor cells, with the longer-term goal of extending exposure into the shallow parenchyma at the CSF–brain interface via perivascular routes [6, 17, 18].

Preclinical translation offers proof of concept. Intrathecal nanoparticle formulations have carried otherwise labile payloads (for example, PARP inhibitors) into CSF with sustained release and measurable antitumor activity in relevant models. As these platforms move toward the clinic, manufacturing must satisfy intrathecal endotoxin and particulate limits, sterilization must preserve carrier integrity and release kinetics, and clinical dosing must be modeled against patient-specific CSF production, posture, and drainage to avoid underexposure through rapid egress [17, 18].

5 Personalization and Multimodal Integration

Personalized intrathecal therapy rests on three pillars (Table 2). Mechanism selection should be driven by contemporaneous CSF profiling: cytology, flow cytometry, and cell-free tumor DNA can reveal CNS-restricted resistance mechanisms invisible in plasma and nominate actionable targets for intrathecal antibodies or small molecules. Exposure control requires adaptive dosing anchored in CSF

Table 2
Practical elements of a personalized intrathecal program

Program element	Operationalization	Primary rationale
Mechanism selection	CSF cytology, flow cytometry, and ctDNA to identify actionable targets	Match agent to CNS resistance biology [2, 3, 5]
Access and distribution	Ommaya reservoir with slow intraventricular dosing; clamp protocols when drains present	Predictable exposure with better tolerability in prolonged courses [2, 23]
Exposure control	Scheduled CSF sampling; model-informed intervals; account for shunt/drain settings	Avoid underexposure and neurotoxic peaks [3, 5, 6]
Response assessment	Composite endpoint across neurology, CSF, and MRI	Earlier discrimination of benefit and futility [2, 3, 5]
Integration with systemic therapy	Coordinate timing with BBB-penetrant regimens and focal radiotherapy	Compartment-appropriate tasking and synergy [2, 3, 5, 21, 22]

sampling and explicit accounting for external ventricular drains or shunts; ventricular access via Ommaya reservoir often enables more consistent distribution and patient comfort during prolonged courses. Finally, response should be gauged with composite endpoints that weight neurological function, CSF molecular remission, and MRI changes across the brain and spine, as endorsed by modern guidelines for LMD management [2, 3, 5, 19-23].

The field continues to refine where intrathecal therapy adds the most value. In diffuse large B-cell lymphoma, contemporary data indicate that high-dose systemic methotrexate reduces CNS relapse in high-risk subsets and may outperform intrathecal prophylaxis, particularly when parenchymal relapse predominates; guidelines increasingly favor systemic CNS-penetrant strategies, reserving intrathecal therapy for specific indications such as documented leptomeningeal involvement or when systemic high-dose regimens are contraindicated [19-22, 24]. Across solid-tumor LMD, intrathecal therapy is evolving from a uniform add-on to a selectively applied, mechanism-matched intervention that complements BBB-penetrant systemic agents and focal radiotherapy according to disease distribution and patient fitness [2, 3, 5, 25].

Conflict of Interest

The authors have no conflicts of interest to declare that are relevant to the content of this chapter.

References

1. Stewart DJ (2006) A critique of the role of the blood–brain barrier in the chemotherapy of human brain tumors. Crit Rev Oncol Hematol 60(3):205–223
2. Le Rhun E, Weller M, van den Bent M et al. (2023) Leptomeningeal metastasis from solid tumours: EANO–ESMO clinical practice guideline for diagnosis, treatment and follow-up. ESMO Open 8(5):101624
3. Wang Y, Chen D, Chen F, Ouyang W, Tang H (2025) A review on intrathecal administration of medications for leptomeningeal metastases in solid tumors. Front Pharmacol 16:1472945
4. Vitanza NA et al. (2024) CAR T-cell therapy for pediatric central nervous system tumors: review of the literature and current North American trials. J Immunother Cancer 12(11):e009845
5. Pan Z, Li L, Zhang W (2023) A review on the efficacy and safety of intrathecal administration of novel medications for leptomeningeal metastases in solid cancers. Curr Cancer Drug Targets 23(6):456–469
6. Lavan DA, Tuan J (2020) Intrathecal drug delivery in the era of nanomedicine. Adv Drug Deliv Rev 160:138–156
7. Li Y, Wang Y, Zhang Y, Wang J (2025) Intrathecal administration of PD-1 inhibitor combined with pemetrexed for leptomeningeal metastases from breast cancer: a case report. Front Immunol 16:1567324
8. Algiraigri AH, Jastaniah W (2025) Intrathecal chemotherapy neurotoxicity: unveiling the challenges of diagnosis, management, and prevention. Pediatr Hematol Oncol 42(3):158-170. Epub 2025 Mar 20. PMID: 40114390. https://doi.org/10.1080/08880018.2025.2471098
9. Delgado EF, Ruggeri RA, Calvo C (2023) Subacute neurotoxicity from methotrexate in onco-hematologic patients: three cases. Arch Argent Pediatr 121(2):e202202649
10. Tomuleasa C et al. (2023) Therapies beyond physiological barriers and drug resistance: a pilot study and review investigating if intrathecal

trastuzumab and new treatment options can improve outcomes in leptomeningeal metastases from HER2-positive breast cancer. Cancers 15(9):2508
11. Kunte S et al. (2023) A systematic review and meta-analysis of intrathecal versus alternate routes of delivery for HER2-targeted therapies in patients with HER2-positive breast cancer leptomeningeal metastases. J Clin Oncol 41(16_suppl):2034
12. Le Rhun E et al (2024) IT-IO: Intrathecal administration of nivolumab and ipilimumab in combination with systemic nivolumab and ipilimumab in newly diagnosed leptomeningeal metastasis from NSCLC or melanoma: a multicentre phase I study [Abstract]. ASCO Annual Meeting
13. Dana-Farber Cancer Institute (2024) Clinical trials show promise in treating central nervous system lymphoma, breast cancer, and glioblastoma. Retrieved December 13, 2025, https://www.dana-farber.org/newsroom/news-releases/2024/clinical-trials-show-promise-in-treating-central-nervous-system-lymphoma-breast-cancer-and-glioblastoma
14. Zugasti I, Espinosa-Aroca L, Fidyt K et al. (2025) CAR-T cell therapy for cancer: current challenges and future directions. Signal Transduct Target Ther 10:210
15. Ferraro D et al. (2024) CAR-T cells for the treatment of central nervous system tumours: known and emerging neurotoxicities. Cells 14(12):1220
16. Bagley SJ et al. (2024) Intrathecal bivalent CAR T cells targeting EGFR and IL13Rα2 in recurrent glioblastoma: phase 1 trial interim results. Nat Med 30(3):764–773.
17. Al-Obaidi M, Zaky W (2024) Advances in intrathecal nanoparticle delivery: targeting the blood–cerebrospinal fluid barrier for enhanced CNS drug delivery. Biomedicines 12(8):2021
18. Bobo D et al. (2024) Intrathecal delivery of nanoparticle PARP inhibitor to the cerebrospinal fluid for treatment of metastatic medulloblastoma. Nat Commun 15:3891
19. Xie J et al. (2021) Systemic vs. intrathecal CNS prophylaxis in primary adrenal/renal diffuse large B-cell lymphoma: a multi-institution retrospective analysis and systematic review. Clin Lymphoma Myeloma Leuk 21(11):e899–e908
20. Grupo Español de Linfomas y Trasplante Autólogo de Médula Ósea (GELTAMO) (2017) Guía de práctica clínica sobre linfoma B difuso de célula grande. GuíaSalud. Retrieved December 12, 2025, https://www.geltamo.com/images/PDF/Guia_LCGBD_v1_1.pdf
21. Horbinski C, Nabors LB, Portnow J et al. (2023) NCCN guidelines® insights: central nervous system cancers, Version 2.2022. J Natl Compr Cancer Netw 21(1):12–20
22. Baskaran AB, Bhatia A, Kumthekar P, Boire A, Lukas RV (2023) Cerebrospinal fluid-administered therapies for leptomeningeal metastases from solid tumors. Future Oncol 19(26):1801-1807. Epub 2023 Sep 22. PMID: 37737023; PMCID: PMC11835020. https://doi.org/10.2217/fon-2022-0926
23. Spirollari E, Vazquez S, Durkin D, Fliegelman J, Carpenter A, Dominguez J, Hanft S (2023) DDEL-08. THE OMMAYA IN LEPTOMENINGEAL DISEASE AND CENTRAL NERVOUS SYSTEM MALIGNANCIES: A RESERVOIR OF THE PAST OR SOURCE OF THE FUTURE? Neuro Oncol 25(Suppl 5):v103. PMCID: PMC10639768. https://doi.org/10.1093/neuonc/noad179.0387
24. Navajas A, Pardo C, Díez B, Sarrión I (2016) Liposomal cytarabine for leptomeningeal dissemination in pediatric CNS tumors. An Pediatr 85(2):98–104
25. Allué MC, Gascón MP, Lomba MLA (2017) Quimioterapia intratecal: revisión de la práctica farmacéutica. Farm Hosp 41(1):105–123

Check for updates

Chapter 10

Intrathecal Therapies for Neurodegenerative Diseases

Manuel Menéndez-González

Abstract

Neurodegenerative diseases (NDDs) pose an immense global health burden, and developing effective treatments is hindered by the blood–brain barrier (BBB). Intrathecal (IT) administration, delivering therapeutics directly into the cerebrospinal fluid (CSF), bypasses the BBB, offering a promising avenue for antisense oligonucleotides (ASOs), gene therapies, antibodies, and stem cells. This review synthesizes the current landscape of IT therapies for Alzheimer's, Parkinson's, Huntington's, and Amyotrophic Lateral Sclerosis, drawing from literature and ClinicalTrials.gov. We highlight key trials and approaches, including the success of ASOs in Spinal Muscular Atrophy and recent progress in other NDDs. However, the efficacy of these novel treatments is often constrained by the limitations of first-generation IT delivery systems, which struggle with uneven distribution, systemic leakage, and the demands of modern biologics. Inspired by recent research and developments, we underscore the critical shortcomings of current devices and detail the innovations shaping next-generation systems: subcutaneous access ports, CSF-flow platforms, AI-driven adaptive dosing, nanoporous membranes, intrathecal pseudodelivery, and hydrogel scaffolds. We conclude by emphasizing the urgent need for these advanced IT drug delivery systems, alongside rigorous comparative assessments, cost-benefit analyses, and clear regulatory pathways, to fully realize the potential of emerging CNS therapies and transform NDD management.

Keywords Neurodegenerative diseases, Intrathecal administration, Drug delivery systems, Alzheimer's disease, Parkinson's disease, Huntington's disease, Amyotrophic lateral sclerosis, Clinical trials, Antisense oligonucleotides, Gene therapy, Intrathecal pumps, Pseudodelivery, CSF dynamics

Abbreviations

AAV	Adeno-associated viruses
AD	Alzheimer's disease
ALS	Amyotrophic lateral sclerosis
ARIA	Amyloid-related imaging abnormalities
Aβ	Amyloid beta
CSF	Cerebrospinal fluid
HD	Huntington disease
IT	Intrathecal

Manuel Menéndez-González (ed.), *Intrathecal Drug Delivery*, Neuromethods, vol. 234,
https://doi.org/10.1007/978-1-0716-5241-1_10,
© The Author(s), under exclusive license to Springer Science+Business Media, LLC, part of Springer Nature 2026

NDD	Neurodegenerative disorder
PD	Parkinson's disease
siRNA	small interfering RNA

1 Introduction

Neurodegenerative diseases (NDDs) encompass a wide array of chronic, progressive disorders characterized by the gradual loss of neurons and represent a growing burden for aging populations worldwide, posing significant challenges for healthcare systems due to their high morbidity, long disease course, and the lack of curative treatments [1]. This heterogeneous group of disorders comprises, among others, Alzheimer's disease (AD), Parkinson's disease (PD), Huntington's disease (HD), and Amyotrophic Lateral Sclerosis (ALS). AD is the most common cause of dementia, affecting over 50 million individuals worldwide [2]. Clinically, it is characterized by progressive memory loss, cognitive dysfunction, and behavioral changes, primarily resulting from the accumulation of extracellular amyloid-beta (Aβ) plaques and intracellular neurofibrillary tangles composed of hyperphosphorylated tau protein [3]. These aggregates disrupt synaptic function, trigger neuroinflammation, and promote neurodegeneration, starting in the hippocampus and ultimately affecting the entire cerebral cortex. PD, the second most frequent NDD, is a progressive movement disorder primarily resulting from the degeneration of dopaminergic neurons in the substantia nigra pars compacta [4]. Clinically, PD is characterized by motor symptoms, including tremor, bradykinesia, rigidity, and postural instability, as well as non-motor symptoms such as cognitive impairment, depression, and autonomic dysfunctions, which significantly impact the patient's quality of life [5]. The pathological hallmarks of PD include the presence of Lewy bodies, composed of misfolded α-synuclein, and extensive loss of dopaminergic neurons [6]. HD is a rare, autosomal dominant NDD caused by a CAG trinucleotide repeat expansion in the HTT gene, leading to the production of mutant huntingtin protein [7]. This toxic protein accumulates in neurons, causing progressive motor dysfunction, psychiatric symptoms, and cognitive decline [8]. Compared to the other NDDs, HD is unique in that the genetic cause is known and consistent across all affected individuals, making it a prime candidate for targeted molecular therapies. Finally, ALS is a rapidly progressive neurodegenerative disorder that affects both upper and lower motor neurons, leading to muscle weakness, atrophy, fasciculations, and ultimately respiratory failure [9]. Most cases are sporadic, but approximately 10% are familial, often linked to mutations in genes such as SOD1, C9orf72, and TARDBP [10].

Despite extensive research, treatment strategies for NDDs remain largely symptomatic, offering only modest improvements in quality of life and limited slowing of disease progression. The complexity of

NDDs arises from the incompletely understood, diverse etiologies, as well as the co-existence of multifactorial pathogenic mechanisms, including protein misfolding and aggregation, mitochondrial dysfunction, oxidative stress, excitotoxicity, impaired axonal transport, and chronic neuroinflammation [11]. Moreover, the blood–brain barrier (BBB), the highly selective semi-permeable border that protects the central nervous system (CNS), simultaneously restricts the entry of potentially therapeutic agents [12–22]. This has significantly hindered the effective delivery of emerging therapies, particularly those based on large or polar molecules such as antisense oligonucleotides (ASOs), monoclonal antibodies, viral vectors for gene therapy, and stem cells.

In this context, intrathecal (IT) drug administration—the delivery of therapeutic agents directly into the CSF—has emerged as a promising strategy to bypass the BBB and enable a more direct access to the CNS [13–37]. IT delivery facilitates higher CNS drug concentrations, reduces systemic exposure, and might open new therapeutic avenues for previously untreatable conditions. Notably, it has gained traction as a feasible route for advanced therapies, including ASOs, gene therapy vectors, immunotherapies, and stem cell-based interventions.

The aim of this chapter is to synthesize the current landscape of IT therapies for the afore-mentioned NDDs, based on the current literature and ClinicalTrials.gov. The most relevant trials and therapeutic approaches are highlighted, including the success of ASOs in Spinal Muscular Atrophy and recent progress in other NDDs. To offer a precise picture of the situation, the limitations of first-generation IT delivery systems are also presented, which struggle with uneven distribution, systemic leakage, and the demands of modern biologics to transform NDD management and improve patients' quality of life and lifespan.

2 Intrathecal Therapies in NDDs: A Clinical Trial Landscape

We have structured this section in a systematic manner to cover the most significant clinical trials conducted in recent years, with a focus on the most relevant NDDs.

2.1 Alzheimer's Disease and other Dementias

While AD remains the most prolific among NDDs in terms of clinical trials of novel therapeutic compounds [23, 24, 38–62], an incompletely explored part is related to IT delivered anti-AD drugs. Based on current literature and ClinicalTrials.gov, three main classes of pharmaceutics are relevant: ASOs targeting Tau protein and Amyloid Precursor Protein (APP), anti-amyloid beta (Aβ) antibodies, and gene therapies.

The development of ASOs that modulate RNA expression represents a transformative approach in the treatment of NDDs. In AD, ASOs target either tau pathology, with BIIB080 a representative compound, or APP. Hyperphosphorylated tau remains one of the main pathological changes, strongly correlated with AD's clinical progression, making it an attractive target for ASO therapy [38–40]. BIIB080 (also known as IONIS-MAPTRx) is a tau-targeted ASO developed through a collaboration between Ionis Pharmaceuticals and Biogen [38–40]. Administered intrathecally, BIIB080 binds to the pre-mRNA of the MAPT gene, promoting its degradation via RNase H-mediated cleavage, thereby reducing tau protein synthesis. Initial clinical evaluation of BIIB080 was the object of a Phase 1b randomized, placebo-controlled trial (NCT03186989) [30–40]. This dose-escalation study enrolled patients with mild AD and demonstrated favorable safety and tolerability up to 60 mg IT every 12 weeks. Notably, CSF total tau and phosphorylated tau levels decreased in a dose-dependent manner, with reductions sustained for several months post-injection. These findings were corroborated by exploratory neuroimaging endpoints, which suggested potential stabilization of disease-related atrophy patterns. The trial's results laid the groundwork for a larger Phase 2 study (NCT04986707), which is currently enrolling participants to assess long-term efficacy and cognitive outcomes [38–40].

Although less developed than tau-targeted approaches, ASOs directed at APP mRNA represent an emerging strategy for reducing Aβ production, as Aβ is a key component of senile plaques. Preclinical models have demonstrated that APP-lowering ASOs can lead to substantial reductions in Aβ42 without perturbing essential physiological functions [38–40]. While no APP-targeted ASO has yet entered late-stage clinical trials for AD, early-phase programs are in development, leveraging the IT route for precise CNS delivery. These programs aim to validate the concept of upstream mRNA intervention, potentially synergizing with or replacing anti-Aβ monoclonal antibodies [40–42]. Still, several challenges remain in the application of ASOs in AD management. Widespread tau expression requires regionally broad CNS distribution, which is currently limited by CSF flow heterogeneity. Dosing schedules involving IT injections every 12–16 weeks may present barriers to compliance in an elderly population [40–42]. Nevertheless, the preliminary promising data from BIIB080 clinical trials suggest that IT ASO therapy is biologically active and well-tolerated in AD, meriting continued investigation.

Monoclonal antibodies targeting Aβ (e.g., aducanumab, lecanemab) have demonstrated modest clinical benefits in early-stage AD when administered intravenously [42–45]. However, systemic delivery often results in limited CNS penetration, with only ~0.1%–0.5% of the injected dose crossing the BBB [43]. This inefficiency not only increases the required dose and costs but also contributes to peripheral adverse effects, such as amyloid-related imaging abnormalities (ARIA) [44–46]. A potential strategy to overcome these limitations could be the IT administration of anti-Aβ antibodies. By delivering antibodies directly into the CSF, higher CNS concentrations may be achieved with lower total doses, potentially enhancing efficacy while reducing systemic exposure. One of the first IT antibody trials in AD is NCT03397506, a Phase 1 study evaluating the safety and pharmacokinetics of an investigational anti-Aβ monoclonal antibody administered via the intrathecal route [45–47]. This open-label, dose-escalation study enrolled subjects with early symptomatic AD and examined various dosing regimens to determine optimal CNS bioavailability. Although detailed results have not yet been published, preliminary data suggest that IT delivery results in significantly higher CSF-to-plasma concentration ratios compared to IV administration, confirming the pharmacological rationale of this approach. With many unknowns still remaining, future trials are expected to assess not only pharmacokinetics but also target engagement (e.g., CSF Aβ reduction, PET imaging) and cognitive outcomes. Even this medication might raise challenges and have associated risks. For example, CSF dynamics and antibody distribution can have individual variability, complicating dosing standardization and requiring tailored therapies. Immunogenicity and ARIA risk remain concerns; however, initial findings suggest that intrathecal administration, by avoiding peripheral accumulation, may reduce the incidence of ARIA.

Finally, an interesting and promising therapeutic avenue in NDDs comprises gene therapy, offering the potential to provide sustained, and in some cases, one-time disease-modifying interventions. Gene therapy in AD aims to either enhance neuroprotection or modulate pathogenic processes, such as amyloidogenesis and tau hyperphosphorylation [46]. The IT route provides a compelling delivery platform for gene therapies, offering direct access to the CNS, particularly for vector-based approaches. Historically, gene therapy in AD focused on targeted intracerebral delivery of neurotrophic factors such as nerve growth factor (NGF) [47] or brain-derived neurotrophic factor (BDNF) [48], based on the observation that basal forebrain cholinergic neurons are particularly vulnerable to degeneration in early AD. However, intracerebral injection is invasive, has limited distribution, and is logistically challenging. The IT route presents an alternative that is both less invasive and potentially capable of achieving more widespread CNS gene expression, particularly when using engineered viral vectors with enhanced neuronal tropism. Among the various vector platforms, adeno-associated virus (AAV) has emerged as the most widely used for CNS-directed gene therapy, due to its favorable safety profile, low immunogenicity, and capacity for long-term gene expression [49]. Specific serotypes, particularly AAV9 and AAVrh10, have demonstrated robust CNS tropism when administered via the IT route [50]. AAV9 has been shown to efficiently transduce neurons, astrocytes, and oligodendrocytes throughout the spinal cord and brain, especially in younger individuals and animal models [51]. AAVrh10, derived from non-human primates, offers similar advantages and may be less susceptible to neutralizing antibodies in the general population [52]. Preclinical studies have demonstrated that IT administration of these vectors can result in widespread CNS gene expression, particularly when administered at the lumbar cistern or via cisterna magna injection. The biodistribution achieved with IT delivery has been shown to encompass hippocampal, cortical, and thalamic regions—areas of high relevance in AD pathology.

The primary therapeutic strategies under investigation for IT gene therapy in AD can be categorized into three main areas: neurotrophic support, modulation of amyloid and tau, and support for synaptic and mitochondrial function [54]. With early-stage AD being characterized by a loss of cholinergic neurons in the basal forebrain, neurotrophic factors such as NGF and BDNF have been investigated for their potential to support these neurons. In early clinical studies (e.g., NCT00087789), NGF gene delivery via stereotactic injections has demonstrated long-term neuronal viability, as evidenced by the upregulation of choline acetyltransferase (ChAT) and persistent transgene expression for several years [54]. However, the limited distribution and surgical risk prompted the exploration of IT delivery as a less invasive, scalable approach. Preclinical IT-AAV-NGF studies have demonstrated successful transgene expression in hippocampal and cortical neurons, accompanied by improvements in memory tasks.

Another promising application is the delivery of genes that either suppress amyloidogenic pathways or promote tau clearance. Preclinical models have utilized AAV-mediated delivery of neprilysin, an enzyme that degrades Aβ, resulting in a reduced plaque burden and improved cognition [55]. Similarly, AAV vectors encoding tau-targeting intrabodies or microRNAs have shown efficacy in reducing tau pathology when delivered via the CSF [56]. More recent strategies aim to enhance synaptic plasticity and metabolic resilience in AD neurons. PGC-1α, a regulator of mitochondrial biogenesis, has been delivered via AAV to improve neuronal energy metabolism and reduce oxidative stress [57]. Similarly, overexpression of synaptic regulators such as PSD95 or Homer1a has been proposed to counteract synaptic failure; however, these remain primarily in early development [58, 59].

While many IT gene therapy strategies for AD are in preclinical or early translational stages, several clinical trials and IND-enabling programs are underway. For example, a Phase 1/2 program (not yet publicly registered) is reportedly investigating IT-AAV-BDNF delivery in early-stage AD patients, building on animal data showing improved memory and synaptogenesis. Other industry-sponsored programs are targeting tau expression modulation using gene therapy vectors encoding tau-lowering RNA interference sequences, administered intrathecally for maximal parenchymal access. Although the Gene Therapy Advisory Committee (GTAC) and regulatory bodies have raised concerns about vector immunogenicity, long-term expression control, and off-target effects, advances in self-limiting vectors, inducible promoters, and immunosuppression regimens are helping mitigate these risks [60–62]. Despite the promise, several challenges limit the near-term translation of IT gene therapy in AD: Nonuniform Vector

Distribution, as even with intrathecal delivery, complete cortical and hippocampal coverage remains difficult, especially in older patients with altered CSF dynamics or brain atrophy; Preexisting Immunity to AAV Serotypes: Up to 60% of the population harbors neutralizing antibodies to common AAVs, which may reduce efficacy or pose safety risks; Dose Escalation Risks: High-dose IT vector administration, as seen in spinal muscular atrophy (SMA) gene therapy, carries risks of aseptic meningitis, liver toxicity, and sensory neuronopathy. The coming years will be critical in determining whether gene therapy, along with the other aforementioned medication administered intrathecally, can transition from theoretical promise to daily clinical practice, improving the prognosis of AD patients. Other genetic dementias may also be targeted intrathecally. For instance the upliFT-D study (Passage Bio) is a multinational Phase 1/2 clinical trial evaluating the safety and efficacy of PBFT02, an intracisternal gene therapy for patients with Frontotemporal Dementia with pathogenic variants in the progranulin or c9orf72 genes [63, 64].

2.2 Parkinson's Disease

When considering PD, although current pharmacotherapies are more diverse than anti-AD medications and provide satisfactory motor symptom relief in most patients, they do not modify disease progression [65–67]. Furthermore, their efficacy diminishes over time, with late-stage PD patients suffering from motor fluctuations, dyskinesias, and the exacerbation of non-motor symptoms [68]. Consequently, disease-modifying therapies that target the underlying neurodegenerative processes are needed. Gene therapy and neurotrophic factor delivery have emerged as promising strategies, particularly when administered via intrathecal or cisternal routes, having advantages over the peripheral delivery method. This section primarily focuses on gene therapy targeting GBA1 mutations and neurotrophic factor delivery, highlighting both their mechanistic rationale and the translational challenges associated with them.

Mutations in the GBA1 gene, which encodes the lysosomal enzyme glucocerebrosidase (GCase), represent the most common genetic risk factor for PD [69–71]. Both homozygous and heterozygous GBA1 mutations reduce GCase activity, resulting in impaired lysosomal function, accumulation of α-synuclein, and increased neuronal vulnerability [72]. GBA1 mutation carriers tend to have earlier PD onset, faster progression, and more severe non-motor symptoms, particularly cognitive impairment [72–74]. Enhancing GCase activity has thus become a therapeutic strategy aimed at restoring lysosomal homeostasis, improving α-synuclein clearance, and slowing disease progression [75–77]. Small-molecule chaperones and enzyme replacement therapies have been tested with limited success due to insufficient CNS penetration. In contrast, AAV-mediated gene therapy provides a means to deliver the GBA1 gene directly to affected neurons, thereby enabling sustained enzymatic activity following a single administration.

PR001 is an investigational AAV9-based gene therapy that delivers a functional copy of the human GBA1 gene under a CAG promoter [78]. Developed by Prevail Therapeutics (now part of Eli Lilly), PR001 is designed to increase GCase expression throughout the CNS following a single intrathecal administration into the cisterna magna. This delivery site, located at the cranial end of the spinal subarachnoid space, is advantageous for distributing viral vectors to both the brain and spinal cord through CSF circulation, with increased access to midbrain structures such as the substantia nigra. Preclinical studies in rodent and non-human primate models of GBA1-associated PD demonstrated that cisterna magna injection of PR001 leads to widespread vector biodistribution, robust transgene expression in cortical and subcortical regions, increased GCase activity, and reductions in α-synuclein accumulation. Behavioral improvements and neuroprotection were also observed in these models, supporting the translational potential of this approach [79].

PR001 is currently under investigation in a Phase 1/2 clinical trial (PROPEL; NCT04127578) in patients with PD and biallelic or heterozygous GBA1 mutations. The trial is an open-label, dose-escalation study evaluating safety, tolerability, and exploratory efficacy of a one-time cisterna magna infusion of PR001. Early data from this trial have shown an acceptable safety profile at low and intermediate doses, with transient fever and headache as the most common adverse events, with dose-dependent increases in CSF GCase activity, sustained for several months post-administration. Preliminary biomarker signals suggesting a reduction in CSF α-synuclein levels. Still, longer-term follow-up and efficacy data, including motor and cognitive assessments, are pending. A parallel natural history study (NCT04128245) is also being conducted to better understand disease progression in GBA1-PD and facilitate the interpretation of endpoints.

Several challenges and aspects should be addressed to improve the outcomes of this therapy. Firstly, patient selection should be more rigorous. While GBA1 mutations are a logical starting point, their clinical heterogeneity necessitates precise stratification. Further studies are needed to evaluate whether non-GBA1 PD patients may benefit from similar lysosomal-targeted therapies [70–72]. A risk of AAV-neutralizing antibodies and cellular immune responses still remains; therefore, pre-screening for anti-AAV9 titers and optimized immunosuppression protocols are essential [73–75]. Finally, vector engineering to enhance cell-specific tropism and optimize spread to the CSF and CNS remains a priority.

A long-standing therapeutic hypothesis in PD is that the exogenous administration of neurotrophic factors could support the survival of dopaminergic neurons, promote regeneration, and slow disease progression [76]. Among the most studied is glial cell line–derived neurotrophic factor (GDNF), which binds to the GFRα1/RET receptor complex to activate intracellular survival pathways in nigrostriatal neurons [77]. In PD patients, endogenous GDNF expression is significantly reduced in the substantia nigra [80]. Preclinical studies in 6-OHDA and MPTP models have demonstrated that GDNF delivery can reverse dopaminergic degeneration and restore motor function. Initial relevant clinical trials of GDNF in PD include an open-label Phase 1 study in which patients received continuous intraputaminal GDNF infusion, resulting in improved motor scores and PET-documented dopaminergic activity [78]. A subsequent randomized, placebo-controlled Phase 2 trial failed to meet primary endpoints, raising questions about efficacy, placebo response, and delivery variability. Considering the limitations associated with intracerebral infusion, alternative delivery routes—including intrathecal and intracisternal administration—have been explored to achieve broader CNS distribution. Preclinical studies have demonstrated that GDNF delivered intrathecally or via the cisterna magna can diffuse through perivascular spaces and reach midbrain targets; however, the concentrations achieved in the putamen are lower than those achieved via direct infusion [79]. To overcome these limitations, gene therapy vectors encoding GDNF have been developed. AAV2-GDNF was tested in Phase 1/2 trials via stereotactic injection into the putamen [81]. These trials demonstrated safety and target expression but were limited by the slow onset of clinical benefit and concerns about retrograde transport to the substantia nigra. Alternatively, several biotechnology companies and academic groups are now investigating next-generation neurotrophic factor therapies, including: encapsulated cell therapies (e.g., NTCELL), which secrete GDNF locally after implantation into the striatum, synthetic mimetics of GDNF that are small enough to cross the BBB and bind GFRα1/RET receptors, and mRNA therapies encoding neurotrophic factors [81].

Recent studies highlight the potential of continuous intracerebroventricular administration of anaerobic dopamine (A-dopamine) as a transformative therapy for PD. Delivered via an abdominal pump and catheter system, A-dopamine bypasses the limitations of traditional oral levodopa, such as its short half-life, inconsistent absorption, and off-target metabolism, which often lead to motor fluctuations and dyskinesias. Unlike current device-aided therapies (DATs), including subcutaneous and intestinal infusions or deep brain stimulation, A-dopamine demonstrated significant advantages in a Phase 1/2 clinical trial conducted from 2020 to 2023. This trial involved 12 patients who received A-dopamine through pump implantation. Phase 1 confirmed the therapy's safety and titration feasibility, while Phase 2 compared A-dopamine with optimized oral treatments. Results indicated a 60% reduction in levodopa equivalent daily dose (LEDD), increased "on" time without dyskinesias, and fewer "off" periods. Notably, A-dopamine did not induce dyskinesias even at high doses. Safety assessments reported no serious adverse events, with only minor, transient issues such as nausea and orthostatic hypotension observed during dose escalation. Over a 3.5-year period, patients maintained clinical benefits, including a 50% reduction in dyskinesia time and a 6.2-h increase in daily "good autonomy" time. The success of this therapy was integral to technological advancements in delivery mechanisms, particularly programmable micro-pumps and refined catheter placement techniques. These innovations minimized complications and established a safe therapeutic index for long-term treatment [82]. Furthermore, A-dopamine's efficacy was unique, outperforming alternative intracerebroventricular approaches that utilized L-dopa or its derivatives. This study aligns with broader innovations in intrathecal and intracerebroventricular drug delivery, which aim to achieve steady-state drug concentrations, reduce fluctuations, and enhance patient outcomes. These findings position A-dopamine as a promising therapeutic paradigm for advanced PD, capable of addressing severe complications unresponsive to existing therapies, and lay the groundwork for large-scale clinical trials.

2.3 Huntington's Disease

In the context of HD, ASOs are used to selectively degrade HTT mRNA, thereby reducing production of both mutant and, in some cases, wild-type huntingtin protein [83–86]. Preclinical studies in HD mouse and non-human primate models have demonstrated that ASO-mediated HTT suppression results in significant reductions in mHTT levels in the brain and CSF, restoration of transcriptional dysregulation, and improvements in motor performance and survival [84–86]. The pioneering molecule in this domain is Tominersen (formerly IONIS-HTTRx or RG6042), a fully phosphorothioate-modified 2′-O-methoxyethyl ASO developed by Ionis Pharmaceuticals and Roche, targeting both wild-type and mutant HTT mRNA transcripts [87, 88]. It was the first HTT-lowering therapy to enter human trials and remains the most extensively studied ASO in HD to date. The phase 1/2a Trial (NCT02519036) was the first-in-human study demonstrating that monthly IT administration of tominersen led to dose-dependent reductions (up to 40%) of mHTT in CSF in early-manifest HD patients [89]. The drug was well tolerated at lower doses, and exploratory cognitive and motor endpoints trended positively. A phase 3 randomized, placebo-controlled trial (GENERATION HD1 trial – NCT03761849) aimed to evaluate the efficacy of tominersen in 791 early-manifest HD patients [83–86, 90]. Participants received either 120 mg every 8 weeks or every 16 weeks via lumbar puncture. However, in March 2021, the study was halted early due to a lack of efficacy and worsening outcomes in the higher-frequency dosing arm. The data showed modest reductions in mHTT, but the every-8-week group experienced more adverse events, including ventricular enlargement, increased neurofilament light chain (NfL) levels, and clinical deterioration [87]. These findings prompted a need for reassessment of HTT-lowering strategies. Still, post hoc analyses suggested that certain subgroups (e.g., younger patients with lower disease burden) may still benefit from therapy. Currently, a new trial (GENERATION HD2, NCT05686551) is ongoing, with tominersen being administered in lower and less frequent doses to reduce toxicity and maintain target engagement [88, 91, 92].

Given the potential risks of lowering wild-type HTT, another interesting alternative approach is based on allele-selective ASOs that preferentially target mutant HTT mRNA. These designs exploit single-nucleotide polymorphisms (SNPs) in linkage disequilibrium with the CAG expansion to selectively degrade the mutant transcript [89]. Programs such as Wave Life Sciences' WVE-120101 and WVE-120102, which targeted specific SNPs (rs362307 and rs362331), initially showed suboptimal efficacy and were discontinued [93]. However, newer backbones and improved chemistry (e.g., PN backbone) have revived interest in next-generation SNP-targeting ASOs. The WVE-003 program (NCT05032196) is currently under evaluation for safety, tolerability, and target engagement after IT administration [89, 93–95].

Unlike ASOs, gene therapy approaches aim to permanently suppress or repair the HTT gene using viral vectors such as AAVs. These vectors can deliver short-hairpin RNAs (shRNAs), artificial microRNAs (miRNAs), or zinc finger repressors directly into the brain, leading to long-lasting mHTT knockdown [78]. However, the large size of the HTT gene (exceeding AAV capacity) and broad CNS involvement of HD make gene therapy delivery particularly challenging. Consequently, many programs have opted for region-specific intracranial delivery, especially targeting the striatum and thalamus. One example is the AMT-130 (NCT04120493) trial. Developed by uniQure, this AAV5-based gene therapy expresses an miRNA targeting exon 1 of the HTT transcript. Administered via bilateral intrastriatal infusion using convection-enhanced delivery (CED), AMT-130 has shown sustained mHTT knockdown in preclinical models [89, 93, 94]. An ongoing Phase 1/2 trial is assessing the long-term safety, reduction in mHTT in CSF, and imaging-based biomarkers. Voyager Therapeutics' VY-HTT01 (now discontinued) utilized an AAV vector with an artificial miRNA delivered directly into the striatum [89, 93]. Despite promising preclinical data, the program was halted due to concerns over delivery complexity and safety. Currently, no AAV-based HTT-lowering therapy has utilized the intrathecal route in clinical trials, primarily due to limitations in vector spread and tropism. However, advances in capsid engineering (e.g., AAV.PHP.B, AAV9 variants) and promoters may enable IT or cisternal delivery of HTT-targeting vectors in future iterations [89, 93–95].

Beyond gene transfer, other RNA-targeted platforms are being explored for IT delivery. For example, similar to ASOs, siRNAs have been investigated for HTT suppression, though challenges related to stability and delivery persist. Preclinical studies have explored the use of CRISPR/Cas9 or CRISPR interference (CRISPRi) to silence or correct the HTT gene, but the delivery of large Cas9 constructs and potential adverse effects remain significant barriers [89, 93–95]. Adenosine deaminase-mediated RNA editing offers a theoretical method to correct CAG expansions or disrupt mHTT translation. These techniques are in early stages but may eventually complement IT delivery strategies.

2.4 Amyotrophic Lateral Sclerosis

Recent therapeutic advances, particularly in ASOs and gene therapy, offer new hope in the therapeutic landscape of ALS. IT delivery is central to their clinical deployment.

ASOs can target several genes. One common target is SOD1, as SOD1 mutations are among the most frequent genetic causes of ALS, accounting for ~20% of familial ALS cases and ~2% of all ALS [96–99]. Tofersen (BIIB067) is a phosphorothioate-modified 2′-O-methoxyethyl (2′-MOE) ASO developed to selectively degrade SOD1 mRNA, thereby reducing toxic protein accumulation [100–102]. The Phase 1/2 Trial (NCT02623699) consisted of the ascending-dose administration of monthly IT injections of tofersen, which were generally well tolerated and resulted in dose-dependent reductions of CSF SOD1 and neurofilament light chain (NfL) [100–104]. Subsequently, in the phase 3 VALOR Trial (NCT02623699), a double-blind, placebo-controlled trial involving 108 participants with confirmed SOD1 mutations, tofersen failed to reach the primary endpoint. However, significant reductions in plasma NfL and CSF SOD1 were observed, and exploratory analyses suggested a potential long-term benefit in early-treated patients [90]. In April 2023, the FDA granted accelerated approval for tofersen based on biomarker evidence of target engagement, marking the first approved therapy targeting a genetic form of ALS [105].

Another interesting target for ASOs is C9orf72, particularly the hexanucleotide (GGGGCC) repeat expansions in C9orf72, which are present in ~40% of familial ALS cases and 5%–10 % of sporadic ALS cases [87]. ASOs targeting C9orf72 mRNA or intronic repeat-containing RNAs are in active development. For example, BIIB078 (Ionis) was designed to selectively suppress the mutant C9orf72 allele. Despite a strong preclinical rationale, a Phase 1 trial (NCT03626012) was discontinued after failing to demonstrate clinical benefit or favorable biomarker profiles [88]. Wave Life Sciences (WVE-004), targeting repeat-containing transcripts, is currently being evaluated in a Phase 1b/2a trial (FOCUS-C9, NCT04931862). Preliminary results show a dose-dependent reduction in poly(GP) DPRs in CSF, suggesting pharmacodynamic engagement, though clinical data are pending [102–104].

Finally, mutations in FUS, an RNA-binding protein, are implicated in juvenile-onset ALS and drive toxic cytoplasmic mislocalization of the protein [100]. A preclinical ASO program by Ionis and

Biogen targeting FUS mRNA has entered IND-enabling studies, with IT delivery planned based on animal efficacy in spinal motor neurons.

In contrast to ASOs, gene therapy aims for more durable, potentially permanent, disease modification. In ALS, AAV vectors, particularly AAV9 and AAVrh10, are suitable for IT delivery. APB-102 is an AAVrh10 vector delivering an artificial microRNA targeting SOD1 mRNA [104–106]. It is administered via intrathecal injection to ensure widespread spinal cord delivery. A Phase 1/2 trial (NCT04833907) is ongoing to assess safety, tolerability, and target engagement. Preclinical studies demonstrated the efficient knockdown of SOD1 and preservation of motor function in murine models [102–106]. AAV9, the vector backbone for Zolgensma (approved for spinal muscular atrophy), is being adapted for ALS due to its demonstrated ability to cross the BBB when delivered systemically or via IT injection [104–106]. Similar regulatory switches could be used in future ALS constructs to fine-tune gene expression and reduce toxicity.

2.5 Spinal Muscular Atrophy (SMA)

SMA, although not a classical NDD, is important to consider when evaluating IT-administered therapies from several perspectives. This devastating autosomal recessive neuromuscular disorder is caused by bi-allelic deletions or mutations in the survival motor neuron 1 (SMN1) gene [107]. The resultant deficiency of the SMN protein leads to progressive degeneration of anterior horn cells in the spinal cord, culminating in muscular weakness, atrophy, and in severe cases, early mortality [96]. Historically, SMA has been one of the leading genetic causes of infant death, and until 2016, SMA's management was purely supportive [97].

The therapeutic revolution began with the approval of nusinersen, followed by onasemnogene abeparvovec (gene therapy) and risdiplam (oral splicing modulator). Among these, nusinersen has been a landmark, administered directly into the CSF via lumbar puncture [98]. As an ASO, nusinersen modulates SMN2 pre-mRNA splicing to promote exon 7 inclusion, increasing functional SMN protein expression predominantly in the CNS [98]. Several reasons support the approach based on intrathecal administration: efficient delivery to spinal motor neurons, minimal systemic exposure, reduced peripheral side effects, and sustained therapeutic concentration in the CSF due to slow clearance.

Moreover, the results from clinical trials demonstrated the compelling efficacy of IT-delivered nusinersen. For example, the ENDEAR trial (2016), which included infantile-onset SMA, demonstrated that nusinersen significantly improved motor milestones and survival compared to the sham control [99]. Another trial, CHERISH (2017), included patients with later-onset (Type II) SMA who received nusinersen, and showed significant gains in motor function

(HFMSE scores) [100]. The SHINE trial and long-term extensions continued to show durable benefits [101]. Additionally, real-world registries (e.g., SMArtCARE, RESTORE) have confirmed these findings across diverse clinical settings and genotypes. Importantly, early initiation—especially in pre-symptomatic individuals—yields the most profound benefits, aligning with the concept of a therapeutic window before irreversible motor neuron loss [108–115].

Despite favorable outcomes, IT nusinersen delivery has also brought technical and logistical challenges. Particularly in infants and young children, sedation and specialized expertise are required. Patients with severe scoliosis or spinal fusion, commonly encountered in Type II and III SMA, pose therapeutic challenges, as standard lumbar access may be impossible. Image-guided (fluoroscopic or CT) intrathecal delivery and implantable reservoirs (e.g., subcutaneous intrathecal catheters) could be potential solutions. Furthermore, studies are investigating modified ASOs with enhanced stability and tissue penetration, potentially allowing less frequent dosing or alternative routes (e.g., intraventricular).

SMA is a success story that exemplifies the triumph of precision medicine in neurology. Intrathecal therapeutic administration has been instrumental in offering a meaningful outcome to a once-fatal disease. Nusinersen, delivered intrathecally, has set a precedent for treating genetic CNS disorders at their source. Still, there are challenges that need to be overcome, detailed in the following section, with current IT delivery techniques needing refinement.

3 Conclusions

With the increasing incidence and prevalence of NDDs worldwide, but still no curative treatment for general use, IT-administered drug therapies offer a promising avenue in this context. The IT route has several advantages when pharmacologically targeting the CNS, including the ability to elude the BBB, but also poses significant challenges to its adequate implementation. Lumbar puncture, intrathecal catheters, and implantable intrathecal infusion pumps are the primary approaches in IT drug delivery, each with its own strengths and drawbacks.

Considering the most relevant NDDs, IT treatments include mainly ASOs, gene therapy using viral vectors such as AAVs, and siRNAs. With a small number of phase 1 and 2 clinical trials currently in progress, it is difficult to draw definitive conclusions yet. The only medication already approved and with a verified positive impact based on real-world data is nusinersen, used in the treatment of SMA. Still, this success story also demonstrates the multiple anatomical, pharmacological, economic, and sometimes ethical challenges that need to be addressed and overcome before realizing similar IT-administered drugs on the market.

With great perspectives, there are a couple of directions that need to be addressed in order to improve the outcomes of IT-delivered drugs. Firstly, there is a pressing need to improve delivery techniques and resolve device-related limitations. Secondly, there is a high need for standardization of administration protocols and drug formulations. Finally, the economic burden related to the healthcare system necessitates rethinking the infrastructure to respond to the needs of both the administration and follow-up.

Conflict of Interest

The authors have no conflicts of interest to declare that are relevant to the content of this chapter.

References

1. Van Schependom J, D'haeseleer M (2023) Advances in neurodegenerative diseases. J Clin Med 12(5):1709. https://doi.org/10.3390/jcm12051709
2. Safiri S, Ghaffari Jolfayi A, Fazlollahi A et al (2024) Alzheimer's disease: a comprehensive review of epidemiology, risk factors, symptoms diagnosis, management, caregiving, advanced treatments and associated challenges. Front Med (Lausanne) 11:1474043. https://doi.org/10.3389/fmed.2024.1474043
3. Kamatham PT, Shukla R, Khatri DK, Vora LK (2024) Pathogenesis, diagnostics, and therapeutics for Alzheimer's disease: breaking the memory barrier. Ageing Res Rev 101:102481. https://doi.org/10.1016/j.arr.2024.102481
4. Zhou ZD, Yi LX, Wang DQ et al (2023) Role of dopamine in the pathophysiology of Parkinson's disease. Transl Neurodegener 12:44. https://doi.org/10.1186/s40035-023-00378-6
5. Váradi C (2020) Clinical features of Parkinson's disease: the evolution of critical symptoms. Biology (Basel) 9(5):103. https://doi.org/10.3390/biology9050103
6. Dickson DW (2018) Neuropathology of Parkinson disease. Parkinsonism Relat Disord 46(Suppl 1):S30–S33. https://doi.org/10.1016/j.parkreldis.2017.07.033
7. Stoker TB, Mason SL, Greenland JC, Holden ST, Santini H, Barker RA (2022) Huntington's disease: diagnosis and management. Pract Neurol 22(1):32–41. https://doi.org/10.1136/practneurol-2021-003074
8. McColgan P, Tabrizi SJ (2018) Huntington's disease: a clinical review. Eur J Neurol 25(1):24–34. https://doi.org/10.1111/ene.13413
9. Segura T, Medrano IH, Collazo S et al (2023) Symptoms timeline and outcomes in amyotrophic lateral sclerosis using artificial intelligence. Sci Rep 13: 702. https://doi.org/10.1038/s41598-023-27863-2
10. Wang H, Guan L, Deng M (2023) Recent progress of the genetics of amyotrophic lateral sclerosis and challenges of gene therapy. Front Neurosci 17:1170996. https://doi.org/10.3389/fnins.2023.1170996
11. Wareham LK, Liddelow SA, Temple S et al (2022) Solving neurodegeneration: common mechanisms and strategies for new treatments. Mol Neurodegeneration 17: 23. https://doi.org/10.1186/s13024-022-00524-0
12. Kadry H, Noorani B, Cucullo L (2020) A blood–brain barrier overview on structure, function, impairment, and biomarkers of integrity. Fluids Barriers CNS 17:69. https://doi.org/10.1186/s12987-020-00230-3
13. De Andres J, Hayek S, Perruchoud C, Lawrence MM, Reina MA, De Andres-Serrano C, Rubio-Haro R, Hunt M, Yaksh TL (2022) Intrathecal drug delivery: advances and applications in the management of chronic pain patient. Front Pain Res (Lausanne) 3:900566. https://doi.org/10.3389/fpain.2022.900566; Erratum in: (2023) 4:1190014. 10.3389/fpain.2023.1190014
14. Wichmann TO, Damkier HH, Pedersen M (2022) A brief overview of the cerebrospinal fluid system and its implications for brain and spinal cord diseases. Front Hum Neurosci

15:737217. https://doi.org/10.3389/fnhum.2021.737217
15. Sakka L, Coll G, Chazal J (2011) Anatomy and physiology of cerebrospinal fluid. Eur Ann Otorhinolaryngol Head Neck Dis 128(6):309–316. https://doi.org/10.1016/j.anorl.2011.03.002
16. Zhao Y, Gan L, Ren L, Lin Y, Ma C, Lin X (2022) Factors influencing the blood-brain barrier permeability. Brain Res 1788:147937. https://doi.org/10.1016/j.brainres.2022.147937
17. Prager J, Deer T, Levy R et al (2014) Best practices for intrathecal drug delivery for pain. Neuromodulation 17(4):354–372. https://doi.org/10.1111/ner.12146
18. Hablitz LM, Plá V, Giannetto M et al (2020) Circadian control of brain glymphatic and lymphatic fluid flow. Nat Commun 11: 4411. https://doi.org/10.1038/s41467-020-18115-2
19. Hornkjøl M, Valnes LM, Ringstad G et al (2022) CSF circulation and dispersion yield rapid clearance from intracranial compartments. Front Bioeng Biotechnol 10:932469. https://doi.org/10.3389/fbioe.2022.932469
20. Stielow M, Witczyńska A, Kubryń N, Fijałkowski Ł, Nowaczyk J, Nowaczyk A (2023) The bioavailability of drugs—The current state of knowledge. Molecules 28:8038. https://doi.org/10.3390/molecules28248038
21. Yamada S, Kelly E (2016) Cerebrospinal fluid dynamics and the pathophysiology of hydrocephalus: new concepts. Semin Ultrasound CT MR 37(2):84–91. https://doi.org/10.1053/j.sult.2016.01.001
22. Burman R, Alperin N (2024) CSF-to-blood toxins clearance is modulated by breathing through cranio-spinal CSF oscillation. J Sleep Res 33(1):e14029. https://doi.org/10.1111/jsr.14029
23. Li Y, Rusinek H, Butler T, Glodzik L, Pirraglia E, Babich J, Mozley PD, Nehmeh S, Pahlajani S, Wang X, Tanzi EB, Zhou L, Strauss S, Carare RO, Theise N, Okamura N, de Leon MJ (2022) Decreased CSF clearance and increased brain amyloid in Alzheimer's disease. Fluids Barriers CNS 19(1):21. https://doi.org/10.1186/s12987-022-00318-y
24. Hay R, Gupta A (2022) Continuous spinal anaesthesia. BJA Educ 22(8):295–297. https://doi.org/10.1016/j.bjae.2022.03.007
25. Rawal N, Tandon B (1985) Epidural and intrathecal morphine in intensive care units. Intensive Care Med 11(3):129–133. https://doi.org/10.1007/BF00258537
26. Dupoiron D, Autier L, Lebrec N, Seegers V, Folliard C, Patsouris A, Campone M, Augereau P (2023) Intrathecal catheter for chemotherapy in leptomeningeal carcinomatosis from HER2-negative metastatic breast cancer. J Breast Cancer 26(6):572–581. https://doi.org/10.4048/jbc.2023.26.e40
27. Qiu J, Wu L, Qu R et al (2022) History of development of the life-saving drug "Nusinersen" in spinal muscular atrophy. Front Cell Neurosci 16:942976. https://doi.org/10.3389/fncel.2022.942976
28. Blair HA (2023) Tofersen: first approval. Drugs 83(11):1039–1043. https://doi.org/10.1007/s40265-023-01904-6
29. Ericson T, Singla P, Kohan L (2022) Intrathecal pumps. Phys Med Rehabil Clin N Am 33(2):409–424. https://doi.org/10.1016/j.pmr.2022.01.004
30. Jane LA, Wray AA (2025) Lumbar puncture. StatPearls [Internet]. StatPearls, Treasure Island (FL). Available from: https://www.ncbi.nlm.nih.gov/books/NBK557553/
31. Hudgins PA, Fountain AJ, Chapman PR, Shah LM (2017) Difficult lumbar puncture: pitfalls and tips from the trenches. AJNR Am J Neuroradiol 38(7):1276–1283. https://doi.org/10.3174/ajnr.A5128
32. Boster AL, Adair RL, Gooch JL et al (2016) Best practices for intrathecal baclofen therapy: dosing and long-term management. Neuromodulation 19(6):623–631. https://doi.org/10.1111/ner.12388
33. Singh E, Gurses ME, Costello MC et al (2024) Intrathecal chemotherapy for leptomeningeal disease in high-grade gliomas: a systematic review [published correction appears in J Neurooncol. 2024 Mar;167(1):49. 10.1007/s11060-024-04619-0.]. J Neurooncol 167(1):39–47. https://doi.org/10.1007/s11060-024-04582-w
34. Binyamin Y, Azem K, Heesen M et al (2023) The effect of placement and management of intrathecal catheters following accidental dural puncture on the incidence of postdural puncture headache and severity: a retrospective real-world study. Anaesthesia 78(10):1256–1261. https://doi.org/10.1111/anae.16088
35. Knight KH, Brand FM, Mchaourab AS, Veneziano G (2007) Implantable intrathecal pumps for chronic pain: highlights and updates. Croat Med J 48(1):22–34
36. Bolash R, Mekhail N (2014) Intrathecal pain pumps: indications, patient selection, techniques, and outcomes. Neurosurg Clin N Am 25(4):735–742. https://doi.org/10.1016/j.nec.2014.06.006
37. Delhaas EM, Huygen FJPM (2020) Complications associated with intrathecal drug delivery systems. BJA Educ 20(2):51–57. https://doi.org/10.1016/j.bjae.2019.11.002

38. Passeri E, Elkhoury K, Morsink M et al (2022) Alzheimer's disease: treatment strategies and their limitations. Int J Mol Sci 23(22):13954. https://doi.org/10.3390/ijms232213954
39. Rawat P, Sehar U, Bisht J, Selman A, Culberson J, Reddy PH (2022) Phosphorylated tau in Alzheimer's disease and other tauopathies. Int J Mol Sci 23(21):12841. https://doi.org/10.3390/ijms232112841
40. Edwards AL, Collins JA, Junge C et al (2023) Exploratory tau biomarker results from a multiple ascending-dose study of BIIB080 in Alzheimer disease: a randomized clinical trial. JAMA Neurol 80(12):1344–1352. https://doi.org/10.1001/jamaneurol.2023.3861
41. Mummery CJ, Börjesson-Hanson A, Blackburn DJ et al (2023) Tau-targeting antisense oligonucleotide MAPTRx in mild Alzheimer's disease: a phase 1b, randomized, placebo-controlled trial. Nat Med 29:1437–1447. https://doi.org/10.1038/s41591-023-02326-3
42. Lozupone M, Dibello V, Sardone R et al (2023) The development of peptide- and oligonucleotide-based drugs to prevent the formation of abnormal tau in tauopathies. Expert Opin Drug Discov 18(5):515–526. https://doi.org/10.1080/17460441.2023.2200245
43. Hung C, Fertan E, Livesey FJ, Klenerman D, Patani R (2024) APP antisense oligonucleotides reduce amyloid-β aggregation and rescue endolysosomal dysfunction in Alzheimer's disease. Brain 147(7):2325–2333. https://doi.org/10.1093/brain/awae092.
44. Klinkovskij A, Shepelev M, Isaakyan Y, Aniskin D, Ulasov I (2023) Advances of genome editing with CRISPR/Cas9 in neurodegeneration: the right path towards therapy. Biomedicines 11:3333. https://doi.org/10.3390/biomedicines11123333
45. Baryakova TH, Pogostin BH, Langer R, McHugh KJ (2023) Overcoming barriers to patient adherence: the case for developing innovative drug delivery systems. Nat Rev Drug Discov 22(5):387–409. https://doi.org/10.1038/s41573-023-00670-0
46. Kim BH, Kim S, Nam Y et al (2025) Second-generation anti-amyloid monoclonal antibodies for Alzheimer's disease: current landscape and future perspectives. Transl Neurodegener 14:6. https://doi.org/10.1186/s40035-025-00465-w
47. Pardridge WM (2020) Blood-brain barrier and delivery of protein and gene therapeutics to brain. Front Aging Neurosci 11:373. https://doi.org/10.3389/fnagi.2019.00373
48. Hampel H, Elhage A, Cho M, Apostolova LG, Nicoll JAR, Atri A (2023) Amyloid-related imaging abnormalities (ARIA): radiological, biological and clinical characteristics. Brain 146(11):4414–4424. https://doi.org/10.1093/brain/awad188
49. Cummings JL (2025) Maximizing the benefit and managing the risk of anti-amyloid monoclonal antibody therapy for Alzheimer's disease: strategies and research directions. Neurotherapeutics 22(3):e00570. https://doi.org/10.1016/j.neurot.2025.e00570
50. Ataei B, Hokmabadi M, Asadi S et al (2024) A review of the advances, insights, and prospects of gene therapy for Alzheimer's disease: A novel target for therapeutic medicine. Gene 912:148368. https://doi.org/10.1016/j.gene.2024.148368
51. Rodrigues BDS, Kanekiyo T, Singh J (2020) Nerve growth factor gene delivery across the blood-brain barrier to reduce beta amyloid accumulation in AD mice. Mol Pharm 17(6):2054–2063. https://doi.org/10.1021/acs.molpharmaceut.0c00218
52. Arora S, Kanekiyo T, Singh J (2022) Functionalized nanoparticles for brain targeted BDNF gene therapy to rescue Alzheimer's disease pathology in transgenic mouse model. Int J Biol Macromol 208:901–911. https://doi.org/10.1016/j.ijbiomac.2022.03.203
53. Wang D, Tai PWL, Gao G (2019) Adeno-associated virus vector as a platform for gene therapy delivery. Nat Rev Drug Discov 18:358–378. https://doi.org/10.1038/s41573-019-0012-9
54. Issa SS, Shaimardanova AA, Solovyeva VV, Rizvanov AA (2023) Various AAV serotypes and their applications in gene therapy: an overview. Cells 12(5):785. https://doi.org/10.3390/cells12050785
55. Pattali R, Mou Y, Li XJ (2019) AAV9 vector: a novel modality in gene therapy for spinal muscular atrophy. Gene Ther 26:287–295. https://doi.org/10.1038/s41434-019-0085-4
56. Bradbury AM, Rafi MA, Bagel JH, Brisson BK, Marshall MS, Pesayco Salvador J, Jiang X, Swain GP, Prociuk ML, ODonnell PA, Fitzgerald C, Ory DS, Bongarzone ER, Shelton GD, Wenger DA, Vite CH (2018) AAVrh10 gene therapy ameliorates central and peripheral nervous system disease in canine globoid cell leukodystrophy (Krabbe disease). Hum Gene Ther 29(7):785–801. https://doi.org/10.1089/hum.2017.151
57. Wang L, Ma L, Gao Z, Wang Y, Qiu J (2025) Significance of gene therapy in neurodegenerative diseases. Front Neurosci 19:1515255. https://doi.org/10.3389/fnins.2025.1515255
58. Amadoro G, Latina V, Balzamino BO et al (2021) Nerve growth factor-based

therapy in Alzheimer's disease and age-related macular degeneration. Front Neurosci 15:735928. https://doi.org/10.3389/fnins.2021.735928

59. Saxena SK, Ansari S, Maurya VK et al. (2024) Neprilysin-mediated amyloid beta clearance and its therapeutic implications in neurodegenerative disorders. ACS Pharmacol Transl Sci 7(12):3645–3657. https://doi.org/10.1021/acsptsci.4c00400
60. Marino M, Holt MG (2022) AAV vector-mediated antibody delivery (A-MAD) in the central nervous system. Front Neurol 13:870799. https://doi.org/10.3389/fneur.2022.870799
61. Shi HZ, Wang YJ, Wang YX et al (2024) The potential benefits of PGC-1α in treating Alzheimer's disease are dependent on the integrity of the LLKYL L3 motif: effect of regulating mitochondrial axonal transportation. Exp Gerontol 194:112514. https://doi.org/10.1016/j.exger.2024.112514
62. Shao CY, Mirra SS, Sait HB, Sacktor TC, Sigurdsson EM (2011) Postsynaptic degeneration as revealed by PSD-95 reduction occurs after advanced Aβ and tau pathology in transgenic mouse models of Alzheimer's disease. Acta Neuropathol 122(3):285–92. https://doi.org/10.1007/s00401-011-0843-x
63. Buccellato FR, D'Anca M, Tartaglia GM, Del Fabbro M, Galimberti D (2024) Frontotemporal dementia: from genetics to therapeutic approaches. Expert Opin Investig Drugs 33(6):561–573. https://doi.org/10.1080/13543784.2024.2349286
64. Sevigny J, Uspenskaya O, Heckman LD et al (2024) Progranulin AAV gene therapy for frontotemporal dementia: translational studies and phase 1/2 trial interim results. Nat Med 30:1406–1415. https://doi.org/10.1038/s41591-024-02973-0
65. Poewe W, Seppi K, Marini K, Mahlknecht P (2020) New hopes for disease modification in Parkinson's Disease. Neuropharmacology 171:108085. https://doi.org/10.1016/j.neuropharm.2020.108085
66. Freitas ME, Hess CW, Fox SH (2017) Motor complications of dopaminergic medications in Parkinson's disease. Semin Neurol 37(2):147–157. https://doi.org/10.1055/s-0037-1602423
67. Huh YE, Usnich T, Scherzer CR, Klein C, Chung SJ (2023) GBA1 variants and Parkinson's disease: paving the way for targeted therapy. J Mov Disord 16(3):261–278. https://doi.org/10.14802/jmd.23023
68. Vieira SRL, Schapira AHV (2022) Glucocerebrosidase mutations and Parkinson disease. J Neural Transm (Vienna) 129(9):1105–1117. https://doi.org/10.1007/s00702-022-02531-3
69. Zhang X, Wu H, Tang B, Guo J (2024) Clinical, mechanistic, biomarker, and therapeutic advances in GBA1-associated Parkinson's disease. Transl Neurodegener 13(1):48. https://doi.org/10.1186/s40035-024-00437-6
70. Fregno I, Pérez-Carmona N, Rudinskiy M et al (2025) Allosteric modulation of GCase enhances lysosomal activity and reduces ER stress in GCase-related disorders. Int J Mol Sci 26(9):4392. https://doi.org/10.3390/ijms26094392
71. Abeliovich A, Hefti F, Sevigny J (2021) Gene therapy for Parkinson's disease associated with GBA1 mutations. J Parkinsons Dis 11(s2):S183–S188. https://doi.org/10.3233/JPD-212739
72. Patel RV, Nanda P, Richardson RM (2024) Neurosurgical gene therapy for central nervous system diseases. Neurotherapeutics 21(4):e00434. https://doi.org/10.1016/j.neurot.2024.e00434
73. Cavallieri F, Cury RG, Guimarães T, Fioravanti V, Grisanti S, Rossi J, Monfrini E, Zedde M, Di Fonzo A, Valzania F, Moro E (2023) Recent advances in the treatment of genetic forms of Parkinson's disease: hype or hope?. Cells 12(5):764. https://doi.org/10.3390/cells12050764
74. Menozzi E, Toffoli M, Schapira AHV (2023) Targeting the GBA1 pathway to slow Parkinson disease: Insights into clinical aspects, pathogenic mechanisms and new therapeutic avenues. Pharmacol Ther 246:108419. https://doi.org/10.1016/j.pharmthera.2023.108419
75. Mendell JR, Connolly AM, Lehman KJ, Griffin DA, Khan SZ, Dharia SD, Quintana-Gallardo L, Rodino-Klapac LR (2022) Testing preexisting antibodies prior to AAV gene transfer therapy: rationale, lessons and future considerations. Mol Ther Methods Clin Dev 25:74–83. https://doi.org/10.1016/j.omtm.2022.02.011
76. El Ouaamari Y, Van den Bos J, Willekens B, Cools N, Wens I (2023) Neurotrophic factors as regenerative therapy for neurodegenerative diseases: current status, challenges and future perspectives. Int J Mol Sci 24(4):3866. https://doi.org/10.3390/ijms24043866
77. Hong M, Mukhida K, Mendez I (2008) GDNF therapy for Parkinson's disease. Expert Rev Neurother 8(7):1125–1139. https://doi.org/10.1586/14737175.8.7.1125
78. Van Laar AD, Christine CW, Phielipp N et al (2025) Intraputaminal delivery of adeno-associated virus Serotype 2-Glial cell line-derived neurotrophic factor in mild or moderate Parkinson's disease. Mov Disord 40(7):1297–1306. https://doi.org/10.1002/mds.30193

79. Pardridge WM (2022) A historical review of brain drug delivery. Pharmaceutics 14(6):1283. https://doi.org/10.3390/pharmaceutics14061283
80. Barker RA, Saarma M, Svendsen CN et al (2024) Neurotrophic factors for Parkinson's disease: current status, progress, and remaining questions. conclusions from a 2023 workshop. J Parkinsons Dis 14(8):1659–1676. https://doi.org/10.1177/1877718X241301041
81. Richardson RM, Kells AP, Rosenbluth KH, Salegio EA, Fiandaca MS, Larson PS, Starr PA, Martin AJ, Lonser RR, Federoff HJ, Forsayeth JR, Bankiewicz KS (2011) Interventional MRI-guided putaminal delivery of AAV2-GDNF for a planned clinical trial in Parkinson's disease. Mol Ther 19(6):1048–57. https://doi.org/10.1038/mt.2011.11
82. Stefani A, Pierantozzi M, Cardarelli S et al (2022) Neurotrophins as therapeutic agents for Parkinson's disease; new chances from focused ultrasound?. Front Neurosci 16:846681. https://doi.org/10.3389/fnins.2022.846681
83. Yao JY, Liu T, Hu XR et al (2024) An insight into allele-selective approaches to lowering mutant huntingtin protein for Huntington's disease treatment. Biomed Pharmacother 180:117557. https://doi.org/10.1016/j.biopha.2024.117557
84. Caron NS, Banos R, Aly AE, Xie Y, Ko S, Potluri N, Anderson C, Black HF, Anderson LM, Gordon B, Southwell AL, Hayden MR (2022) Cerebrospinal fluid mutant huntingtin is a biomarker for huntingtin lowering in the striatum of Huntington disease mice. Neurobiol Dis 166:105652. https://doi.org/10.1016/j.nbd.2022.105652
85. McColgan P, Thobhani A, Boak L et al (2023) Tominersen in adults with manifest Huntington's disease. N Engl J Med 389(23):2203–2205. https://doi.org/10.1056/NEJMc2300400
86. Tabrizi SJ, Estevez-Fraga C, van Roon-Mom WMC, Flower MD, Scahill RI, Wild EJ, Muñoz-Sanjuan I, Sampaio C, Rosser AE, Leavitt BR (2022) Potential disease-modifying therapies for Huntington's disease: lessons learned and future opportunities. Lancet Neurol 21(7):645–658. https://doi.org/10.1016/S1474-4422(22)00121-1
87. Liddy V, West D, Nguyen M et al (2021) F14 Collaborating with the community to conduct clinical trials in Huntington's disease: lessons from the tominersen phase III generation HD1 study. J Neurol Neurosurg Psychiatry 92:A24–A25
88. Estevez-Fraga C, Tabrizi SJ, Wild EJ (2023) Huntington's disease clinical trials corner: August 2023. J Huntingtons Dis 12(2):169–185. https://doi.org/10.3233/JHD-239001
89. Estevez-Fraga C, Tabrizi SJ, Wild EJ (2023) Huntington's disease clinical trials corner: August 2023. J Huntingtons Dis 12(2):169–185. https://doi.org/10.3233/JHD-239001
90. Boareto M, Yamamoto Y, Long JD, Sampaio C, McColgan P, Diack C, Ducray PS (2025) Modeling disease progression and placebo response in Huntington disease: insights from Enroll-HD and GENERATION HD1 cohorts. Neurology 104(10):e213646. https://doi.org/10.1212/WNL.0000000000213646
91. Claassen DO, Corey-Bloom J, Dorsey ER, Edmondson M, Kostyk SK, LeDoux MS, Reilmann R, Rosas HD, Walker F, Wheelock V, Svrzikapa N, Longo KA, Goyal J, Hung S, Panzara MA (2020) Genotyping single nucleotide polymorphisms for allele-selective therapy in Huntington disease. Neurol Genet 6(3):e430. https://doi.org/10.1212/NXG.0000000000000430
92. Svrzikapa N, Longo KA, Prasad N et al (2020) Investigational assay for haplotype phasing of the huntingtin gene. Mol Ther Methods Clin Dev 19:162–173. https://doi.org/10.1016/j.omtm.2020.09.003
93. Byun S, Lee M, Kim M (2022) Gene therapy for Huntington's disease: the final strategy for a cure?. J Mov Disord 15(1):15–20. https://doi.org/10.14802/jmd.21006
94. Kotowska-Zimmer A, Przybyl L, Pewinska M, Suszynska-Zajczyk J, Wronka D, Figiel M, Olejniczak M (2022) A CAG repeat-targeting artificial miRNA lowers the mutant huntingtin level in the YAC128 model of Huntington's disease. Mol Ther Nucleic Acids 28:702–715. https://doi.org/10.1016/j.omtn.2022.04.031
95. Ferguson MW, Kennedy CJ, Palpagama TH, Waldvogel HJ, Faull RLM, Kwakowsky A (2022) Current and possible future therapeutic options for Huntington's disease. J Cent Nerv Syst Dis 14:11795735221092517. https://doi.org/10.1177/11795735221092517
96. Berdyn´ski M, Miszta P, Safranow K et al (2022) SOD1 mutations associated with amyotrophic lateral sclerosis analysis of variant severity. Sci Rep 12: 103. https://doi.org/10.1038/s41598-021-03891-8
97. Saini A, Chawla PA (2024) Breaking barriers with tofersen: Enhancing therapeutic opportunities in amyotrophic lateral sclerosis. Eur J Neurol 31(2):e16140. https://doi.org/10.1111/ene.16140
98. Miller T, Cudkowicz M, Shaw PJ et al (2020) Phase 1-2 trial of antisense oligonucleotide tofersen for SOD1 ALS. N Engl J Med 383(2):109–119. https://doi.org/10.1056/NEJMoa2003715

99. Miller TM, Cudkowicz ME, Genge A et al (2022) Trial of antisense oligonucleotide tofersen for SOD1 ALS. N Engl J Med 387(12):1099–1110. https://doi.org/10.1056/NEJMoa2204705
100. Smeyers J, Banchi EG, Latouche M (2021) C9ORF72: what it is, what it does, and why it matters. Front Cell Neurosci 15:661447. https://doi.org/10.3389/fncel.2021.661447
101. van den Berg LH, Rothstein JD, Shaw PJ et al (2024) Safety, tolerability, and pharmacokinetics of antisense oligonucleotide BIIB078 in adults with C9orf72-associated amyotrophic lateral sclerosis: a phase 1, randomised, double blinded, placebo-controlled, multiple ascending dose study. Lancet Neurol 23(9):901–912. https://doi.org/10.1016/S1474-4422(24)00216-3
102. Chamakioti M, Karantzelis N, Taraviras S (2022) Advanced gene-targeting therapies for motor neuron diseases and muscular dystrophies. Int J Mol Sci 23:4824. https://doi.org/10.3390/ijms23094824
103. Xiao X, Li M, Ye Z, He X, Wei J, Zha Y (2024) FUS gene mutation in amyotrophic lateral sclerosis: a new case report and systematic review. Amyotroph Lateral Scler Front Degener 25(1-2):1–15. https://doi.org/10.1080/21678421.2023.2272170
104. Meijboom KE, Brown RH (2022) Approaches to gene modulation therapy for ALS. Neurotherapeutics 19(4):1159–1179. https://doi.org/10.1007/s13311-022-01285-w
105. Zhu L, Li S, Li XJ, Yin P (2023) Pathological insights from amyotrophic lateral sclerosis animal models: comparisons, limitations, and challenges. Transl Neurodegener 12(1):46. https://doi.org/10.1186/s40035-023-00377-7
106. Lejman J, Panuciak K, Nowicka E, Mastalerczyk A, Wojciechowska K, Lejman M (2023) Gene therapy in ALS and SMA: advances, challenges and perspectives. Int J Mol Sci 24(2):1130. https://doi.org/10.3390/ijms24021130
107. Sharma G, Sharma AR, Bhattacharya M, Lee SS, Chakraborty C (2021) CRISPR-Cas9: A preclinical and clinical perspective for the treatment of human diseases. Mol Ther 29(2):571–586. https://doi.org/10.1016/j.ymthe.2020.09.028
108. Keinath MC, Prior DE, Prior TW (2021) Spinal muscular atrophy: mutations, testing, and clinical relevance. Appl Clin Genet 14:11–25. https://doi.org/10.2147/TACG.S239603
109. Cancès C, Richelme C, Barnerias C, Espil C (2020) Clinical features of spinal muscular atrophy (SMA) type 2. Arch Pediatr 27(7S):7S18–7S22. https://doi.org/10.1016/S0929-693X(20)30272-4
110. Boardman FK, Young PJ, Griffiths FE (2017) Newborn screening for spinal muscular atrophy: the views of affected families and adults. Am J Med Genet A 173(6):1546–1561. https://doi.org/10.1002/ajmg.a.38220
111. Günther R, Wurster CD, Brakemeier S et al (2024) Long-term efficacy and safety of nusinersen in adults with 5q spinal muscular atrophy: a prospective European multinational observational study. Lancet Reg Health Eur 39:100862. https://doi.org/10.1016/j.lanepe.2024.100862
112. Finkel RS, Mercuri E, Darras BT et al (2017) Nusinersen versus sham control in infantile-onset spinal muscular atrophy. N Engl J Med 377(18):1723–1732. https://doi.org/10.1056/NEJMoa1702752
113. Mercuri E, Darras BT, Chiriboga CA et al (2018) Nusinersen versus sham control in later-onset spinal muscular atrophy. N Engl J Med 378(7):625–635. https://doi.org/10.1056/NEJMoa1710504
114. Pearson SD, Thokala P, Stevenson M, Rind D (2019) The effectiveness and value of treatments for spinal muscular atrophy. J Manag Care Spec Pharm 25(12):1300–1306. https://doi.org/10.18553/jmcp.2019.25.12.1300
115. Abbas KS, Eltaras MM, El-Shahat NA, Abdelazeem B, Shaqfeh M, Brašić JR (2022) The safety and efficacy of nusinersen in the treatment of spinal muscular atrophy: a systematic review and meta-analysis of randomized controlled trials. Medicina (Kaunas) 58(2):213. https://doi.org/10.3390/medicina58020213
116. Li L, Vasan L, Kartono B, Clifford K, Attarpour A, Sharma R, Mandrozos M, Kim A, Zhao W, Belotserkovsky A, Verkuyl C, Schmitt-Ulms G (2023) Advances in recombinant adeno-associated virus vectors for neurodegenerative diseases. Biomedicines 11(10):2725. https://doi.org/10.3390/biomedicines11102725

Check for updates

Chapter 11

Intrathecal Delivery for Leptomeningeal Disease

Victor M. Lu and Ashish H. Shah

Abstract

Leptomeningeal disease (LMD) is a rare but devastating development of many cancers with an extremely poor prognosis. Barriers to improving outcomes are related to the inability of many traditional therapies to effectively reach the cerebrospinal fluid (CSF) space within the central nervous system. Intrathecal treatment, therefore, is the most logical and direct route to accessing this space. However, treatment options are still lacking given the need for appropriate LMD models. We highlight how our group is able to model LMD and the current techniques available to verify these models.

Keywords CSF, Cerebrospinal fluid, LMD, Leptomeningeal disease, Filtration

1 Background

Leptomeningeal disease (LMD, also known as leptomeningeal metastases or carcinomatous meningitis) is a rare but devastating sequela of advanced solid tumors where tumor cells spread and seed into the leptomeningeal space of the central nervous system (CNS) via the cerebrospinal fluid (CSF). It is estimated that 5%–10% of patients with any solid cancer, and 10%–15% of patients with hematological cancer, will develop LMD [1, 2]. The most common solid cancers are breast and lung primary, and the most common hematological cancer is lymphoma [3].

Prognosis is extremely poor at the time of diagnosis, irrespective of the primary. Expected survival is a matter of months, which is partly due to the absence of any effective treatments known to date [4, 5]. Current systemic treatment options, such as chemotherapy, are limited both by the difficulty to penetrate the blood–brain barrier to reach the CSF space, as well as neurotoxicity concerns [6].

LMD itself is characterized by the simultaneous multifocal tumor involvement of the leptomeninges throughout the CNS. It is within this space that the CSF circulates, providing a route for tumor cells to disseminate throughout the leptomeninges, travel to distant sites, settle, and grow into LMD [7]. There are multiple proposed

Manuel Menéndez-González (ed.), *Intrathecal Drug Delivery*, Neuromethods, vol. 234,
https://doi.org/10.1007/978-1-0716-5241-1_11,
© The Author(s), under exclusive license to Springer Science+Business Media, LLC, part of Springer Nature 2026

mechanisms by which tumor cells access the CSF space to enter the leptomeninges and precipitate LMD. Most simply, tumor cells located within the CNS and adjacent to CNS structures can invade directly into the space [8]. Tumor cells can also invade via the circulation. Within the arterial circulation, they can enter the CSF via the fenestrated epithelium of the choroid plexus in the ventricular system [8]. Within the venous system, tumor cells can pass through various venous plexi, including Batson's plexus, at draining sites along the skull and vertebrae [9].

A reason this issue is pertinent today is that the incidence of LMD is increasing due to improved survival rates secondary to the improved systemic control of the disease, and better imaging and other diagnostic modalities [10]. The longer patients live with systemic cancer, the higher the chances of tumor spread and seeding of the leptomeninges in the CSF space. Hence, there is a parallel growing need to investigate novel therapeutic approaches in the setting of LMD. Intrathecal delivery offers a direct route to the primary space where LMD exists and proliferates. As more is discovered regarding intrathecal delivery and markers of progression, this route is poised to significantly impact LMD prognosis in the future.

2 Current Intrathecal Treatments

Intrathecal (IT) therapy has emerged as a cornerstone of treatment, aiming to deliver chemotherapy directly into the CSF to bypass the blood–brain barrier. The most commonly used intrathecal agents are methotrexate, cytarabine (especially its sustained-release liposomal form), and thiotepa [11]. As seen in Chap. 8, these drugs are typically administered via lumbar puncture or an implanted reservoir, such as an Ommaya reservoir, allowing for repeated dosing with less discomfort (Fig. 1). Methotrexate and cytarabine are often used in rotation or in combination due to their differing mechanisms of action, although neurotoxicity can be a limiting factor.

Recent advances have expanded IT options to include targeted therapies and immunotherapeutics. For example, trastuzumab, a monoclonal antibody targeting HER2, has been administered intrathecally in HER2-positive breast cancer patients with LMD, with some reports showing improved CSF cytology and survival [12]. Similarly, intrathecal rituximab has been explored for central nervous system involvement in CD20-positive lymphomas [13]. These targeted approaches aim to personalize therapy based on tumor biology and have shown promise in small clinical trials and case series. However, challenges remain in drug delivery, toxicity management, and achieving sustained responses.

Despite these advancements, intrathecal therapy for LMD is generally palliative, and outcomes remain poor. Multimodal approaches, combining IT therapy with systemic treatment and radiation, are often

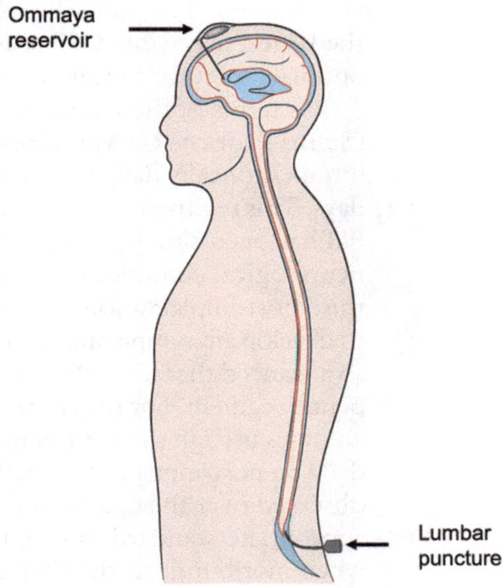

Fig. 1 Location of the most common access to the intrathecal space for therapy, including the Ommaya reservoir in the brain and the lumbar puncture in the spine. The cerebrospinal fluid (CSF, blue) surrounds all the brain and spine, which are covered by thin leptomeninges (red)

employed to optimize disease control [14]. Ongoing clinical trials are investigating newer agents and delivery methods, including intrathecal immune checkpoint inhibitors [15] and nanoparticles [16]. These efforts reflect the growing recognition of LMD as a distinct clinical entity requiring innovative strategies for effective management.

Our lab is interested in the targeted treatment of leptomeningeal disease secondary to glioblastoma (GBM). Glioblastoma (GBM) is an incurable malignant primary brain tumor that has a median overall survival of 14 months despite maximum treatment [17]. Estimates suggest LMD occurs in approximately 4% of all GBM diagnoses, making it a very rare disease to study [18]. We herein describe the material and methods we utilize to both model LMD of GBM and how to evaluate a potential target (pro-inflammatory TLR-4 pathway in our case) and measure its effect.

3 Materials and Methods to Investigate Intrathecal Treatments

3.1 Mouse Model and Treatment

Immunocompetent C57BL/6 female mice were obtained from The Jackson Laboratory Black aged 6. After 1 week, mice underwent stereotactic implantation of immunofluorescent 5×10^4 SB28 cells resuspended in 2 μL of PBS into the cisterna magnum under microscopic visualization of the dura (Fig. 2A). The cisterna magnum represents the largest accessible CSF space along the neuraxis, making it the optimal target for cancer cell implantation.

In terms of the treatment group, TLR-4 antagonist TAK-242 (Sigma-Aldrich, USA) was administered 3 mg/kg in PBS solution intraperitoneally daily from post-implantation day 3 for a total of 5 days. This regimen was selected as its safety and translational potential has been shown in multiple other mouse models of different neurological conditions [19–21]. The timing of treatment on the third post-implantation day was chosen to allow time for the LMD to develop after implantation, but not to allow the disease to become so advanced that it would not be affected by our exploratory therapeutic regimen. For the control group, equal volumes of PBS to the volumes used in the treatment group per mouse were administered daily on post-implantation day 3 for a total of 5 days. All mice were observed over the course of 3 months after implantation using the appropriate standard of care techniques, and they were sacrificed when moribund. At the time of sacrifice, the brain and spine were extracted and then processed (Fig. 2B).

3.2 Confocal Microscopy

To confirm the overlap of immunofluorescent GBM cancer cells along the ependyma of the ventricle (i.e., the leptomeninges), we utilized confocal microscopy to evaluate the brain and spine of specimens at the time of sacrifice. Extracted specimens were fixed in paraformaldehyde (PFA) and prepared on histological slides for evaluation. The slides were dried at 37°C overnight and baked at 59°C for 1 h. Then they were deparaffinized and rehydrated using xylene and graded ethanol (100%, 95%, 70%), mounted using DAPI. Finally, the slides were imaged using a Dragonfly confocal microscope (Oxford Instruments, USA). The immunofluorescent nature of the GFP-tagged SB28 cells meant tumor cells were interrogated at excitation wavelength 488 nm, and compared to DAPI fluorescent biological cells at excitation wavelength 405 nm to ensure there was no shine-through artifact (Fig. 3).

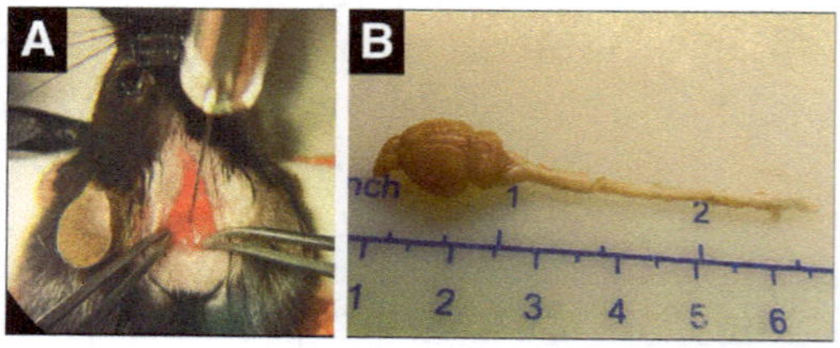

Fig. 2 Our novel LMD animal model results. (**A**) Development of our GBM LMD mouse model using SB28 cell line implantation by stereotactic injection into the cisterna magna to generate our LMD mouse model. (**B**) Perfusion protocol allows us to extract the brain and spine en bloc for processing

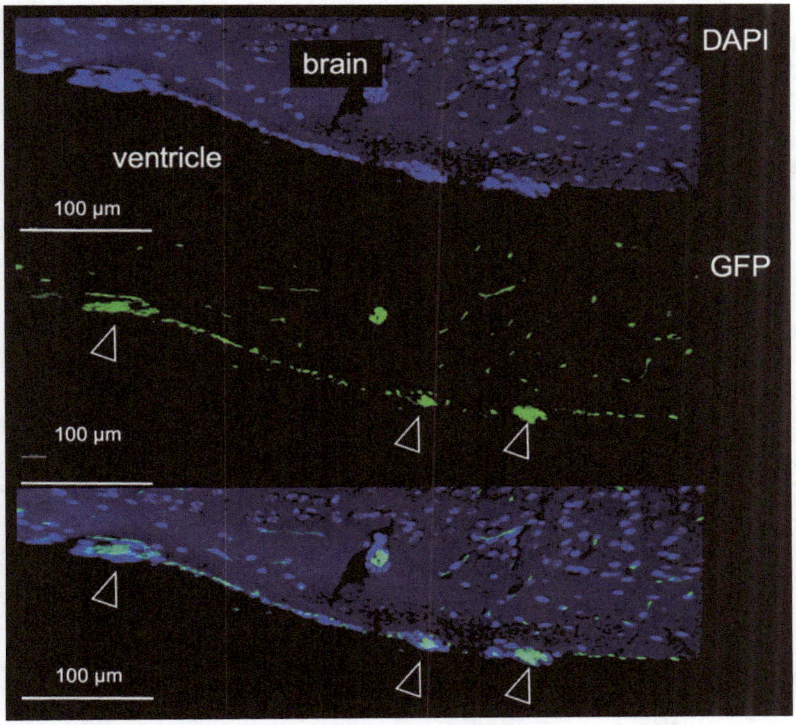

Fig. 3 Confocal microscopy allows confirmation of target tumor cells along the ependymal lining of CSF spaces of the leptomeninges. All samples were submitted to confocal imaging to confirm the presence of fluorescent (GFP) murine GBM SB28 cell deposits (black triangles) along the ventricle lining with corresponding DAPI overlay

3.3 Clearing Protocol

To evaluate the 3D distribution of SB28 cancer cells throughout the CSF space along the brain and spine, we used a clearing technique of the specimen brain and spine, as described herein. Whole brains with spinal cords were dissected and fixed overnight in 4% PFA at 4°C and washed three times in PBS. The brain with the spinal cord was cleared as previously described [22–24]. Briefly, whole mouse brains were incubated on a rotating shaker at room temperature in 50%, 80%, and twice with 100% peroxide-free tetrahydrofuran (THF; Sigma-Aldrich, USA) for 12 h each for a total of 2 days. Samples were transferred to BABB solution (1:2 ratio of benzyl alcohol, Sigma-Aldrich, USA; benzyl benzoate, Sigma-Aldrich, USA) for 3 h. After clearing, whole brains and spinal cords were imaged the same day using light sheet Ultramicroscope microscopy (LaVision BioTec, Germany). The fluorescent SB28 cells were interrogated for using excitation wavelength 488 nm to demonstrate 3D distribution of cells along the specimens (Fig. 4). Post-acquisition, images were stitched and merged using 2D/3D montage tools in the SlideBook digital microscopy software (3i, Denver, CO, USA).

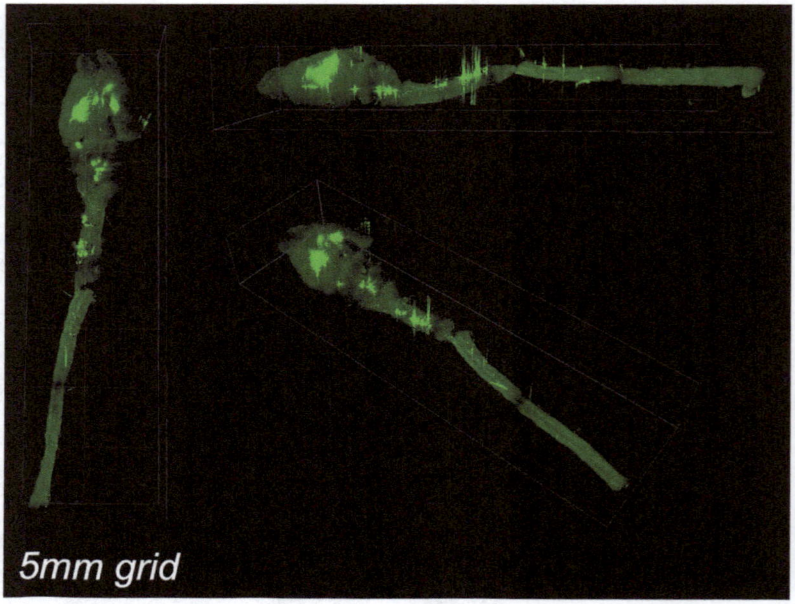

Fig. 4 LMD model can be verified using a clearing protocol to demonstrate tumor fluorescence along the brain and spine axis as a 3D model along all axes

3.4 Cerebrospinal Fluid Collection

The primary component of LMD analysis is the analysis of the CSF within the intrathecal space. At the time of sampling, be it under anesthesia or at sacrifice, specimens can undergo a puncture at the foramen magnum to extract CSF from the model using a light microscope. The CSF is then maintained at 4°C on ice and directly transported to the laboratory. Each specimen is centrifuged for 5 min at 1200 RPM. Then, the supernatant is removed, and the remaining CSF can be stored at −80°C until analysis such as western blotting and ELISA (Fig. 5).

3.5 Western Blotting

Protein can be extracted from all CSF samples using RIPA buffer (Sigma-Aldrich) with protease/phosphatase inhibitor (Cell Signaling). Protein concentration of each sample was then measured using the BCA assay (Thermo Scientific; Rockford, IL). For CSF samples, western blotting was performed using 100 μg protein loading control, which has been shown to be an effective way of controlling CSF content [25]. Total protein can then be stained for using a number of stains, including Coomassie blue stain [26]. Western blotting was performed as previously described [27]. In brief, proteins were denatured using Laemmli denaturing buffer (Bio-Rad Laboratories) supplemented with β-mercaptoethanol and separated on a 12% polyacrylamide SDS-PAGE gel. Proteins were transferred to a PVDF membrane and immunoblotted using antibodies against the desired target. Membranes were developed using Clarity Western

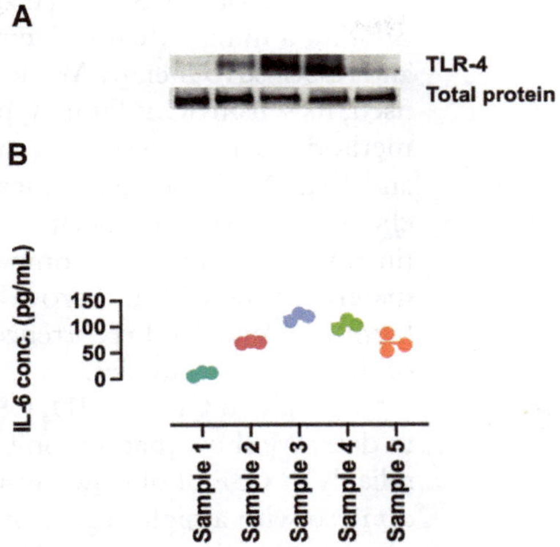

Fig. 5 Example cerebrospinal fluid results across five samples using (**A**) Western blotting for TLR-4 antibody and total protein by Coomassie blue stain and (**B**) ELISA analysis for IL-6 content in CSF

ECL Substrate Biorad Chemiluminescence system (BioRAD #170-506). Blots were imaged using the Western Blot Imaging System 600 (Azure Biosystems, United States) at appropriate exposure times.

3.6 Enzyme-Linked Immunosorbent Assay (ELISA)

Sandwich ELISA against desired targets can be performed with CSF samples. The analysis was performed according to the manufacturer's instructions, which utilizes the protein content as determined in the western blotting methods. Absorbance in each well is read at 450 nm using a Spectramax M5 automated reader (Molecular Devices, San Jose, CA), and evaluated using system-associated Softmax pro software v4.8.

4 Discussion

Intrathecal (IT) drug delivery plays a central role in LMD treatment, as it allows therapeutic agents to bypass the blood–brain barrier and achieve higher concentrations directly within the CSF. However, effective intrathecal therapy depends on a detailed understanding of CSF dynamics, drug pharmacokinetics, and distribution patterns within the brain and spine. Modeling LMD and analyzing CSF pharmacodynamics helps identify optimal dosing schedules, delivery methods (e.g., Ommaya reservoir versus lumbar puncture), and potential toxicities. Combining disease-specific models with detailed CSF analytics provides a foundation for developing rational intrathecal therapies tailored to the unique challenges of treating malignancies within the central nervous system.

Analysis of CSF is a vital diagnostic and monitoring tool in LMD, offering a minimally invasive method to track disease progression and response to therapy. While conventional cytology has long been used, its sensitivity is limited, prompting the development of other methods that can reveal specific targets, such as western blotting and ELISA. These techniques allow for more accurate molecular characterization and dynamic assessment of tumor burden in real time. Understanding the composition and behavior of tumor-derived material in the CSF also provides insights into mechanisms of treatment resistance and recurrence, guiding the rational development of therapeutic strategies.

Specific to GBM LMD, CSF content has not been well defined to date. We show that not only can the TLR-4 pathway be detected reliably in CSF of our own institutional patients, but it can also be done so with a high degree of sensitivity and specificity from CSF of primary brain tumors without LMD as well as CSF controls. The high degree of sensitivity of TLR-4 in LMD CSF versus control CSF was already implicated by our analysis of the publicly available LMD CSF data [28]. One rationale for the observation that GBM LMD shows higher levels of TLR pathway expression compared to other brain tumors without LMD is the presence of circulating tumor cells in the CSF spaces, with LMD being the medium through which CSF TLR-4 pathway is expressed. This would align with the theory that the CSF in GBM LMD exists in a pro-inflammatory state to promote leptomeningeal spread.

There have not been many studies that have evaluated the CSF results of a GBM LMD model, making our methods somewhat novel. One study used a similar implantation target of the cisterna magna to generate mouse lung and breast LMD models—they then studied the CSF primarily using the ELISA technique we described and found that complement component 3 (C3) was upregulated in their leptomeningeal metastatic models and proved necessary for cancer growth within the leptomeningeal space [29]. Another study used immunofluorescence techniques similar to our confocal microscopy and clearing techniques to show dura-derived macrophages migrate into the CSF in a matrix metalloproteinase 14 (MMP14)-dependent manner to promote their mouse LMD model [30]. Finally, western blotting has been utilized to study the CSF content of other LMD models too, making this methodology also one that should be considered when studying the CSF of LMD models [31].

5 Conclusions

Modeling leptomeningeal disease (LMD) is critical to advancing our understanding of this aggressive manifestation of metastatic cancer. LMD involves the dissemination of malignant cells into the cerebrospinal fluid (CSF) and leptomeninges, a compartment that is biologically distinct from the systemic circulation. Traditional systemic cancer models often fail to replicate the unique environment of the CSF. Therefore, specialized in vivo models, such as our intracisternal tumor cell injections in mice methods, are necessary to study LMD-specific tumor growth, immune responses, and treatment effects. These models help elucidate how tumor cells adapt to the nutrient-poor, immunologically active CSF and how they resist therapies designed for solid parenchymal tumors.

Disclosures

All authors report no funding sources concerning the materials or methods used in this study or the findings specified in this chapter.

Conflict of Interest

The authors have no conflicts of interest to declare that are relevant to the content of this chapter.

References

1. Jeyapalan SA, Batchelor TT (2000) Diagnostic evaluation of neurologic metastases. Cancer Invest 18(4):381–394. https://doi.org/10.3109/07357900009012181
2. Glass JP, Melamed M, Chernik NL, Posner JB (2011) Malignant cells in cerebrospinal fluid (CSF): the meaning of a positive CSF cytology. Neurology 77(20):1793. https://doi.org/10.1212/01.wnl.0000407892.25373.7b
3. Kaplan JG, DeSouza TG, Farkash A, Shafran B, Pack D, Rehman F et al (1990) Leptomeningeal metastases: comparison of clinical features and laboratory data of solid tumors, lymphomas and leukemias. J Neurooncol 9(3):225–229. https://doi.org/10.1007/bf02341153
4. Hitchins RN, Bell DR, Woods RL, Levi JA (1987) A prospective randomized trial of single-agent versus combination chemotherapy in meningeal carcinomatosis. J Clin Oncol 5(10):1655–1662. https://doi.org/10.1200/jco.1987.5.10.1655
5. Glantz MJ, Jaeckle KA, Chamberlain MC, Phuphanich S, Recht L, Swinnen LJ et al (1999) A randomized controlled trial comparing intrathecal sustained-release cytarabine (DepoCyt) to intrathecal methotrexate in patients with neoplastic meningitis from solid tumors. Clin Cancer Res 5(11):3394–3402
6. Bokstein F, Lossos A, Siegal T (1998) Leptomeningeal metastases from solid tumors. Cancer 82(9):1756–1763. https://doi.org/10.1002/(SICI)1097-0142(19980501)82:9<1764::AID-CNCR24>3.0.CO;2-1
7. Kesari S, Batchelor TT (2003) Leptomeningeal metastases. Neurol Clin 21(1):25–66. https://doi.org/10.1016/s0733-8619(02)00032-4
8. Kokkoris CP (1983) Leptomeningeal carcinomatosis. How does cancer reach the pia-arachnoid? Cancer 51(1):154–160. https://doi.org/10.1002/1097-0142(19830101)51:1<154::aid-cncr2820510130gt;3.0.co;2-k
9. Glover RL, Brook AL, Welch MR (2014) Teaching neuroimages: leptomeningeal lung carcinoma. Neurology 82(21):e183–e184. https://doi.org/10.1212/wnl.0000000000000455
10. El Shafie RA, Böhm K, Weber D, Lang K, Schlaich F, Adeberg S et al (2019) Outcome and prognostic factors following palliative craniospinal irradiation for leptomeningeal carcinomatosis. Cancer Manag Res 11:789–801. https://doi.org/10.2147/cmar.S182154
11. Perego G, Longobardo G, Viganò L, Locatelli A, Veneziano C, Fazio M (2022) Stability of a standardized preparation of methotrexate, cytarabine, and methylprednisolone hemisuccinate for intrathecal use. J Oncol Pharm Pract. 2022:10781552221117228. https://doi.org/10.1177/10781552221117228.

12. Lazaratos AM, Maritan SM, Quaiattini A, Darlix A, Ratosa I, Ferraro E et al (2023) Intrathecal trastuzumab versus alternate routes of delivery for HER2-targeted therapies in patients with HER2+ breast cancer leptomeningeal metastases. Breast 69:451–468. https://doi.org/10.1016/j.breast.2023.04.008
13. Vitolo U, Chiappella A, Ferreri AJ, Martelli M, Baldi I, Balzarotti M et al (2011) First-line treatment for primary testicular diffuse large B-cell lymphoma with rituximab-CHOP, CNS prophylaxis, and contralateral testis irradiation: final results of an international phase II trial. J Clin Oncol 29(20):2766–2772. https://doi.org/10.1200/jco.2010.31.4187
14. Berg SL, Chamberlain MC (2005) Current treatment of leptomeningeal metastases: systemic chemotherapy, intrathecal chemotherapy and symptom management. Cancer Treat Res 125:121–146. https://doi.org/10.1007/0-387-24199-x_8
15. Mills MN, Uno A, Li P, Liveringhouse C, Kim Y, Oliver DE et al (2024) Clinical outcomes of patients with non-small cell lung cancer leptomeningeal disease following receipt of EGFR-targeted therapy, immune-checkpoint blockade, intrathecal chemotherapy, or radiation therapy alone. Clin Lung Cancer 25(5):417–423.e1. https://doi.org/10.1016/j.cllc.2024.04.005
16. Khang M, Lee JH, Lee T, Suh HW, Lee S, Cavaliere A et al (2023) Intrathecal delivery of nanoparticle PARP inhibitor to the cerebrospinal fluid for the treatment of metastatic medulloblastoma. Sci Transl Med 15(720):eadi1617. https://doi.org/10.1126/scitranslmed.adi1617
17. Stupp R, Mason WP, van den Bent MJ, Weller M, Fisher B, Taphoorn MJ et al (2005) Radiotherapy plus concomitant and adjuvant temozolomide for glioblastoma. N Engl J Med 352(10):987–996. https://doi.org/10.1056/NEJMoa043330
18. Alatakis S, Malham GM, Thien C (2001) Spinal leptomeningeal metastasis from cerebral glioblastoma multiforme presenting with radicular pain: case report and literature review. Surg Neurol 56(1):33–37. https://doi.org/10.1016/s0090-3019(01)00459-1; discussion 7–8
19. Fellner A, Barhum Y, Angel A, Perets N, Steiner I, Offen D et al (2017) Toll-like receptor-4 inhibitor TAK-242 attenuates motor dysfunction and spinal cord pathology in an amyotrophic lateral sclerosis mouse model. Int J Mol Sci 18(8):1666. https://doi.org/10.3390/ijms18081666
20. Wang YC, Wang PF, Fang H, Chen J, Xiong XY, Yang QW (2013) Toll-like receptor 4 antagonist attenuates intracerebral hemorrhage-induced brain injury. Stroke 44(9):2545–2552. https://doi.org/10.1161/strokeaha.113.001038
21. Hua F, Tang H, Wang J, Prunty MC, Hua X, Sayeed I et al (2015) TAK-242, an antagonist for toll-like receptor 4, protects against acute cerebral ischemia/reperfusion injury in mice. J Cereb Blood Flow Metab 35(4):536–542. https://doi.org/10.1038/jcbfm.2014.240
22. Ertürk A, Becker K, Jährling N, Mauch CP, Hojer CD, Egen JG et al (2012) Three-dimensional imaging of solvent-cleared organs using 3DISCO. Nat Protoc 7(11):1983–1995. https://doi.org/10.1038/nprot.2012.119
23. Wang Z, Maunze B, Wang Y, Tsoulfas P, Blackmore MG (2018) Global connectivity and function of descending spinal input revealed by 3D microscopy and retrograde transduction. J Neurosci 38(49):10566–10581. https://doi.org/10.1523/jneurosci.1196-18.2018
24. Soderblom C, Lee D-H, Dawood A, Carballosa M, Santamaria AJ, Benavides FD, et al. (2015) 3D imaging of axons in transparent spinal cords from rodents and nonhuman primates. eNeuro 2(2):ENEURO.0001-15.2015. https://doi.org/10.1523/eneuro.0001-15.2015
25. Collins MA, An J, Peller D, Bowser R (2015) Total protein is an effective loading control for cerebrospinal fluid western blots. J Neurosci Methods 251:72–82. https://doi.org/10.1016/j.jneumeth.2015.05.011
26. Collins MA, An J, Peller D, Bowser R (2015) Total protein is an effective loading control for cerebrospinal fluid western blots. J Neurosci Methods 251:72–82. https://doi.org/10.1016/j.jneumeth.2015.05.011
27. Lu VM, Jue TR, McDonald KL (2020) Cytotoxic lanthanum oxide nanoparticles sensitize glioblastoma cells to radiation therapy and temozolomide: an in vitro rationale for translational studies. Sci Rep 10(1):18156. https://doi.org/10.1038/s41598-020-75372-3
28. Ruan H, Zhou Y, Shen J, Zhai Y, Xu Y, Pi L et al (2020) Circulating tumor cell characterization of lung cancer brain metastases in the cerebrospinal fluid through single-cell transcriptome analysis. Clin Transl Med 10(8):e246. https://doi.org/10.1002/ctm2.246
29. Boire A, Zou Y, Shieh J, Macalinao DG, Pentsova E, Massagué J (2017) Complement component 3 adapts the cerebrospinal fluid for leptomeningeal metastasis. Cell 168(6):1101–1113.e13. https://doi.org/10.1016/j.cell.2017.02.025
30. Zhao J, Zeng R, Li X, Lu Y, Wang Z, Peng H et al (2024) Dura immunity configures leptomeningeal metastasis immunosuppression for cerebrospinal fluid barrier invasion. Nat Cancer 5(12):1940–1961. https://doi.org/10.1038/s43018-024-00858-2
31. Grausam KB, Dooyema SDR, Bihannic L, Premathilake H, Morrissy AS, Forget A et al (2017) ATOH1 promotes leptomeningeal dissemination and metastasis of sonic hedgehog subgroup medulloblastomas. Cancer Res 77(14):3766–3777. https://doi.org/10.1158/0008-5472.Can-16-1836

Chapter 12

Design, Preparation, and Optimization of mRNA Nanoemulsions for CNS Delivery

Philipp Lapuhs, Inés Rubio-Prego, Laura Pineiro-Alonso and María José Alonso

Abstract

Intrathecal delivery of RNA therapeutics is an increasingly credible route to treat central nervous system (CNS) disorders, yet the dense, heterogeneous brain extracellular matrix (ECM) imposes a formidable post-barrier obstacle to nanoscale transport. This chapter addresses that gap by presenting the design rationale, manufacturing workflow, analytical characterization, and in vivo evaluation of ionizable nanoemulsions (iNEs) engineered for enhanced diffusion and functional RNA delivery within brain parenchyma. iNEs are oil-in-water droplets stabilized by biocompatible surfactants and ionizable lipids that complex RNA at low pH and promote endosomal escape after uptake. Four physicochemical levers—hydrodynamic size (<100 nm, prioritized near 70 nm), near-neutral surface charge (−5–0 mV), dense PEGylation to minimize adhesive interactions, and a deformable oily core to improve interstitial adaptability—are tuned in concert. Microfluidic assembly yields monodisperse particles with high RNA association; complementary gel electrophoresis and RiboGreen assays confirm post-purification association efficiencies exceeding 80%. A standardized preclinical pipeline is detailed for intraparenchymal administration and quantitative diffusion mapping. In rats receiving intrathalamic injections of iNEs encapsulating GFP mRNA (3 μL at 0.5 μL/min), confocal microscopy of serial sections enables whole-brain mosaics, distance/area/volume metrics, and cell-type colocalization. Under these conditions, functional reporter expression is observed across neuronal populations with diffusion distances up to ~3.4 mm between the edges of GFP-positive regions—exceeding the ~1–1.5 mm spread commonly reported for conventional LNPs after local dosing. The chapter also discusses sources of variability (injection site, volume and rate, CSF egress, labeling strategy), proposes harmonized readouts to compare formulations, and highlights opportunities for design-of-experiments and machine learning to accelerate multi-parameter optimization. Collectively, the methods and data provide a practical blueprint to rationally tailor iNEs for CNS use and to rigorously quantify their parenchymal dispersion and transfection, advancing the translational readiness of RNA nanomedicines for neurological disease.

Keywords mRNA, Ionizable nanoemulsions, Intraparenchymal, CNS drug delivery, Nanotechnology

The authors Philipp Lapuhs and Inés Rubio-Prego contributed equally.

Manuel Menéndez-González (ed.), *Intrathecal Drug Delivery*, Neuromethods, vol. 234,
https://doi.org/10.1007/978-1-0716-5241-1_12,
© The Author(s), under exclusive license to Springer Science+Business Media, LLC, part of Springer Nature 2026

1 Introduction

RNA-based therapeutics are reshaping the landscape of modern medicine by enabling precise modulation of gene expression in a wide range of diseases [1]. Central to the success of these therapies is nanotechnology [2], with lipid nanoparticles (LNPs) emerging as the most clinically advanced delivery platforms for RNA molecules. Their promise is highlighted by the approval of the first siRNA-based therapy, Onpattro® (Alnylam) in 2018 for hereditary transthyretin-mediated amyloidosis [3], and the global deployment of mRNA vaccines Comirnaty® (Pfizer-BioNTech) and Spikevax® (Moderna) during the COVID-19 pandemic [4]. These milestones underscore the clinical relevance and translational capacity of LNPs in safely delivering nucleic acids in vivo, positioning them as pivotal tools in the advancement of genetic medicine.

The scope of RNA therapeutics has increasingly extended to the treatment of central nervous system (CNS) disorders, an area of high unmet clinical need. So far, ten antisense oligonucleotides (ASOs) and four mRNA-based therapeutics [5] have been approved for CNS-related conditions, and nearly 50 RNA-based brain therapies are currently in clinical trials [6]. Yet, despite these advances, achieving effective delivery of RNA therapeutics to the brain remains a formidable challenge, to which nanomedicine offers a solution [7].

Notably, the presence of the blood-brain barrier (BBB), a highly selective and protective membrane formed by tightly connected endothelial cells, astrocytes, pericytes, and a supportive basement membrane, severely restricts macromolecule access to the brain from systemic circulation [8]. Although numerous strategies are being explored to enhance BBB permeability, delivery efficiency remains suboptimal [9], and alternative administration routes are being considered.

Among these, intrathecal (IT) administration, which involves direct delivery into the cerebrospinal fluid (CSF), has gained traction as a clinically viable strategy to bypass the BBB [10]. This route is already validated by the recent FDA approval of the ASO Qalsody® (tofersen) for amyotrophic lateral sclerosis [11], and is under clinical and preclinical investigation for numerous CNS-targeted therapies [12].

However, even after direct access to the CNS, another significant barrier remains: the brain extracellular matrix (ECM). The ECM is a dense and complex network primarily composed of hyaluronic acid, proteoglycans, and collagen that severely limits the diffusion of macromolecules such as LNPs to effectively reach their target sites, thereby reducing the efficacy of genetic therapeutics [13]. Intracranial injections of LNPs carrying fluorescently labeled mRNA [14] or siRNA [15] have demonstrated limited diffusion ranges restricted to 1–1.5 mm from the injection site, underscoring the critical need to enhance nanosystems brain diffusion for effective therapeutic delivery.

To address this limitation, ionizable nanoemulsions (iNEs) have emerged as a promising alternative to conventional LNPs. These nanosystems consist of oil-in-water droplets stabilized by biocompatible surfactants [16, 17] and incorporate ionizable lipids, whose pH-responsive behaviour facilitates efficient RNA complexation and promotes endosomal escape upon cellular uptake, enhancing cytoplasmic delivery of the RNA payload [18, 19]. In this line, our group has previously demonstrated that iNEs exhibit improved diffusion across the brain parenchyma and achieve enhanced neuronal transfection after local administration [20].

Furthermore, the successful development of iNEs involves the fine-tuning of several physicochemical parameters, including particle size, surface charge, surface modification, and core composition. Furthermore, tailoring nanoparticles to target specific brain regions or cell types may enhance the precision and therapeutic efficacy of RNA-based therapies for CNS diseases [21].

Despite growing interest in CNS-targeted nanomedicine, the field still lacks standardized methods to assess nanoparticles diffusion in the brain. Many studies report overall nucleic acid content in the CNS without providing their specific localization across the tissue [22, 23], and those that provide information about the potential of nanosystems to achieve widespread distribution within the brain often provide not clear evidence. In this line, some reports suggest broad distribution following intracerebroventricular (ICV) administration with poor understanding of the diffusion mechanism, such as in case of oligonucleotide-iron oxide nanoparticles visualized by MRI [24], or fluorescently labelled PEI/siRNA complexes showing neuronal uptake across various brain regions in a Parkinson's disease mouse model [25]. Consequently, advanced imaging and tracking techniques are essential for elucidating the diffusion patterns and cellular targeting of nanosystems in the brain.

In this chapter, we present the design principles, key considerations, and experimental methodologies applied to develop iNEs optimized for enhanced diffusion in the CNS following intraparenchymal administration and, consequently, RNA therapeutic delivery. Special emphasis is given to in vivo assessment of nanocarriers' diffusion across brain parenchyma, presenting fluorescence microscopy–based tracking techniques for evaluating distribution and transfection within the brain. Through this work, we aim to guide the rational development of nanoparticles tailored for effective RNA-based therapies targeting neurological disorders.

2 Materials

2.1 iNE Preparation

All reagents and materials used for iNEs formulation were of pharmaceutical or analytical grade. C12–200, HCl salt (1, 1′-((2-(4-(2-((2-(bis(2-hydroxydodecyl)amino) ethyl)(2-hydroxydodecyl)amino)ethyl)piperazin-1-yl)ethyl)azanediyl) bis(dodecan-2-ol)) and DMG-PEG2000 ((*R*)-methoxy-polyethyleneglycol- 2000-carbamoyl-di-O-myristyl-sn-glyceride) were obtained from Alnylam Pharmaceuticals (MA, USA); while DOPE (1,2-dioleoyl-sn-glycerol-3-phosphoethanolamine) and Vitamin E (D, L-α-tocopherol) were purchased from Avanti Polar Lipids (AL, USA) and BASF (Mannheim, Germany), respectively. mGFP, CleanCap green fluorescent protein (GFP) mRNA, was purchased from TrikLink Biotechnologies (CA, USA).

To prepare the iNEs, an organic and aqueous phase were mixed using a microfluidic device Nano Assemblr™ bench-top instrument, Precision NanoSystems Inc.). The organic phase prepared with the compound dissolved in ethanol was composed of 12.8 mg/mL of C12–200, 3.7 mg/mL of DOPE, 6 mg/mL of Vitamin E, and 2 mg/mL of DMG-PEG2000. On the other hand, the aqueous phase was based on RNA solution dissolved in citrate buffer at 10 mM, pH 4. Both phases were pushed into the microfluidic device using a 5 mL syringe (Becton Dickinson Syringes), and mixed at a flow ratio of 12 mL/min.

2.2 iNE Characterization

The physicochemical properties of the produced nanoemulsions (NEs) were evaluated by dynamic light scattering (Zetasizer® Nano ZS, Malvern Instruments, Malvern, UK). Particle size and polydispersity index (PDI) were measured after diluting the nanoemulsion samples ten-fold in PBS, while zeta potential was assessed after a 20-fold dilution in RNase-free water.

The association efficiency (AE%) of RNA with the NEs was determined using two complementary assays.

First, agarose gel electrophoresis was performed by loading 3 μg of siRNA onto a 1% (w/v) agarose gel prepared in Tris-Acetate-EDTA (TAE) buffer (Sigma-Aldrich, MO, USA), both before and after incubation with Triton X-100. Samples were mixed with an equal volume of loading buffer containing 1× SYBR® Gold nucleic acid stain (Invitrogen, CA, USA). Free siRNA (3 μg) was included as a control. Electrophoresis was carried out for 20 min at 70 V using a Sub-Cell GT cell 96/192 system (Bio-Rad Laboratories, CA, USA), and bands were visualized using a UV transilluminator imaging system (Molecular Imager® Gel Doc™ XR, Bio-Rad Laboratories, CA, USA).

Second, AE% was quantified using the Quant-iT RiboGreen RNA assay. Two standard curves were prepared with RNA concentrations ranging from 0.1 to 1 μg/mL, in the presence or absence of Triton X-100. All samples were diluted 40-fold in 1× TE buffer and then further diluted two-fold with either Triton X-100 (to disrupt NEs) or PBS, followed by incubation at 37°C for 30 min. Subsequently, 100 μL of each sample and standard were pipetted in triplicate into a black 96-well polystyrene microplate (Corning®, NY, USA). The RiboGreen reagent was diluted 200-fold in 1× TE buffer, and 100 μ L was added to each well, yielding a final volume of 200 μL per well. Fluorescence intensity was measured using a microplate reader (Synergy H1, BioTek Instruments, VT, USA) with excitation at 485 nm and emission at 530 nm. siRNA concentrations were calculated using standard curves with an $r^2 \geq 0.99$.

To gain further insight into the nanostructures, additional characterization techniques can be employed. For example, both particle size and nanoparticle concentration can be analyzed by Nanoparticle Tracking Analysis (NTA) using a NanoSight NS3000 (Malvern Instruments, Amesbury, UK). For accurate measurements, samples typically require dilution between 1000- to 10,000-fold in the appropriate buffer.

For morphological analysis, transmission electron microscopy (TEM) can be used to visualize the structure of the nanosystems (JEOL JEM-2010, 200 kV). Samples should be diluted in a suitable buffer and mixed 1:1 with an aqueous solution of phosphotungstic acid (2% w/v) for negative staining. Then, 1 μL of the mixture is applied onto copper grids with carbon films (CF400-CU, Electron Microscopy Sciences, Hatfield, PA, USA) and allowed to dry.

Finally, Cryo-TEM can provide additional morphological insights. Sample vitrification may be performed using the Automatic Plunge Freezer EM GP2 (Leica), and prepared using a Vitrobot Mark IV (FEI – Thermo Fisher Scientific) under conditions of 70%–80% humidity at room temperature. Data acquisition can be carried out using the JEM-2200FS/CR (JEOL Ltd.) 200 kV TEM, equipped with a K2 Summit direct detection camera (GATAN). Images can be acquired at ×2000 magnification for overview ("aerial") views, and at ×25,000–×30,000 for high-magnification micrographs. Size distribution histograms can be generated by manually measuring individual nanoparticles.

2.3 Animal Studies, Intrastriatal Injection, and NP Diffusion Measurement In Vivo

Surgical Procedures—Administration, Tissue Collection, and Processing

Administration of iNEs was performed in male Sprague-Dawley rats (three animals in the iNE-mGFP group and one animal in the PBS control group) deeply anesthetized with ketamine (50 mg/kg) and medetomidine (0.4 mg/kg). Anesthetized animals were mounted in a stereotaxic frame (Kopf Instruments, CA, USA) and were injected with 3 μg of mGFP encapsulated into iNEs, in a total volume of 3 μL. The solution was injected using a 5 uL Hamilton syringe, coupled to a motorized injection (Stoelting), at a rate of 0.5 μL/min. Stereotaxic coordinates were A/P: 0.8 mm, M/L: 3.0 mm and D/V: 5.0 mm. The needle used for the administration was left in place for an additional 5 min before withdrawal to avoid iNEs reflux. After 24 h, animals received an anesthetic overdose, and they were sacrificed by decapitation; their brains were rapidly removed, cryoprotected, and cut into coronal tissue sections using a sliding microtome. Sections were processed for immunofluorescence labeling as follows.

Sample Preparation and Imaging of Reporter mRNA GFP Delivery to Determine the Diffusion

Free-floating tissue sections were pre-incubated in KPBS-1% BSA with 5% normal donkey serum (Sigma-Aldrich, MO, USA) and 0.03% Triton X-100 for 60 min at room temperature. Immunofluorescence was performed to identify cells capable of expressing GFP after mGFP transfection. Different brain cell types were labeled with glial fibrillary acidic protein (GFAP, 1:500, MAB360, Merk Millipore, MA, USA) as an astrocytic marker, or ionized calcium-binding adaptor molecule 1 (Iba- 1, 1:500, 019–19,741, Wako Chemicals, Neuss, Germany) as a microglial marker. In the case of neurons, considering the auto-green fluorescence depicted by them [26], a double immunofluorescence against GFP (anti-GFP, 1:1000, C10362, Molecular Probes, OR, USA) and β III-tubulin, followed by the incubation with fluorescence secondary antibodies Alexa Fluor 568-conjugated donkey anti-rabbit IgG (1:200, Molecular Probes, OR, USA) and Alexa Fluor 488-conjugated donkey anti-mouse IgG (1:200, Molecular Probes, OR, USA), respectively, was performed. Cell nuclei were marked with the DNA-binding dye Hoechst 33342 (1:2000, 62,249, Sigma-Aldrich, MO, USA). Immunoreaction was visualized with the fluorescence secondary antibody Alexa Fluor 568- conjugated donkey anti-rabbit IgG (1:200, Molecular Probes, OR, USA) or Alexa Fluor 488-conjugated donkey anti-mouse IgG (1:200, Molecular Probes, OR, USA). Sections were mounted on gelatin-coated slides and cover-slipped with Immumount (Thermo-Shandon). Co-localization of these markers was assessed by confocal laser microscopy (AOBS-SP5X; Leica Microsystems Heidelberg GmbH, Mannheim, Germany). For whole-brain slide imaging, several confocal images were stitched together, and the extent of diffusion was determined by measuring the distance between transfected cells. The extent of diffusion was measured in confocal images as the distance between the most distal transfected areas.

3 Methods

3.1 iNE Design

iNEs were designed for effective diffusion within the brain parenchyma by carefully modulating four critical features: particle size, surface charge, surface composition, and core composition, as illustrated in Fig. 1.

Particle Size

iNEs were formulated with a particle size below 100 nm, as it is generally recommended to reduce steric entrapment within the dense and mesh-like architecture of the brain ECM, therefore facilitating interstitial transport in the CNS [27, 28]. In this line, formulations around 70 nm were prioritized [20] based on prior studies demonstrating a gradually spread profile through the rat brain tissue with 70 nm PEG-PLGA NPs radiolabeled with 18F following convection-enhanced delivery (CED) in the striatum, compared to larger NPs (150 nm) particles that showed irregular and localized distribution [29].

Surface Charge

Surface charge was tuned to achieve near-neutral to slightly negative values, between −5 and 0 mV [20]. Neutral or negative surface charges are generally recommended to improve nanocarriers' diffusion potential as this charge minimizes nonspecific electrostatic interactions with the anionic ECM components, which are known to impede the movement of positively charged nanoparticles and cause entrapment near the injection site [30, 31].

Surface Composition

PEGylation was achieved by incorporating a lipid containing polyethylene glycol (PEG) into the formulation to reduce opsonization and nonspecific adhesive interactions with the ECM, which improved

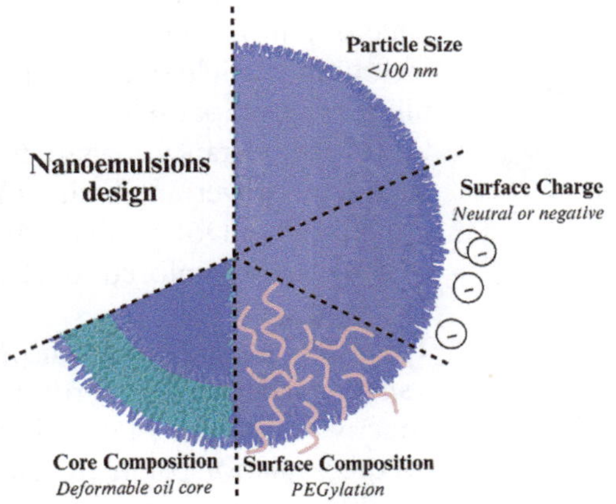

Fig. 1 Design considerations for enhancing iNEs diffusion within the brain, focusing on particle size, surface charge, surface composition, and core composition

iNEs mobility through the brain parenchyma [32–34]. Additionally, PEGylation density should be optimized as it has been described that high-density PEG coatings were used to create a sterically stabilized "stealth" surface and, therefore, enhance brain diffusion [35, 36]. The importance of both PEG presence and surface density has been underscored in multiple studies, with enhanced brain diffusion observed in these PEGylated iNEs administered via CED in rodent models [37, 38].

Core Composition

To enhance deformability and facilitate mechanical adaptability within brain tissue, conventional LNPs—composed of an ionizable lipid, cholesterol, a helper lipid, and a PEGylated lipid [18]—can be structurally modified. Their dense and rigid internal core that limits their ability to adapt to the tight interstitial spaces of the brain [39], as observed in other solid-core nanosystems [40, 41], can be modified for alternative LNP designs, such as the iNEs described in this chapter, featuring more fluid and deformable cores. In these systems, cholesterol is replaced with biocompatible oils as core materials, resulting in more flexible nanosystems that have shown superior diffusion profiles following intraparenchymal administration, outperforming conventional LNPs in mice studies [20].

Integrated Optimization: Experimental and Computational Approaches

To develop nanocarrier formulations with improved diffusion across the brain parenchyma, a formulation strategy was implemented that integrated multiple key design parameters into a single nanosystem. Rather than optimizing each factor independently, particle size, surface charge, and surface and core composition were considered simultaneously during the iNE design to generate a synergistically optimized formulation.

Moreover, to streamline and accelerate this multifactorial optimization process, Design of Experiments (DoE), using Design-Expert® software (Stat-Ease Inc., MN, USA), could have been employed as a robust statistical methodology. DoE enable to evaluate the influence of nanoparticle components on essential response variables, including particle size, zeta potential, encapsulation efficiency, stability, cytocompatibility, and transfection in CNS-relevant cell lines. Feature importance analyses further highlighted the most influential parameters governing brain diffusion and transfection efficiency. The integration of DoE enables the precise and accelerated development of nanocarriers tailored for RNA brain therapies [42].

3.2 iNE Preparation

To obtain a highly monodisperse population of nanosystems with a small particle size, microfluidics has emerged as one of the most effective strategies [43, 44]. This approach allows precise control over formulation parameters in real time, enabling reproducible and scalable production of nanocarriers that often outperform those generated by conventional bulk mixing methods. One of the key advantages of microfluidic-assisted synthesis is its ability to achieve rapid and homogeneous mixing, which minimizes particle growth and aggregation.

This is particularly relevant in the context of brain delivery, where, as mentioned before, particle size plays a critical role not only in crossing the blood–brain barrier, but also in ensuring effective diffusion through the brain parenchyma [45, 46]. In fact, smaller and more homogeneous NPs are less prone to being sequestered by the extracellular matrix and exhibit enhanced mobility within the dense and heterogeneous brain tissue [47, 48], ultimately improving their ability to reach target cells deep within the central nervous system.

3.3 iNE Characterization

The amount of RNA that successfully reaches the cytoplasm after systemic administration is remarkably low, with studies reporting less than 2% efficiency [49, 50]. This highlights the importance of optimizing the RNA (AE%) to ensure that the highest amount of RNA is delivered to the target cells, thereby increasing the therapeutic outcome. In this context, evaluating the AE% during both the synthesis and purification steps was essential, which confirmed that over 80% of the RNA remained associated with the NP structure post-purification, ensuring a high payload retention prior to in vivo application.

3.4 iNE Diffusion Studies In Vivo

Understanding how nanoparticles distribute within brain tissue following direct administration is crucial for developing effective RNA-based therapeutics and other nanomedicines targeting brain disorders. Fluorescence-based methods have emerged as powerful tools for visualizing and quantifying nanoparticle diffusion, cellular uptake, and functional RNA delivery in the brain. This section will put particular emphasis on direct brain administration of iNEs containing GFP-encoding mRNA via intraparenchymal injection and the assessment of nanocarrier distribution via confocal microscopy [20], while carefully considering and discussing the most relevant factors for the design of in vivo diffusion studies of nanoparticles in a broader scope.

mRNA Encoding GFP as Reporter System

Prior to being able to evaluate iNEs for their capacity to diffuse across the brain, a reporter system must be selected.

In principle, there are two conceptually distinct labeling approaches to track nanoparticle distribution. The first is based on direct fluorescence labeling, which involves the covalent attachment of fluorescent moieties either to the RNA cargo [14, 51] or nanoparticle components [52, 53], or the incorporation of lipophilic dyes in the hydrophobic sections of the particles [15, 54]. This approach enables real-time visualization of particle biodistribution, cellular internalization, and tissue penetration patterns; however, it may affect the particle behavior due to the necessary interventions to achieve labeling [55]. Attachment of dyes to the nanoparticle surface can change surface properties, thereby augmenting the biodistribution profile [56], and dye leakage, on the other hand, may lead to misinterpretation of particle distribution [57].

On the other hand, considering the biological barriers implied in RNA delivery, opting for mRNA-encoded reporter-based tracking offers certain advantages. This involves the assessment of functional mRNA delivery by monitoring the expression of fluorescent reporter proteins following successful cellular transfection [22, 58–60]. This methodology provides direct evidence of particle distribution, cellular uptake, and endosomal escape of RNA prior to functional translation, offering an assessment of the entire delivery process rather than simply tracking physical particle distribution. Most commonly utilized reporter mRNAs are encoding for GFP (mGFP) [20], Luciferase (mLuc) [58, 61], Cre recombinase (mCre) [60], or Cherry (mCherry) [59].

Reporter-based strategies are not without limitations, as the requirement of successful transfection may lead to underestimation of physical diffusion behavior. Additionally, the delay between injection and detectable reporter expression (typically 6–24 h) prevents real-time tracking of initial particle distribution patterns and must be considered during study design. Study endpoints must be defined carefully, as peak expression depends not only on the delivery vehicle and RNA construct, but also on the route of administration, as described by Weissman and colleagues [62]. In the case of intraparenchymal mGFP delivery via iNEs, 24 h post-administration is suitable, whereas mLuc or mCre delivery is evaluated after around 6–24 h [58] or several days [60] after injection, respectively.

Considerations During Intraparenchymal Administration

Upon design, development and characterization of suitable nanocarriers, as the herein described iNEs loaded with mRNA GFP, they were administered via intraparenchymal, more specifically intrathalamic injection in rats. Rats were selected for their technical and anatomical advantages over mice, facilitated by their larger brain sizes. Experimental details can be found in Section:Surgical procedures – administration, tissue collection and processing.

To ensure reproducibility and unbiased results, execution of the injection is of utmost importance for the experimental outcome [10] and will be discussed in the following paragraphs.

Both injection volume and speed can significantly affect the experimental outcome. Large volumes may increase the pressure and therefore the distribution through the tissue, can however cause significant damage around the injection site. Injection volumes must be adapted with the animal model and the therapeutic purpose in mind [63]. Volumes of about 0.5–3 μL and 1–10 μL in mice and rats respectively, have been reported [15, 60, 64].

The speed of injection, not unlike the total volume, may increase pressure and cause tissue damage when too high. Rates of around 0.1–0.5 μL/min [22, 53] in small rodents not only ensure sufficiently slow injection to avoid tissue damage by allowing the fluid to flow off, but it can also avoid backflow [65]. Recent approaches for the delivery of therapeutics via intracranial administration by Hanes and Suk described how CED can enable widespread delivery [66]. After the administration, the needle is to be kept in place for around 1–5 min to avoid additional backflow of the mRNA NPs along the needle path, which can potentially reduce the effective dose at the intended site and artificially amplify the fluorescence signal [67].

The injection site plays a role in the observed diffusion and needs to be considered. Generally, ICV injection gives direct access to the CSF, whereby nanoparticles spread quickly over large areas of the brain and spinal cord surfaces, but penetration into the brain parenchyma is limited. Intraparenchymal delivery is more localized, and while it gives access to possible target cells at the injection site, it exposes the particles to a penetration barrier of dense tissue, demanding highly diffusive particles [10].

In the specific example of iNE-mediated delivery of mGFP, rats were injected with 3 μL iNE-mGFP at a speed of 0.5 μL/min and sacrificed 24 h post-injection, after which brains were processed to allow for brain delivery analysis. Specifics thereof are described in Section:Sample preparation and imaging of reporter mRNA GFP deliver to determine the diffusion.

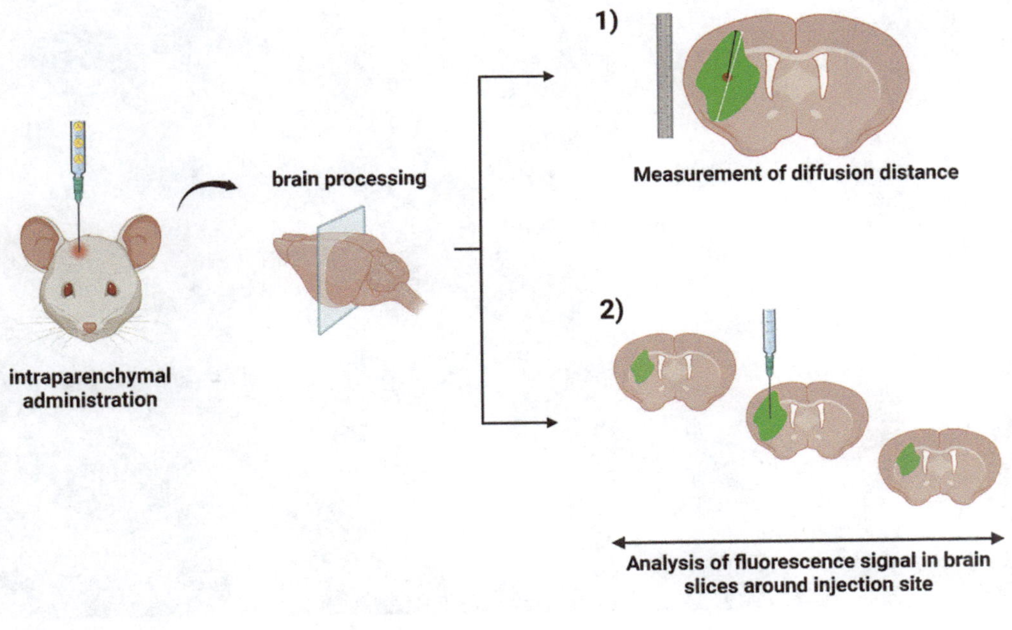

Fig. 2 Analysis scheme to evaluate brain diffusivity ex vivo. Upon intraparenchymal injection, brains are processed and sliced (coronal) and analyzed. (1) The diffusion is analyzed either via measurement of the extension from the injection site (red dot) to the distal points (black line) or via measurement between distal points (white line). (2) The analysis may involve consecutive brain slices to add another dimension, where the fluorescence signal in brain slices may decrease with increasing distance from the central injection plane

Analysis of Nanoparticle mRNA Delivery and Diffusion

The brain slices, prepared for confocal microscopy, were analyzed to determine mGFP transfection and thereby assess the diffusivity of injected nanoparticles. Confocal microscopy provides superior spatial resolution for visualizing nanoparticle distribution within brain tissue sections. By using a pinhole aperture to eliminate out-of-focus light, confocal microscopy enables optical sectioning and three-dimensional reconstruction of nanoparticle or reporter protein localization [68].

Ex vivo analysis of nanoparticle diffusion based on brain slice confocal images can be performed and is illustrated in Fig. 2. A region of interest can be defined, based on for example distance between injection site and the most distal transfected point, or the overall distance between edges of transfected area (Fig. 2 (1)). A clear definition of the boundary of nanoparticle penetration is crucial and may be based on a threshold fluorescence intensity with respect to the maximum or background intensity.

Image processing software allows for the measurement of distances in different dimensions, based on which the area can be calculated. Upon analysis of different brain slices with increasing distance

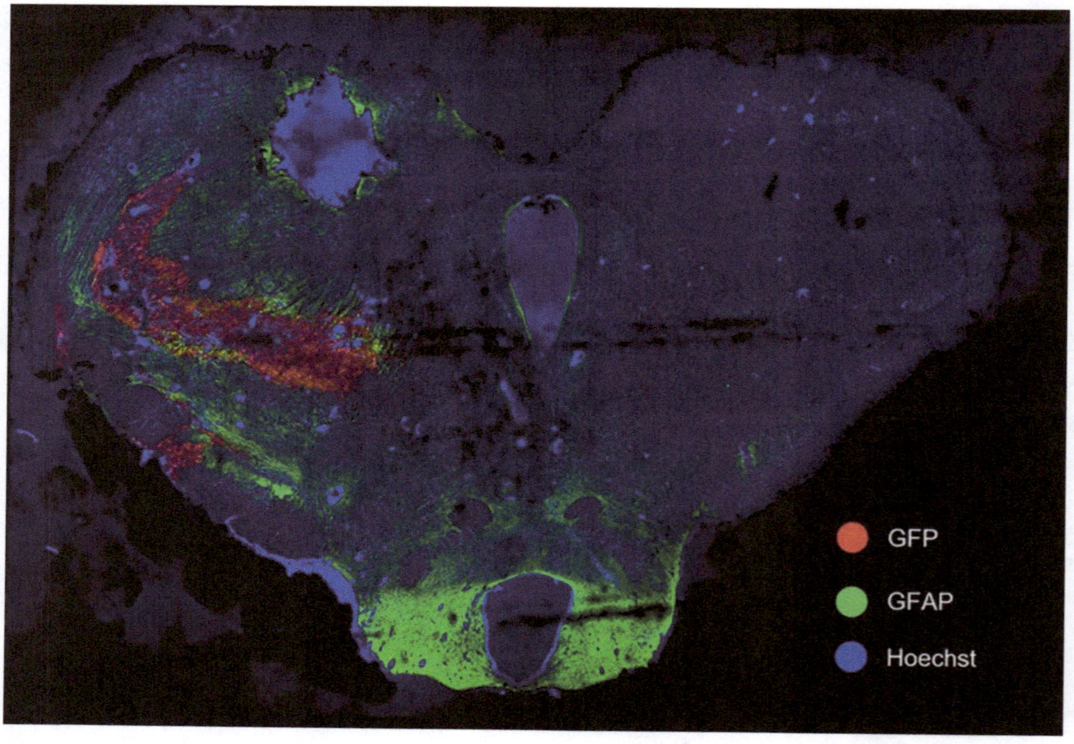

Fig. 3 Confocal brain map upon intraparenchymal delivery of iNE-mGFP. Double immunolabelling in the whole brain for GFP (red) and astrocytes (labeled with GFAP, green). Hoechst was used as nuclei marker (blue). Distance measurement between edges of GFP-positive area revealed diffusion and functional mRNA delivery of around 3.4 mm. GFAP glial fibrillary acidic protein, GFP green fluorescence protein, iNE ionizable nanoemulsion, mGFP mRNA encoding GFP. (Reproduced with permission from [20])

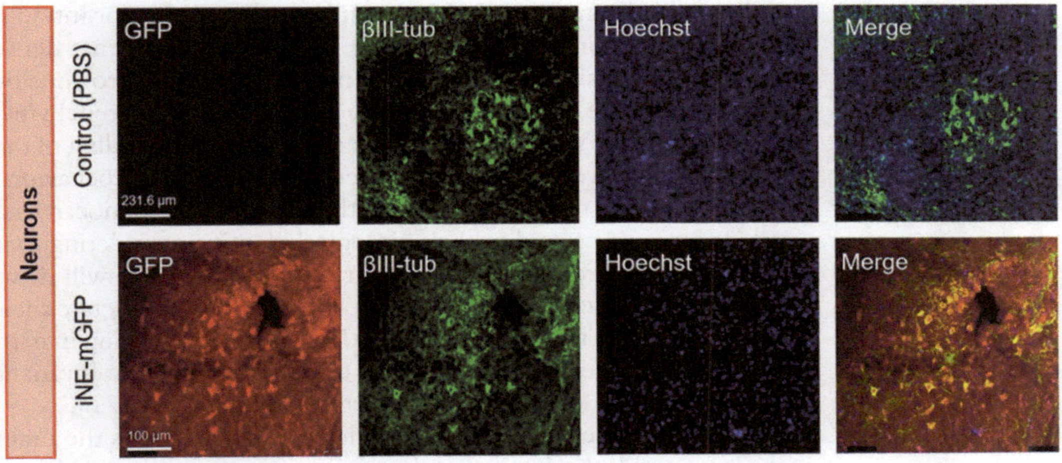

Fig. 4 Neuron-specific delivery of mGFP to neurons in vivo. Immunolabelling for GFP in neurons (labeled with βIII-tubulin), in rat brain slides after intraparenchymal administration of control (PBS) or iNE-mGFP. Hoechst was used as a nucleus marker. βIII-tubulin class III β-tubulin, GFP green fluorescence protein, iNE ionizable nanoemulsion, mGFP mRNA encoding GFP. (Reproduced and adapted with permission from [20])

from the injection site, information in a third plane is obtained, allowing for calculation of the transfected volume (Fig. 2 (2)).

The literature, however, mostly reports the distance measured between injection site and distal points in a 2D space, generally reporting between 1 and 2 mm in function of injection site, volume, and mRNA dose [14, 15, 60]. In the case of the described iNEs, distance measurement in confocal brain maps revealed GFP-expressing cells (Fig. 3) up to approximately 3.4 mm apart, confirming proper particle design for diffusive RNA delivery systems.

In addition to the acquisition of brain maps for diffusion distance, area, or volume determination, confocal microscopy can at the same time be used for colocalization studies to determine whether nanoparticles are associated with specific cell types, such as neurons, astrocytes, and micrgolia. This increases the value of the information by adding another layer of complexity, equally relevant and determinant of therapeutic success. In the case of the iNEs, neuron-specific mRNA delivery was observed (Fig. 4), information that, in combination with the diffusive character, guides their therapeutic exploitation [20].

4 General Conclusions

In nanocarrier-mediated brain delivery, several biological barriers must be overcome, with diffusion across the brain parenchyma being one of the least studied. This chapter has aimed at guiding nanocarrier design and the study of their diffusion based on reporter mRNA delivery, as exemplified by our group's work on iNEs for intraparenchymal administration. While we and others found that dense PEGylation, small particle size, and enhanced flexibility provided by the oily core are key contributors to improved brain parenchymal diffusion, future studies need to aim at a better understanding of the correlation between particle characteristics and diffusive behavior.

In addition to studying the synthetic identity of nanocarriers, attention must be paid to their biological identity, considering that once administered, their exposure to biological media will drive adsorption of biomolecules, mainly proteins, that may dictate their behavior [69]. The employment of AI and machine learning in nanoparticle design offers great potential to tackle complex, multivariant problems such as nanoparticle delivery to the brain [70, 71].

Available literature on nanoparticle diffusivity across the brain is scarce and lacks detailed and quantitative information. It is not surprising that drawing solid conclusions is difficult, not only due to the complexity of the question, but also the variability in the approach itself. Be it in the relevance the administration itself plays, specifically in terms of location, volume, or speed, but also nanoparticle design, labelling strategy, and the way distance or volume are measured or quantified.

Fluorescence-based techniques have significantly advanced the study of nanoparticle diffusion and delivery. Direct labeling approaches enable immediate visualization of physical distribution, while reporter-based systems provide functional confirmation of successful RNA delivery and protein expression. Both strategies offer complementary insights, and their selection should be guided by the specific experimental objectives, distinguishing physical spread from biologically active delivery.

The effectiveness of fluorescence-based tracking depends not only on the labeling strategy but also on the properties of the fluorescent probe itself. Key parameters such as tissue autofluorescence, spatial and temporal resolution, penetration depth, and photostability can substantially influence signal interpretation. As such, the choice of fluorophore is as critical as the method of incorporation into the nanoparticle or RNA cargo [55].

The development of advanced imaging modalities, including confocal and two-photon microscopy as well as label-free approaches such as stimulated Raman scattering microscopy, has further expanded the ability to visualize nanoparticle behavior across different resolutions and depths. However, each technique presents trade-offs in terms of resolution, imaging speed, and tissue compatibility [68, 72, 73].

Notably, recent advances in tissue clearing methods have transformed fluorescence imaging by enabling three-dimensional visualization of nanoparticle distribution in intact brain tissue. By rendering the tissue optically transparent while preserving anatomical architecture, clearing approaches circumvent the need for physical sectioning and allow for volumetric analysis of nanoparticle biodistribution on a whole-organ 3D scale. This development opens new possibilities for comprehensive diffusion mapping and cell-type colocalization in complex brain regions [74, 75].

As fluorescence imaging technologies continue to advance, we can expect further improvements in sensitivity, resolution, and quantitative analysis capabilities. These developments, coupled with innovations in nanoparticle design and labeling strategies, will enhance our understanding of nanoparticle-brain delivery and accelerate the development of effective and diffusive RNA-based therapeutics for neurological disorders.

Acknowledgments

This work was supported by Competitive Reference Groups, Consellería de Educación e Ordenación Universitaria, Xunta de Galicia, Ref: ED431C 2021/17. Financial support from the Xunta de Galicia (Centro singular de investigación de Galicia accreditation 2019–2022) and the European Union (European Regional Development Fund - ERDF), is gratefully acknowledged. REF: ED431G 2019/02. The Grant PID2020-119368RB-I00 funded by MICIU/ AEI/ 10.13039/501100011033 and by “ERDF/EU” is gratefully acknowledged. Philipp Lapuhs acknowledges a grant (ED481A 2022/237), Programa de ayudas a la etapa predoctoral 2022 de la Xunta de Galicia (Consellería de Cultura, Educación, Formación Profesional y Universidades o de la Agencia Gallega de Innovación). Inés Rubio-Prego acknowledges a grant (IFI23/00035) from Instituto de Salud Carlos III “Contratos i-PFIS: Doctorados ISS-empresa en Ciencias y Tecnologías de la Salud”. Laura Pineiro-Alonso acknowledges a grant (ED481A 2022/237) from Xunta de Galicia “Axudas de apoio á etapa predoutoral 2021”. Figures have been created with BioRender.com.

Conflict of Interest The authors have no conflicts of interest to declare that are relevant to the content of this chapter.

References

1. Chancellor D, Barrett D, Nguyen-Jatkoe L, Millington S, Eckhardt F (2023) The state of cell and gene therapy in 2023. Mol Ther 31:3376–3388. https://doi.org/10.1016/j.ymthe.2023.11.001
2. López-Estévez AM, Lapuhs P, Pineiro-Alonso L, Alonso MJ (2024) Personalized cancer nanomedicine: overcoming biological barriers for intracellular delivery of biopharmaceuticals. Adv Mater 36(14):e2309355. https://doi.org/10.1002/adma.202309355
3. Hoy SM (2018) Patisiran: first global approval. Drugs 78:1625–1631. https://doi.org/10.1007/s40265-018-0983-6
4. Fiolet T, Kherabi Y, MacDonald C-J, Ghosn J, Peiffer-Smadja N (2022) Comparing COVID-19 vaccines for their characteristics, efficacy and effectiveness against SARS-CoV-2 and variants of concern: a narrative review. Clin Microbiol Infect 28:202–221. https://doi.org/10.1016/j.cmi.2021.10.005

5. Curreri A, Sankholkar D, Mitragotri S, Zhao Z (2023) RNA therapeutics in the clinic. Bioeng Transl Med 8(1):e10374. https://doi.org/10.1002/btm2.10374
6. Search for: cns OR Brain Cancer or Alzheimer | Other terms: RNA therapeutics OR mRNA OR siRNA OR ASO OR aptamers OR miRNA | Not yet recruiting, Recruiting, Active, not recruiting studies | Phase: Early Phase 1, 1, 2, 3, 4 | Interventional studies, (n.d.). https://clinicaltrials.gov/search?cond=cns OR Brain Cancer or Alzheimer&aggFilters=phase:0 1 2 3 4,status:rec act not,studyType:int&limit=100&term=RNA therapeutics OR mRNA OR siRNA OR ASO OR aptamers OR miRNA. Accessed 21 May, 2025
7. Gao J, Xia Z, Gunasekar S, Jiang C, Karp JM, Joshi N (2024) Precision drug delivery to the central nervous system using engineered nanoparticles. Nat Rev Mater 9:567–588. https://doi.org/10.1038/s41578-024-00695-w
8. Abbott NJ (2013) Blood–brain barrier structure and function and the challenges for CNS drug delivery. J Inherit Metab Dis 36:437–449. https://doi.org/10.1007/s10545-013-9608-0
9. Pineiro-Alonso L, Rubio-Prego I, Lobyntseva A, González-Freire E, Langer R, Alonso MJ (2025) Nanomedicine for targeting brain neurodegeneration: critical barriers and circadian rhythm Considerations. Adv Drug Deliv Rev 222:115606. https://doi.org/10.1016/j.addr.2025.115606
10. Fowler MJ, Cotter JD, Knight BE, Sevick-Muraca EM, Sandberg DI, Sirianni RW (2020) Intrathecal drug delivery in the era of nanomedicine. Adv Drug Deliv Rev 165–166:77–95. https://doi.org/10.1016/j.addr.2020.02.006
11. Blair HA (2023) Tofersen: first approval. Drugs 83:1039–1043. https://doi.org/10.1007/s40265-023-01904-6
12. Ajeeb R, Clegg JR (2023) Intrathecal delivery of macromolecules: clinical status and emerging technologies. Adv Drug Deliv Rev 199:114949. https://doi.org/10.1016/j.addr.2023.114949
13. Zimmermann DR, Dours-Zimmermann MT (2008) Extracellular matrix of the central nervous system: from neglect to challenge. Histochem Cell Biol 130:635–653. https://doi.org/10.1007/s00418-008-0485-9
14. Waggoner LE, Miyasaki KF, Kwon EJ (2023) Analysis of PEG-lipid anchor length on lipid nanoparticle pharmacokinetics and activity in a mouse model of traumatic brain injury. Biomater Sci 11:4238–4253. https://doi.org/10.1039/D2BM01846B
15. Rungta RL, Choi HB, Lin PJ, Ko RW, Ashby D, Nair J, Manoharan M, Cullis PR, MacVicar BA (2013) Lipid nanoparticle delivery of siRNA to silence neuronal gene expression in the brain. Mol Ther Acids 2:e136. https://doi.org/10.1038/mtna.2013.65
16. Patel MR, Patel RB, Thakore SD (2018) Nanoemulsion in drug delivery. In: Applications of nanocomposite materials in drug delivery. Elsevier, Amsterdam, pp 667–700. https://doi.org/10.1016/B978-0-12-813741-3.00030-3
17. Sutradhar KB, Amin ML (2013) Nanoemulsions: increasing possibilities in drug delivery. Eur J Nanomed 5(2):97. https://doi.org/10.1515/ejnm-2013-0001
18. Cullis PR, Hope MJ (2017) Lipid nanoparticle systems for enabling gene therapies. Mol Ther 25:1467–1475. https://doi.org/10.1016/j.ymthe.2017.03.013
19. Patel S, Kim J, Herrera M, Mukherjee A, Kabanov AV, Sahay G (2019) Brief update on endocytosis of nanomedicines. Adv Drug Deliv Rev 144:90–111. https://doi.org/10.1016/j.addr.2019.08.004
20. Borrajo ML, Quijano A, Lapuhs P, Rodriguez-Perez AI, Anthiya S, Labandeira-Garcia JL, Valenzuela R, Alonso MJ (2024) Ionizable nanoemulsions for RNA delivery into the central nervous system – importance of diffusivity. J Control Release 372:295–303. https://doi.org/10.1016/j.jconrel.2024.06.051
21. Durán-Lobato M, López-Estévez AM, Cordeiro AS, Dacoba TG, Crecente-Campo J, Torres D, Alonso MJ (2021) Nanotechnologies for the delivery of biologicals: Historical perspective and current landscape. Adv Drug Deliv Rev 176:113899. https://doi.org/10.1016/j.addr.2021.113899
22. Tanaka H, Nakatani T, Furihata T, Tange K, Nakai Y, Yoshioka H, Harashima H, Akita H (2018) In vivo introduction of mRNA encapsulated in lipid nanoparticles to brain neuronal cells and astrocytes via intracerebroventricular administration. Mol Pharm 15:2060–2067. https://doi.org/10.1021/acs.molpharmaceut.7b01084
23. Liu S, Liu J, Li H, Mao K, Wang H, Meng X, Wang J, Wu C, Chen H, Wang X, Cong X, Hou Y, Wang Y, Wang M, Yang Y-G, Sun T (2022) An optimized ionizable cationic lipid for brain tumor-targeted siRNA delivery and glioblastoma immunotherapy. Biomaterials 287:121645. https://doi.org/10.1016/j.biomaterials.2022.121645
24. Liu CH, Ren JQ, Yang J, Liu C, Mandeville JB, Rosen BR, Bhide PG, Yanagawa Y, Liu PK (2009) DNA-based MRI probes for specific detection of chronic exposure to amphetamine in living brains. J Neurosci 29:10663–10670. https://doi.org/10.1523/JNEUROSCI.2167-09.2009

25. Helmschrodt C, Höbel S, Schöniger S, Bauer A, Bonicelli J, Gringmuth M, Fietz SA, Aigner A, Richter A, Richter F (2017) Polyethylenimine nanoparticle-mediated siRNA delivery to reduce α-synuclein expression in a model of Parkinson's disease. Mol Ther - Nucleic Acids 9:57–68. https://doi.org/10.1016/j.omtn.2017.08.013
26. Spitzer N, Sammons GS, Price EM (2011) Autofluorescent cells in rat brain can be convincing impostors in green fluorescent reporter studies. J Neurosci Methods 197:48–55. https://doi.org/10.1016/j.jneumeth.2011.01.029
27. Yuan T, Gao L, Zhan W, Dini D (2022) Effect of particle size and surface charge on nanoparticles diffusion in the brain white matter. Pharm Res 39:767–781. https://doi.org/10.1007/s11095-022-03222-0
28. MacKay JA, Deen DF, Szoka FC (2005) Distribution in brain of liposomes after convection enhanced delivery; modulation by particle charge, particle diameter, and presence of steric coating. Brain Res 1035:139–153. https://doi.org/10.1016/j.brainres.2004.12.007
29. Sirianni RW, Zheng M-Q, Patel TR, Shafbauer T, Zhou J, Saltzman WM, Carson RE, Huang Y (2014) Radiolabeling of Poly(lactic- co -glycolic acid) (PLGA) nanoparticles with biotinylated F-18 prosthetic groups and imaging of their delivery to the brain with positron emission tomography. Bioconjug Chem 25:2157–2165. https://doi.org/10.1021/bc500315j
30. Yuan T, Gao L, Zhan W, Dini D (2022) Effect of particle size and surface charge on nanoparticles diffusion in the brain white matter. Pharm Res 39:767–781. https://doi.org/10.1007/s11095-022-03222-0
31. MacKay JA, Deen DF, Szoka FC (2005) Distribution in brain of liposomes after convection enhanced delivery; modulation by particle charge, particle diameter, and presence of steric coating. Brain Res 1035:139–153. https://doi.org/10.1016/j.brainres.2004.12.007
32. Negron K, Zhu C, Chen S-W, Shahab S, Rao D, Raabe EH, Eberhart CG, Hanes J, Suk JS (2020) Non-adhesive and highly stable biodegradable nanoparticles that provide widespread and safe transgene expression in orthotopic brain tumors. Drug Deliv Transl Res 10:572–581. https://doi.org/10.1007/s13346-020-00759-8
33. López-Estevez AM, Gref R, Alonso MJ (2024) A journey through the history of PEGylated drug delivery nanocarriers. Drug Deliv Transl Res 14:2026–2031. https://doi.org/10.1007/s13346-024-01608-8
34. Berrecoso G, Bravo SB, Arriaga I, Abrescia N, Crecente-Campo J, Alonso MJ (2024) Controlling the protein corona of polymeric nanocapsules: effect of polymer shell on protein adsorption. Drug Deliv Transl Res 14:918–933. https://doi.org/10.1007/s13346-023-01441-5
35. Mastorakos P, Zhang C, Berry S, Oh Y, Lee S, Eberhart CG, Woodworth GF, Suk JS, Hanes J (2015) Highly PEGylated DNA nanoparticles provide uniform and widespread gene transfer in the brain. Adv Healthc Mater 4:1023–1033. https://doi.org/10.1002/adhm.201400800
36. Nance EA, Woodworth GF, Sailor KA, Shih T-Y, Xu Q, Swaminathan G, Xiang D, Eberhart C, Hanes J (2012) A dense Poly(Ethylene Glycol) coating improves penetration of large polymeric nanoparticles within brain tissue. Sci Transl Med 4(149):149ra119. https://doi.org/10.1126/scitranslmed.3003594
37. Pickering AJ, Lamson NG, Marand MH, Hwang W, Straehla JP, Hammond PT (2023) Layer-by-layer polymer functionalization improves nanoparticle penetration and glioblastoma targeting in the brain. ACS Nano 17:24154–24169. https://doi.org/10.1021/acsnano.3c09273
38. Séhédic D, Chourpa I, Tétaud C, Griveau A, Loussouarn C, Avril S, Legendre C, Lepareur N, Wion D, Hindré F, Davodeau F, Garcion E (2017) Locoregional confinement and major clinical benefit of 188 re-loaded CXCR4-targeted nanocarriers in an orthotopic human to mouse model of glioblastoma. Theranostics 7: 4517–4536. https://doi.org/10.7150/thno.19403
39. Eygeris Y, Patel S, Jozic A, Sahay G (2020) Deconvoluting lipid nanoparticle structure for messenger RNA delivery. Nano Lett 20:4543–4549. https://doi.org/10.1021/acs.nanolett.0c01386
40. Etame AB, Diaz RJ, O'Reilly MA, Smith CA, Mainprize TG, Hynynen K, Rutka JT (2012) Enhanced delivery of gold nanoparticles with therapeutic potential into the brain using MRI-guided focused ultrasound. Nanomed Nanotechnol Biol Med 8:1133–1142. https://doi.org/10.1016/j.nano.2012.02.003
41. Pan S, Yang PH, DeFreitas D, Ramagiri S, Bayguinov PO, Hacker CD, Snyder AZ, Wilborn J, Huang H, Koller GM, Raval DK, Halupnik GL, Sviben S, Achilefu S, Tang R, Haller G, Quirk JD, Fitzpatrick JAJ, Esakky P, Strahle JM (2023) Gold nanoparticle-enhanced X-ray microtomography of the rodent reveals region-specific cerebrospinal fluid circulation in the brain. Nat Commun 14: 453. https://doi.org/10.1038/s41467-023-36083-1
42. Rampado R, Peer D (2023) Design of experiments in the optimization of nanoparticle-based drug delivery systems. J Control Release

358:398–419. https://doi.org/10.1016/j.jconrel.2023.05.001
43. Liu Y, Yang G, Hui Y, Ranaweera S, Zhao C (2022) Microfluidic nanoparticles for drug delivery. Small 18:2106580. https://doi.org/10.1002/smll.202106580
44. Gimondi S, Ferreira H, Reis RL, Neves NM (2023) Microfluidic devices: a tool for nanoparticle synthesis and performance evaluation. ACS Nano 17:14205–14228. https://doi.org/10.1021/acsnano.3c01117
45. Krol S (2012) Challenges in drug delivery to the brain: nature is against us. J Control Release 164:145–155. https://doi.org/10.1016/j.jconrel.2012.04.044
46. Abbott NJ (2013) Blood–brain barrier structure and function and the challenges for CNS drug delivery. J Inherit Metab Dis 36:437–449. https://doi.org/10.1007/s10545-013-9608-0
47. Thorne RG, Nicholson C (2006) In vivo diffusion analysis with quantum dots and dextrans predicts the width of brain extracellular space. Proc Natl Acad Sci 103:5567–5572. https://doi.org/10.1073/pnas.0509425103
48. Yuan T, Gao L, Zhan W, Dini D (2022) Effect of particle size and surface charge on nanoparticles diffusion in the brain white matter. Pharm Res 39:767–781. https://doi.org/10.1007/s11095-022-03222-0
49. Chen P, Yang W, Hong T, Miyazaki T, Dirisala A, Kataoka K, Cabral H (2022) Nanocarriers escaping from hyperacidified endo/lysosomes in cancer cells allow tumor-targeted intracellular delivery of antibodies to therapeutically inhibit c-MYC. Biomaterials 288:121748. https://doi.org/10.1016/j.biomaterials.2022.121748
50. Gilleron J, Querbes W, Zeigerer A, Borodovsky A, Marsico G, Schubert U, Manygoats K, Seifert S, Andree C, Stöter M, Epstein-Barash H, Zhang L, Koteliansky V, Fitzgerald K, Fava E, Bickle M, Kalaidzidis Y, Akinc A, Maier M, Zerial M (2013) Image-based analysis of lipid nanoparticle-mediated siRNA delivery, intracellular trafficking and endosomal escape. Nat Biotechnol 31:638–646. https://doi.org/10.1038/nbt.2612
51. Yang YL, Zhang XY, Wu SW, Zhang R, Zhou BL, Zhang XY, Tang L, Tian Y, Men K, Yang L (2022) Enhanced nose-to-brain delivery of siRNA using hyaluronan-enveloped nanomicelles for glioma therapy. J Control Release 342:66–80. https://doi.org/10.1016/j.jconrel.2021.12.034
52. Reimold I, Domke D, Bender J, Seyfried CA, Radunz HE, Fricker G (2008) Delivery of nanoparticles to the brain detected by fluorescence microscopy. Eur J Pharm Biopharm 70:627–632. https://doi.org/10.1016/j.ejpb.2008.05.007
53. Householder KT, Dharmaraj S, Sandberg DI (2019) Fate of nanoparticles in the central nervous system after intrathecal injection in healthy mice. Sci Rep 9(1):12587. https://doi.org/10.1038/s41598-019-49028-w
54. Akita H, Nakatani T, Kuroki K, Maenaka K, Tange K, Nakai Y, Harashima H (2015) Effect of hydrophobic scaffold on the cellular uptake and gene transfection activities of DNA-encapsulating liposomal nanoparticles via intracerebroventricular administration. Int J Pharm 490:142–145. https://doi.org/10.1016/j.ijpharm.2015.05.043
55. Chen H, Hu Q, Li W, Cai X, Mao L, Li R (2023) Approaches to nanoparticle labeling: a review of fluorescent, radiological, and metallic techniques. Environ Heal 1:75–89. https://doi.org/10.1021/envhealth.3c00034
56. Álamo P, Pallarès V, Céspedes MV, Falgàs A, Sanchez JM, Serna N, Sánchez-garcía L, Voltà-duràn E, Morris GA, Sánchez-chardi A, Casanova I, Mangues R, Vazquez E, Villaverde A, Unzueta U (2020) Fluorescent dye labeling changes the biodistribution of tumor-targeted nanoparticles. Pharmaceutics 12:1–18. https://doi.org/10.3390/pharmaceutics12111004
57. Zhukova V, Osipova N, Semyonkin A, Malinovskaya J, Melnikov P, Valikhov M, Porozov Y, Solovev Y, Kuliaev P, Zhang E, Sabel BA, Chekhonin V, Abakumov M, Majouga A, Kreuter J, Henrich-Noack P, Gelperina S, Maksimenko O (2021) Fluorescently labeled PLGA nanoparticles for visualization in vitro and in vivo: The importance of dye properties. Pharmaceutics 13(8):1145. https://doi.org/10.3390/pharmaceutics13081145
58. Nabhan JF, Wood KM, Rao VP, Morin J, Bhamidipaty S, Labranche TP, Gooch RL, Bozal F, Bulawa CE, Guild BC (2016) Intrathecal delivery of frataxin mRNA encapsulated in lipid nanoparticles to dorsal root ganglia as a potential therapeutic for Friedreich's ataxia. Sci Rep 6:1–10. https://doi.org/10.1038/srep20019
59. Tamaru M, Akita H, Nakatani T, Kajimoto K, Sato Y, Hatakeyama H, Harashima H (2014) Application of apolipoprotein E-modified liposomal nanoparticles as a carrier for delivering DNA and nucleic acid in the brain. Int J Nanomed 9:4267–4276. https://doi.org/10.2147/IJN.S65402
60. Tuma J, Chen YJ, Collins MG, Paul A, Li J, Han H, Sharma R, Murthy N, Lee HY (2023) Lipid nanoparticles deliver mRNA to the brain after an intracerebral injection. Biochemistry 62:3533–3547. https://doi.org/10.1021/acs.biochem.3c00371

61. Buckley SMK, Delhove JMKM, Perocheau DP, Karda R, Rahim AA, Howe SJ, Ward NJ, Birrell MA, Belvisi MG, Arbuthnot P, Johnson MR, Waddington SN, McKay TR (2015) In vivo bioimaging with tissue-specific transcription factor activated luciferase reporters. Sci Rep 5:11842. https://doi.org/10.1038/srep11842
62. Pardi N, Tuyishime S, Muramatsu H, Kariko K, Mui BL, Tam YK, Madden TD, Hope MJ, Weissman D (2015) Expression kinetics of nucleoside-modified mRNA delivered in lipid nanoparticles to mice by various routes. J Control Release 217:345–351. https://doi.org/10.1016/j.jconrel.2015.08.007
63. Alamouti AK, Raouf I, Zahabi S, Salimibani M (2024) Numerical study of magnetic nanoparticles injection into a brain tumor considering the effects of injection volume and location on the termination of cancerous cells. Biointerphases 19: 61005. https://doi.org/10.1116/6.0003814
64. Guo S, Cázarez-Márquez F, Jiao H, Foppen E, Korpel NL, Grootemaat AE, Liv N, Gao Y, Van Der Wel N, Zhou B, Nie G, Yi CX (2022) Specific silencing of microglial gene expression in the rat brain by nanoparticle-based small interfering RNA delivery. ACS Appl Mater Interfaces 14:5066–5079. https://doi.org/10.1021/acsami.1c22434
65. Casanova F, Carney PR, Sarntinoranont M (2014) Effect of needle insertion speed on tissue injury, stress, and backflow distribution for convection-enhanced delivery in the rat brain. PLoS One 9(4):e94919. https://doi.org/10.1371/journal.pone.0094919
66. Rao D, Kwak G, Wang H, Eberhart CG, Hanes J, Suk JS (2023) Bioreducible gene delivery platform that promotes intracellular payload release and widespread brain dispersion. ACS Biomater Sci Eng 9:4567–4572. https://doi.org/10.1021/acsbiomaterials.3c00799
67. Bartoletti S, Raimondi F, Garofani BC, Ren E, Ciarpella F, Capodiferro A, Palazzolo G, Vilella A, Leo G, Zoli M, Decimo I, Curia G (2025) Two minimally invasive strategies to implant guide cannulas for multiple injections in deep brain areas. Methods 238:27–39. https://doi.org/10.1016/j.ymeth.2025.03.005
68. Del Bonis-O'Donnell JT, Chio L, Dorlhiac GF, McFarlane IR, Landry MP (2018) Advances in nanomaterials for brain microscopy. Nano Res 11:5144–5172. https://doi.org/10.1007/s12274-018-2145-2
69. Wang Y, Zhang H, Xiao W, Liu Y, Zhou Y, He X, Xia X, Gong T, Wang L, Gao H (2021) Unmasking CSF protein corona: effect on targeting capacity of nanoparticles. J Control Release 333:352–361. https://doi.org/10.1016/j.jconrel.2021.04.001
70. Yousfan A, Al Rahwanji MJ, Hanano A, Al-Obaidi H (2024) A comprehensive study on nanoparticle drug delivery to the brain: application of machine learning techniques. Mol Pharm 21:333–345. https://doi.org/10.1021/acs.molpharmaceut.3c00880
71. Dipankar P, Salazar D, Dennard E, Mohiyuddin S, Nguyen QC (2025) Artificial intelligence based advancements in nanomedicine for brain disorder management: an updated narrative review. Front Med 12:1599340. https://doi.org/10.3389/fmed.2025.1599340
72. Ban W, You Y, Yang Z (2022) Imaging Technologies for cerebral pharmacokinetic studies: progress and perspectives. Biomedicines 10(10):2447. https://doi.org/10.3390/biomedicines10102447
73. Wei M, Qian N, Gao X, Lang X, Song D, Min W (2024) Single-particle imaging of nanomedicine entering the brain. Proc Natl Acad Sci U S A 121(5):e2309811121. https://doi.org/10.1073/pnas.2309811121
74. Epp JR, Niibori Y, Hsiang HLL, Mercaldo V, Deisseroth K, Josselyn SA, Frankland PW (2015) Optimization of CLARITY for clearing whole-brain and other intact organs. ENeuro 2(3):ENEURO.0022-15.2015. https://doi.org/10.1523/ENEURO.0022-15.2015
75. Ueda HR, Ertürk A, Chung K, Gradinaru V, Chédotal A, Tomancak P, Keller PJ (2020) Tissue clearing and its applications in neuroscience. Nat Rev Neurosci 21:61–79. https://doi.org/10.1038/s41583-019-0250-1

Part IV

The Need for New Intrathecal Drug Delivery System

Chapter 13

The New Wave of Intrathecal Therapies and Related Challenges in Intrathecal Delivery Systems

Manuel Menéndez-González

Abstract

Intrathecal drug delivery (ITDD) has traditionally served as a niche solution for chronic pain and spasticity, utilizing first-generation programmable pumps designed for small-molecule infusion. However, the emergence of a "new wave" of sophisticated biologics—including antisense oligonucleotides (ASOs), adeno-associated virus (AAV) vectors, and enzyme replacement therapies—has exposed a critical mismatch between modern therapeutic potential and legacy delivery hardware. This chapter explores the multi-dimensional challenges of ITDD in the era of precision medicine, detailing the anatomical, pharmacological, device-related, and socio-economic barriers that hinder the effective treatment of complex neurological disorders. A comprehensive review of current clinical pipelines, pharmacokinetic data, and procedural outcomes was conducted to identify the "innovation gap" in central nervous system (CNS) drug delivery. Analysis reveals that conventional methods, such as repeated lumbar punctures and static-tip catheters, fail to achieve uniform drug distribution, often resulting in a "distribution-toxicity paradox." This phenomenon is characterized by potentially toxic local drug pooling at the injection site and sub-therapeutic concentrations in deep or rostral brain structures. Furthermore, anatomical obstacles (e.g., degenerative spinal changes), altered cerebrospinal fluid (CSF) dynamics in neurodegenerative states, and significant economic burdens associated with specialized infrastructure and maintenance create substantial barriers to health equity and patient access. In conclusion, the transition from palliative care to disease-modifying genetic therapies necessitates a paradigm shift in ITDD technology. Incremental improvements to existing devices are insufficient; rather, a fundamental re-engineering of delivery platforms is required to incorporate active mixing and spatiotemporal control. Addressing these challenges through multidisciplinary innovation is essential to realize the promise of next-generation biologics for patients with devastating neurological diseases.

Keywords Intrathecal drug delivery, CNS barriers, Antisense oligonucleotides, Gene therapy, Neurodegeneration, Cerebrospinal fluid dynamics, Medical device innovation, Novel intrathecal therapies, Delivery system challenges, Device optimization, Emerging therapeutics, CSF dynamics

1 Introduction to Intrathecal Drug Delivery and Central Nervous System Barriers

Intrathecal drug delivery (ITDD) emerged in the 1970s as a pragmatic solution for refractory cancer pain. This approach, designed to administer pharmaceutical agents directly into the cerebrospinal fluid

Manuel Menéndez-González (ed.), *Intrathecal Drug Delivery*, Neuromethods, vol. 234,
https://doi.org/10.1007/978-1-0716-5241-1_13,
© The Author(s), under exclusive license to Springer Science+Business Media, LLC, part of Springer Nature 2026

(CSF), aims to circumvent the formidable physiological barriers that typically restrict systemic drug access to the central nervous system (CNS). The CNS is protected by highly selective interfaces, primarily the blood-brain barrier (BBB) and the blood–CSF barrier (B-CSFB), which together prevent approximately 98% of small-molecule drugs and nearly all large-molecule biologics from entering the brain and spinal cord from the bloodstream [1, 2] By delivering medications directly into the subarachnoid space, ITDD bypasses these barriers, enabling therapeutic effects at the target site with significantly lower dosages compared to oral or parenteral routes, thereby minimizing systemic side effects. This fundamental advantage has made ITDD a well-established modality for the management of chronic non-cancer pain, cancer-related pain, and severe spasticity, particularly in patient populations where conventional systemic therapies have proven inadequate or cause intolerable side effects.

The initial success and subsequent evolution of ITDD were driven by the clinical need to manage these specific conditions. Consequently, the hardware—implantable pumps and catheters—co-evolved to solve the problems posed by chronic pain and spasticity. These devices, such as the Medtronic SynchroMed pump, were optimized for the continuous, low-volume, low-pressure infusion of stable, small-molecule drugs like morphine and baclofen [3]. This path-dependent development, while successful for its original purpose, has inadvertently created a significant "innovation gap." Half a century after its inception, ITDD now stands at a new inflection point. The advent of novel therapeutics such as recombinant enzymes, single-dose adeno-associated virus (AAV) vectors, splice-switching antisense oligonucleotides (ASOs), and CRISPR-Cas editors presents dosing paradigms that the classical programmable pump was never designed to handle [4, 5]. These next-generation biologics are often large, fragile, administered intermittently in large volumes, and possess entirely different pharmacokinetic and safety profiles.

The very hardware that pioneered the field is now a primary obstacle to its future advancement. The challenge is no longer one of simply improving existing devices but of fundamentally re-inventing the delivery platform to match a new class of medicines. This chapter briefly reviews the new wave of intrathecal therapies and surveys the multidimensional gap between first-generation devices and these next-generation therapeutics, detailing the current anatomical, pharmacological, device-related, and systemic challenges that must be addressed to advance ITDD for complex neurological disorders.

2 The New Wave of Intrathecal Therapies

Today, an increasing number of therapies for CNS conditions intended for intrathecal administration are under development. This pipeline reflects a major shift in neurology towards targeting the genetic and molecular root causes of disease. However, the clinical development path is fraught with challenges, as evidenced by the high rate of discontinuation even for promising candidates. A clear trend of "high-risk, high-reward" is evident, with a significant number of ASO therapies for major neurodegenerative diseases like Huntington's disease and C9orf72-associated amyotrophic lateral sclerosis (ALS) being discontinued despite promising mechanisms [6–8]. In several cases, the drug successfully engaged its molecular target, but this did not translate into clinical benefit, suggesting a fundamental gap in our understanding of the relationship between target engagement and functional outcomes [9]. This high failure rate underscores the complexity of treating established neurodegenerative disease and points to the urgent need for better biomarkers that are more predictive of clinical efficacy than simple target protein reduction.

Table 1 provides a comprehensive, up-to-date summary of the most recently approved and investigational intrathecal therapies, categorized by their indication, molecule type, and current development stage. This snapshot is intended to provide clinicians and researchers with an accurate overview of the evolving therapeutic landscape.

3 Anatomical and Procedural Challenges in ITDD

From an anatomical and procedural perspective, the main challenges in ITDD are related to accessing the intrathecal space, complications arising from repeated dosing, and the inherent biological barriers of the human body, which are often altered by the underlying disease.

3.1 Access to the Intrathecal Space

Lumbar puncture (LP) is the basic procedure for IT drug administration. However, what is a routine procedure in a healthy young adult becomes increasingly difficult in patients with CNS diseases. The target demographic for many new therapies consists of older adults who frequently present with age-related degenerative changes of the spine, such as lumbar spondylosis, which affects over 80% of individuals older than 40 [10]. This condition involves disc desiccation, loss of disc height, and osteophyte formation, which can obscure anatomical landmarks and narrow the interlaminar space available for needle entry [10].

Furthermore, disease-specific complications can present formidable barriers. Patients with advanced ALS or spinal muscular atrophy (SMA) often develop severe degenerative scoliosis, which distorts the normal spinal anatomy [11–14]. Previous spinal fusion surgeries, common in this population, create osseous barriers that make a standard LP impossible [15]. In such cases, fluoroscopy-guided or CT-guided access becomes necessary [15, 16]. While effective, these image-guided procedures significantly increase costs, radiation exposure to the patient, and procedural time, and they require scheduling in specialized radiology suites, which can create access delays [17].

Table 1
Recently approved and investigational intrathecal therapies

Therapy Name	Indication	Molecule type	Mechanism	Delivery	Dosing regimen	Approval status	Development stage (Trial ID)
Tofersen (Qalsody™)	ALS (SOD1 mutation)	Antisense Oligonucleotide	Degrades SOD1 mRNA	IT Bolus	Three loading doses, then every 3 months	Approved (accelerated)	Phase III (ATLAS: NCT04856982)
BIIB078	ALS (C9orf72 expansion)	Antisense Oligonucleotide	Reduces toxic RNA/protein aggregates	IT Bolus	Monthly	Discontinued	Phase I (terminated)
WVE-004	ALS & FTD (C9orf72)	Antisense Oligonucleotide	Targets mutant C9orf72 transcripts	IT Bolus	Q4W or Q12W	Discontinued	Phase I/IIa (FOCUS-C9) [9]
Tominersen (RG6042)	Huntington's disease	Antisense Oligonucleotide	Suppresses huntingtin (HTT) mRNA	IT Bolus	Every 16 weeks	Investigational	Phase II (GENERATION HD2: NCT05686551) [46]
Branaplam (LMI070)	Huntington's disease	Small molecule splicing modifier	Modifies HTT mRNA splicing	Oral	Once weekly	Discontinued	Phase II (VIBRANT-HD) [47]
Onasemnogene abeparvovec (Zolgensma/ OAV101 IT)	Spinal muscular atrophy	Gene therapy (AAV9)	Delivers functional SMN1 gene	IT Bolus	Single dose	Investigational	Phase III (STEER: NCT05089656) [37]
TSHA-102	Rett syndrome	Gene therapy (AAV9)	Delivers shortened MECP2 gene with regulation	IT Bolus	Single dose	Investigational	Phase I/II (REVEAL: NCT05606614) [34]

(continued)

Table 1
continued

Therapy Name	Indication	Molecule type	Mechanism	Delivery	Dosing regimen	Approval status	Development stage (Trial ID)
NGN-401	Rett syndrome	Gene therapy (AAV9)	Delivers full-length MECP2 gene with regulation	Intraventricular	Single dose	Investigational	Phase I/II (NCT05898620) [34]
VAL-1221	Pompe Disease	Modified enzyme replacement	Replaces defective GAA enzyme	IV Infusion	Every other week	Investigational	Phase I/II completed [40]
AXO-AAV-GM2	Tay-Sachs, Sandhoff	Gene therapy (AAVrh8)	Corrects HEXA/HEXB enzyme deficiency	Intraparenchymal & IT	Single dose	Investigational	Phase I/II (NCT04669535) [20]
AAV9-GLB1	GM1 Gangliosidosis	Gene therapy (AAV9)	Corrects GLB1 enzyme deficiency	IV infusion	Single dose	Investigational	Phase I/II (recruiting) (NCT03952637) [20]
Anti-NGF monoclonal antibodies	Chronic pain	Monoclonal antibody	Blocks nerve growth factor (NGF)	Systemic	Under investigation	Investigational	Preclinical/clinical [48]
SP-Saporin	Chronic pain	Targeted toxin	Selectively destroys NK1-expressing neurons	IT Bolus	Single injection	Investigational	Preclinical (Canine) [49]
Resiniferatoxin (RTX)	Chronic pain (cancer)	Targeted toxin	Ablates nociceptive neurons via TRPV1 receptors	IT Bolus	Single dose	Investigational	Phase I (NCT00804154) [8]
Dopamine/stem cells	Parkinson's disease	Small molecule/cell therapy	Replaces dopamine/dopaminergic neurons	IT/ICV/Intraparenchymal	Continuous/single transplant	Experimental	Clinical (STEM-PD: NCT05635409) [29]

3.2 Complications from Repeated Dosing

The challenge of access is magnified by the chronicity of many new therapeutic regimens. Most IT-administered ASOs require chronic administration; for example, nusinersen for SMA and the approved regimen for tofersen for SOD1-ALS involve repeated injections over a patient's lifetime [18–20]. This necessitates repetitive LPs, with each procedure introducing a new set of risks. Post-dural puncture headache (PDPH) is a common and often debilitating complication, affecting approximately 32%–40% of patients [12, 21]. It is caused by persistent CSF leakage from the puncture site and can be severe enough to immobilize the patient for days, with potential for serious sequelae such as subdural hematoma, cranial nerve palsies, and seizures [12, 22–26].

Beyond PDPH, each puncture carries risks of infection (meningitis), bleeding (spinal epidural hematoma), direct nerve root irritation causing radicular pain, and CSF leaks [12]. This cumulative procedural burden significantly decreases patient compliance and quality of life. The need for repeated procedures, often under sedation or even general anesthesia in pediatric or anxious patients, places a heavy strain on patients, their families, and the healthcare system. This fundamental conflict between the chronic nature of the therapy and the acute, invasive nature of the delivery method is a powerful driver for the development of more reliable, long-term access solutions or "one-and-done" gene therapies.

3.3 Altered Biological Barriers and CSF Dynamics

The CNS environment is not a passive receptacle for drugs. Disease-specific conditions can significantly alter the already highly selective biological barriers, CSF dynamics, and local immune responses. Many neurodegenerative diseases, such as the development of hydrocephalus ex vacuo in late-stage Alzheimer's disease (AD) or the narrowing of the spinal canal in ALS, directly modify CSF volume, flow patterns, and composition [27]. These alterations can profoundly affect drug distribution, absorption, and clearance, making pharmacokinetic modeling highly unpredictable. Disruptions in CSF clearance pathways, such as the glymphatic system, are now understood to be a key feature of neurodegeneration, contributing to the accumulation of toxic proteins like amyloid-beta and tau [27]. An IT-administered drug is therefore injected into a dynamic and pathologically altered system, not a static one.

Furthermore, the CNS is immune-privileged but not immune-ignorant. Repeated IT administration of large, complex biologics or oligonucleotides can provoke local inflammatory responses. This can manifest as CSF pleocytosis (an increase in white blood cells in the CSF) or, in more severe cases, drug-induced aseptic meningitis or choroid plexus inflammation [23]. For instance, cases of aseptic meningoencephalitis, a serious inflammatory condition, have been reported in patients receiving long-term nusinersen therapy, highlighting the potential for the immune system to react to chronic exposure to these novel agents within the intrathecal space [24].

4 Pharmacological Challenges

Beyond the physical hurdles of delivery, several critical pharmacological aspects must be addressed to improve the efficacy and safety of IT-administered therapies. These challenges revolve around the fundamental biophysics of drug movement within the complex architecture of the CNS.

4.1 Uneven Drug Distribution

Uneven drug distribution in the CSF remains a persistent and fundamental problem for ITDD. Following a standard bolus injection via LP, the spread of the drug is governed primarily by passive diffusion and the slow, complex patterns of bulk CSF flow. This process is remarkably inefficient. The result is a steep and rapid exponential decline in drug concentration with increasing distance from the injection site. Medications tend to achieve very high, potentially toxic peak concentrations in the cauda equina and lumbar region while failing to reach therapeutic levels in more rostral CNS regions like the thoracic spinal cord, brainstem, and cerebral cortex.

This phenomenon, sometimes termed the "diffusion deficit," is particularly problematic for diseases like Huntington's disease, Rett syndrome, or Alzheimer's disease, where widespread, supraspinal targeting is essential for therapeutic effect. The inherent limitations of passive diffusion mean that drugs often spread only a few millimeters from the injection site, compromising efficacy at distant targets while simultaneously creating local "hot spots" of high drug exposure that increase the risk of localized neurotoxicity or inflammatory reactions like granuloma formation. This non-uniformity is influenced by numerous factors, including CSF pulsatility driven by cardiac and respiratory cycles, patient posture, and the physicochemical properties of the drug itself, such as its molecular weight and charge [10, 11, 27–32].

4.2 Limited Parenchymal Penetration

Achieving adequate concentration in the CSF is only the first step; the drug must then penetrate the brain and spinal cord tissue (parenchyma) to reach its cellular targets. This represents another significant barrier, particularly for large molecules such as ASOs, enzymes, and antibodies [21]. To enter the parenchyma from the subarachnoid space, a drug must cross the pial surface and the underlying glia limitans, a dense network of astrocytic foot processes.

While it was once thought that this barrier was nearly impermeable to large molecules, more recent research suggests that active transport mechanisms and specific anatomical pathways, such as translocation along perivascular (Virchow-Robin) spaces and white matter tracts, facilitate some degree of entry [25]. However, these pathways are inefficient and result in a distribution pattern that is often limited to periventricular regions or the superficial layers of the CNS, with poor penetration into deeper structures like the basal ganglia or brainstem nuclei [25]. This limited penetration means that even with high CSF concentrations, the actual dose reaching deep target neurons may be insufficient, representing a major obstacle for treating a wide range of neurodegenerative disorders.

4.3 Pharmacokinetic Contrasts with Modern Biologics

The demands placed on ITDD systems have escalated dramatically with the shift from traditional small molecules to complex biologics. The pharmacokinetic chasm between the archetypal IT drug of the 1980s, morphine, and the new wave of therapeutics is vast and highlights the inadequacy of existing delivery technologies. Table 2 provides a quantitative comparison that illustrates this paradigm shift.

This table starkly illustrates the scale of the challenge. The molecular mass increases by orders of magnitude, from hundreds of daltons for morphine to millions for an AAV vector. The clearance dynamics shift from hours to months, or even become permanent in the case of gene therapy. Most critically, the therapeutic window collapses. While morphine dosing can be adjusted across a wide range, ASOs like nusinersen have a narrow safety margin, with risks of toxicity (e.g., thrombocytopenia, proteinuria) emerging at concentrations not far above therapeutic levels [33]. For AAVs, the safety profile is not one of simple dose-toxicity but is dominated by the risk of a potent, and potentially fatal, immune response [34].

This combination of factors—a massive leap in molecular size, a dramatic extension of half-life, and a collapse of the safety margin—creates a "distribution-toxicity paradox." The very properties that make these biologics potent and long-lasting also cause them to pool dangerously at the injection site while failing to reach distant targets. Simply increasing the dose is not a viable solution, as it would exacerbate local toxicity without solving the distribution problem. This fundamental biophysical conflict demands a new delivery strategy that can achieve widespread, uniform, and controlled distribution, a capability that current hardware does not possess.

Table 2
Pharmacokinetic contrasts between traditional small molecules and modern biologics for ITDD

Attribute	Morphine (1980s Archetype)	Nusinersen (ASO)	Onasemnogene abeparvovec (AAV9)
Molecular size	285 Da [42]	~7.5 kDa [50]	~5 MDa
Target CSF Cmax	~0.3 μM	~0.1 μM	~2 × 109 vg/mL
Half-life in CSF	1.5–6 h (biphasic) [42]	135–177 days [33]	One-off transduction
Therapeutic window	Wide [42]	Narrow (neurotoxicity above ~3×) [33]	Cell-specific, immunogenic [51]

5 Device-Related Limitations

The technological tools used for ITDD, largely developed and optimized for delivering small-molecule analgesics and antispasmodics, are fundamentally ill-equipped to handle the complex demands of modern biologics. This mismatch between hardware capability and therapeutic need creates significant limitations and risks.

5.1 First-Generation Hardware Limitations

The devices that pioneered IT delivery were not designed for the precision, safety, and distribution requirements of therapies for neurodegenerative diseases.

- **Lumbar Puncture (LP):** While being the simplest and most common method for IT administration, allowing for the delivery of approved ASOs like nusinersen and tofersen, it is a crude tool for precision medicine. As discussed, repeated LPs are burdensome for patients and clinicians, causing procedural pain, a high incidence of PDPH, and significant technical challenges in patients with abnormal spinal anatomy [15]. It offers no control over drug distribution, leading to the pharmacological challenges of uneven spread and local toxicity.
- **Intraventricular Ports (e.g., Ommaya Reservoir):** An Ommaya reservoir, a subcutaneously implanted silicone dome connected to a catheter surgically placed into a lateral ventricle, provides repeated, direct access to ventricular CSF for drug injection or sampling. While this bypasses the difficulties of repeated LPs, it requires a neurosurgical procedure for placement and carries substantial risks. The most significant is infection, which can track along the catheter and lead to life-threatening meningitis or epidural abscess, often necessitating device explantation. Mechanical complications, including catheter malpositioning (reported in up to 6% of placements), kinking, or disconnection, are also common and frequently require surgical revision. Critically, like bolus injections via LP, Ommaya reservoirs still rely entirely on passive diffusion from a single point source for drug dissemination, failing to solve the fundamental problem of achieving widespread and uniform drug distribution throughout the CNS.
- **Implantable Programmable Pumps:** These devices, most commonly implanted in the abdominal wall with a catheter tunneled subcutaneously to the lumbar intrathecal space, represent the standard of care for delivering IT baclofen for severe spasticity and IT morphine for chronic pain [3]. They offer the advantage of providing continuous or complex, user-defined infusion patterns. However, they require major surgical implantation and periodic refilling via percutaneous injection

[3]. When considered for modern biologics, their design reveals several critical flaws:

- Static Tip Location: The catheter tip remains in a fixed location, which, for lipophilic or high-binding biologics, leads to the creation of steep cranio-caudal concentration gradients and local drug pooling. Pharmacokinetic studies of IT baclofen, a small molecule, have demonstrated a 4:1 lumbar-to-cisternal concentration gradient and significant interpatient variability, a problem that is magnified with much larger molecules [35].
- Shear Stress: The mechanical action of positive-pressure pumps can generate shear stresses that are damaging to the complex structures of viral capsids, immunoglobulins, and cell-based therapies.
- Lack of Active Mixing: Simple, continuous infusion from a silicone catheter without any active mixing mechanism can create "dead zones" behind nerve-root sleeves and other anatomical niches, contributing to as much as a five-fold variability in dose-to-effect between subjects [2, 16].
- Telemetry and Safety: Before the advent of modern telemetry in devices like the SynchroMed III, significant latency in communication meant that flow-checks performed only every 24 h could miss a critical event like catheter occlusion or a life-threatening runaway delivery.

This portfolio of existing devices forces clinicians into a dangerous trade-off between invasiveness and control. The choice is between a minimally invasive but completely uncontrolled method (LP) and highly invasive methods (Ommaya, pumps) that offer some temporal control but fail to address the primary challenge of spatial distribution and introduce their own significant risks. The ideal tool—one that is both minimally invasive and provides spatiotemporal control—does not yet exist in clinical practice.

5.2 Device-Related Complications

Beyond their functional limitations, implanted ITDD hardware is susceptible to a range of serious complications. Mechanical failure of the pump, catheter occlusion, kinking, or migration out of the intrathecal space are known issues that can lead to a sudden cessation of therapy (and risk of withdrawal syndrome) or delivery of drug into the wrong compartment, often requiring imaging and surgical revision [5, 18, 36]. Post-surgical infections are a constant threat and can be devastating in the CNS context, potentially causing meningitis or abscesses that require device removal [5, 18, 36].

A particularly concerning complication of long-term infusion is the formation of an intrathecal catheter-tip granuloma, an inflammatory mass that can develop in response to high local drug concentrations [34, 37, 38]. First reported in 1991, this complication is a known risk with chronic opioid infusions but is theoretically possible with any irritating substance delivered at high concentration [39, 40]. These masses can grow to compress the spinal cord, leading to progressive and potentially permanent neurological deficits, including paralysis and loss of bowel and bladder control [34, 36–38]. The incidence increases with the duration of therapy, and the risk is thought to be higher with greater drug concentrations and dosages [39, 40]. This highlights a direct link between the poor distribution from a static catheter tip (creating a "hot spot" of high concentration) and a severe, device-related complication. For patients with neurodegenerative diseases who require periodic MRI for disease monitoring, device-related imaging artifacts remain a significant concern that can obscure important clinical findings [39, 40].

5.3 Lack of Standardization

The field of ITDD is marked by a significant lack of standardization in clinical practice, which complicates the interpretation of clinical trial data, hinders the development of best practices, and contributes to inconsistent patient outcomes. A 2022 survey of practitioners by the American Society of Pain and Neuroscience (ASPN) revealed substantial variation in nearly every aspect of care [36]. This included differences in the training background of implanting physicians, the choice of medications used for pre-implantation trials (with morphine and fentanyl being the most common), the surgical location for pump implantation (e.g., abdomen vs. gluteal region), and the protocols for patient-controlled bolus dosing [36].

This variability extends to fundamental parameters like drug formulation (concentration), delivery volume, and infusion rate. Such inconsistencies make it difficult to compare outcomes across studies and institutions, complicating the efforts of regulatory agencies to establish clear guidelines for safety and efficacy. This lack of a standardized approach hinders the widespread adoption of ITDD and prevents patients from receiving consistent, optimized therapeutic benefit.

6 Logistical and Economic Challenges

Even if the anatomical, pharmacological, and device-related challenges of ITDD were solved, significant system-level hurdles related to healthcare infrastructure and economics would remain. These logistical and financial burdens represent critical limitations to the broad and equitable implementation of advanced intrathecal therapies.

6.1 Health System Limitations

The administration of advanced intrathecal therapies is a resource-intensive endeavor that requires a sophisticated and well-coordinated infrastructure. These resources are predominantly concentrated in large, tertiary academic medical centers, creating a major bottleneck for patient access [36]. A successful ITDD program requires:

- **A Multidisciplinary Team:** This includes highly skilled proceduralists (neurologists, neurosurgeons, anesthesiologists, interventional radiologists), as well as specialized pharmacists, infusion nurses, physical therapists, program coordinators, and billing experts [36].
- **Specialized Facilities:** Procedures often require access to sterile operating rooms, fluoroscopy or CT imaging suites for guided placement, and dedicated infusion centers for drug administration and pump refills [3, 16].
- **Long-Term Follow-Up Clinics:** Patients with implanted devices require lifelong management, including regularly scheduled appointments for pump refills, programming adjustments, and monitoring for complications [3].

The necessity of this complex infrastructure means that access to these potentially life-altering therapies is severely limited for patients living outside of major urban areas or in regions and countries with less developed healthcare systems [27]. This creates a "treatment-access paradox": the revolutionary therapies are being developed for devastating diseases that affect patients globally, yet the systems required to deliver them are available to only a privileged few. This disparity raises profound questions of health equity and highlights the need to develop not only new technologies but also scalable and accessible models of care.

6.2 Economic Burden

The economic burden of ITDD is substantial and multifaceted, extending far beyond the initial acquisition cost of the drug or device. The total cost of care includes the high upfront expense of the implantable hardware, surgical implantation fees, hospital stays, frequent imaging studies, and the ongoing costs of drug refills and complication management [27, 38]. For example, in one analysis, the implantation cost of an IT pump was over $15,000, with ongoing maintenance costs.

While several health-economic analyses have suggested that ITDD can be cost-effective in the long run for conditions like chronic pain and spasticity when compared to the costs of conventional medical management (including hospitalizations and alternative procedures), the high initial financial outlay remains a significant barrier for many health systems and payers [27, 38]. These costs are magnified in the context of chronic progressive diseases that require lifelong therapy, creating major sustainability concerns. As new, high-cost gene therapies and biologics enter the market, the economic pressures on healthcare budgets will only intensify, making the demonstration of clear cost-effectiveness and long-term value essential for reimbursement and patient access.

7 Ethical and Psychosocial Considerations

Finally, the implementation of advanced ITDD involves critical ethical and psychosocial dimensions that can act as limiting factors if not properly addressed. These considerations are central to patient-centered care and have a direct impact on treatment outcomes.

7.1 Patient-Specific Considerations

The decision to initiate ITDD is not merely a clinical calculation but a profound life choice that requires careful and continuous deliberation. Informed consent procedures must be especially robust, tailored to the specific patient population. When treating pediatric patients, this involves navigating complex consent and assent processes with parents and guardians. For patients with neurodegenerative diseases that may be associated with cognitive decline, assessing decisional capacity is paramount. All patients and their families must receive a clear and comprehensive explanation of the procedural risks, including the risks of sedation or anesthesia, the potential for complications, and the realistic expectations for therapeutic benefit [17].

For therapies requiring chronic administration, the balance between the potential benefits of treatment and the cumulative burden of the procedures must be regularly reassessed. This is not a one-time decision made at implantation but an ongoing dialogue. The patient's identity often shifts to one of being "device-dependent," a life tethered to the maintenance schedules, alarms, and potential malfunctions of an implanted machine [41]. This can create a significant psychological burden of technological dependence on top of the underlying chronic illness. Therefore, a successful ITDD program must include strong psychosocial support to help patients and caregivers manage the logistical and emotional challenges of living with an implanted device. Addressing these factors is highly relevant, as it significantly impacts treatment adherence, improves patients' quality of life, and ultimately leads to better long-term outcomes [11].

8 Limitations of Conventional ITDD Methods

Despite the inherent advantage of bypassing systemic barriers, conventional ITDD methods, including bolus injections via lumbar puncture (LP), Ommaya reservoirs, or external ventricular drains (EVDs), are associated with significant limitations, primarily concerning drug distribution and safety.

8.1 Bolus Injections

When drugs are administered as boluses, their spread within the CSF relies predominantly on passive diffusion. This passive mechanism inherently restricts drug distribution to a few millimeters from the injection site, characterized by a rapid exponential decline in concentration with increasing distance. Such non-uniformity can result in inadequate drug concentrations at distant target sites, potentially compromising therapeutic efficacy, while simultaneously leading to excessively high and potentially toxic peak concentrations in the immediate vicinity of the injection site. Furthermore, the intermittent nature of bolus dosing causes fluctuating drug concentrations, which can result in periods of sub-therapeutic levels or transient overdose. For instance, intrathecal morphine, while providing effective analgesia, is associated with dose-related adverse effects such as pruritus, urinary retention, sedation, and delayed respiratory depression due to its rostral spread to the brainstem, necessitating prolonged patient monitoring [42]. Similarly, intrathecal baclofen boluses achieve peak spasmolytic effects approximately four hours post-administration, with effects lasting four to eight hours, and require vigilant monitoring for potential CNS depression, cardiovascular collapse, and respiratory failure [35, 43].

8.2 Ommaya Reservoirs

These devices facilitate repetitive drug delivery into the CSF; however, they carry a notable risk of infection. Infections can track along the catheter, potentially leading to meningitis or epidural abscess, often requiring device explantation. Minimizing this risk necessitates strict adherence to aseptic conditions and the involvement of specially trained personnel. Mechanical complications, such as catheter malpositioning (reported in up to 6% of placements), kinking, or disconnection, can also occur, frequently requiring surgical revision. Like bolus injections, Ommaya reservoirs primarily depend on passive diffusion for drug dissemination within the CSF, leading to similar challenges in achieving widespread and uniform drug distribution.

8.3 External Ventricular Drains (EVDs)

The administration of intrathecal medications through EVDs inherently requires opening a sterile system, which elevates the risk of infection. As with other bolus methods, EVDs deliver medication in bolus form, necessitating clamping after delivery and contributing to non-uniform drug distribution and fluctuating drug levels. A significant gap exists in the understanding of pharmacokinetics for continuous IT medication delivery, as current protocols are largely based on intermittent bolus instillation. This highlights a critical need for prospective studies to establish optimal dosing standards and safety profiles for continuous infusions.

8.4 Lumbar Punctures (LPs)

Repeated lumbar punctures are associated with considerable risks, including neurological injury, dislodgement of device components, infection, and drug-related complications. A common adverse event is post-lumbar puncture headache (PDPH), affecting approximately 32% of patients. These headaches can be severe and immobilizing, with potential for serious sequelae such as subdural hematoma and seizures [12]. Catheter migration outside the intrathecal space has also been documented, causing pain and often requiring imaging and potential surgical revision. Furthermore, while drugs injected into the CSF via LP are rapidly transported into the bloodstream through bulk flow, their penetration into brain tissue remains poor due to the inherent limitations of diffusion.

The consistent observation across these conventional ITDD methods is a fundamental inadequacy in achieving widespread therapeutic exposure for many CNS disorders, particularly those with diffuse pathology like leptomeningeal metastases or widespread neurodegenerative conditions. The inherent biophysical limitations of passive diffusion within the complex CNS environment mean that for many conditions, this approach is fundamentally insufficient, creating a significant clinical need for technologies that can overcome this "diffusion deficit." The persistent challenges associated with conventional ITDD, including infection risks, mechanical complications, and systemic side effects, despite their long-standing use, underscore that while these methods are established, they are far from optimal. The current "standard of care" for some conditions is recognized as insufficient, particularly regarding the uniform spread of drugs or the effective clearance of pathological contaminants.

9 Regulatory and Economic Landscape Challenges

The regulatory and economic landscape presents additional significant challenges for the development and adoption of next-generation ITDD systems.

9.1 Comparative Trials and Evidence Generation

Regulatory pathways, such as the FDA's Center for Devices and Radiological Health (CDRH) Breakthrough Device pathway, often mandate head-to-head comparative trials against the current standard of care. This requirement can be resource-intensive and time-consuming, slowing down the translation of innovative devices to clinical practice. While adaptive platform designs (e.g., I-SPY ITDD) can accelerate evidence generation, these are still complex undertakings that require significant investment and coordination. Generating the robust evidence needed to prove superiority over an established, albeit flawed, standard is a high bar for device developers.

9.2 Health Economics and Cost Burden

The health-economic burden of novel ITDD therapies extends beyond the cost of the drug itself. A Markov cost-utility model of AAV9 SMA therapy delivered by ThecaFlex™ versus repeated LP showed an incremental cost of US$38,000 per quality-adjusted life-year (QALY), which was still below the orphan-drug threshold. However, the overall economic impact, including procedural costs, imaging, and monitoring requirements, especially for chronic progressive diseases, raises significant sustainability concerns for health systems and payers [27]. The high upfront costs associated with implantable systems can be a major deterrent to adoption, even if long-term cost-effectiveness can be demonstrated [27].

9.3 Post-Market Surveillance and Cybersecurity

Post-market surveillance for advanced ITDD systems requires robust mechanisms to ensure long-term safety and performance. International standards such as ISO 14708-3:2023 for active implantable medical devices now mandate features like remote log retrieval to monitor device function over time. For a new generation of "smart" pumps that may incorporate artificial intelligence or be connected to external networks, cybersecurity becomes a paramount concern. Regulatory standards like IEC 81001-5-1 require rigorous cybersecurity risk mitigation measures to be integrated into the device design and lifecycle management. These regulations, while essential for patient safety and data integrity in an increasingly interconnected healthcare environment, add significant layers of complexity, cost, and time to the device development and maintenance process.

10 Conclusions

The therapeutic revolution in neurology, characterized by the emergence of sophisticated biologics targeting the molecular underpinnings of disease, will undoubtedly face significant hurdles if hardware delivery systems do not evolve in parallel. The current generation of ITDD devices, while foundational to the field, is ill-equipped to meet the escalating demands of precise dosing, widespread distribution, and long-term safety required for these novel therapeutics. The very systems that were optimized for the chronic, low-volume infusion of small molecules for pain and spasticity are now revealing their fundamental limitations when tasked with delivering large, fragile, and intermittently dosed biologics for neurodegeneration.

Addressing the multifaceted challenges outlined in this chapter—anatomical, pharmacological, device-related, logistical, and ethical—is paramount to unlocking the full potential of ITDD. An incremental approach of merely refining existing technologies will be insufficient. What is required is a paradigm shift in device engineering, driven by a clear understanding of the existing limitations and a strategic roadmap for innovation. This necessitates a concerted, multidisciplinary effort involving neurologists, biomedical engineers, neuropharmacologists, and regulatory scientists to design and validate next-generation ITDD systems from the ground up. These future systems must move beyond the limitations of single-point, passive diffusion and incorporate technologies for active, controlled, and widespread drug distribution. This is not merely a technical problem to be solved; it is a critical step to ensure that the profound promise of the new wave of neurological therapies can be safely, effectively, and equitably realized for patients suffering from these devastating disorders [44, 45].

Conflict of Interest

The authors have no conflicts of interest to declare that are relevant to the content of this chapter.

References

1. Hao D, Di Capua CA, Dumas C et al (2025) Real-world prescription data for intrathecal drug delivery: a retrospective analysis in 32,784 patients. Neuromodulation 28(7):1054–1058. https://doi.org/10.1016/j.neurom.2024.07.007
2. De Andres J, Hayek S, Perruchoud C et al (2022) Intrathecal drug delivery: advances and applications. Front Pain Res 3:900566
3. Magill E, Demartis S, Gavini E et al (2023) Solid implantable devices for sustained drug delivery. Adv Drug Deliv Rev 199:114950
4. Ailawadhi P, Sung VW (2021) Management of adult degenerative scoliosis. PM R 13(9):1042–1050
5. Rigo F, Hua Y, Krainer AR, Bennett CF (2012) Antisense-based therapies for the treatment of central nervous system diseases. J Cell Biol 199(1):21–25
6. Burla GKR, Shrestha D, Bowen M et al (2024) Evaluating the effect of injection protocols on intrathecal solute dispersion in non-human primates. Fluids Barriers CNS 21:61
7. Roche (2021) Roche provides update on tominersen programme in manifest Huntington's disease. Press Release. March 22
8. Novartis (2025) New Novartis Phase III data demonstrate meaningful efficacy and safety results of intrathecal onasemnogene abeparvovec in broad patient population with SMA. Press Release. March 19
9. Muccio M, Sun Z, Chu D et al. (2024) The impact of body position on neurofluid dynamics. Front Aging Neurosci 16:1454282
10. National Tay-Sachs & Allied Diseases Association. Gene therapy. [Internet]. [cited 2024 Nov 20]. Available from: https://ntsad.org/gene-therapy-general/
11. Brown DL, Intoccia PA, Iadarola MJ et al (2013) Intrathecal substance-p saporin in the dog: efficacy in bone cancer pain. Anesthesiology 119(5):1178–1185
12. Biogen (2016) SPINRAZA® (nusinersen) injection, for intrathecal use. In: Prescribing information. Biogen, Cambridge, MA
13. Novartis Gene Therapies, Inc. (2023) ZOLGENSMA® (onasemnogene abeparvovec-xioi) suspension, for intravenous infusion. In: Prescribing Information. Novartis Gene Therapies, Inc., Bannockburn, IL
14. North RB, Cutchis PN, Epstein JA, Long DM (1991) Catheter-tip granuloma in the spinal subarachnoid space: a new complication of implanted drug infusion systems. J Neurosurg 75(6):940–943
15. Obeso JA, Stamelou M, Goetz CG et al (2017) Past, present, and future of Parkinson's disease: a special essay on the 200th anniversary of the shaking palsy. Mov Disord 32(9):1264–1310
16. Eskander MS, Freeman AL, Sturycz CA et al (2016) A review of fluoroscopy-guided lumbar puncture. AJR Am J Roentgenol 206(3):479–488
17. Eskander JP, Lai RP, Yamani A et al (2021) Intrathecal and intracerebroventricular dopamine for Parkinson's disease. Clin Neurol Neurosurg 200:106374
18. IRRAS AB (2023) Pre-clinical evaluation of active irrigation CSF systems. White paper
19. Biogen (2023) QALSODY™ (tofersen) injection, for intrathecal use. In: Prescribing information. Biogen, Cambridge, MA
20. Biogen (2022) Update on BIIB078 (C9orf72) ALS Program
21. Manuel MG, Tamba BI, Leclere M et al (2023) Intrathecal pseudodelivery of drugs in the therapy of neurodegenerative diseases. Pharmaceutics 15:768
22. Deer TR, Pope JE, Lamer TJ et al (2017) The polyanalgesic consensus conference (PACC): recommendations for intrathecal drug delivery: guidance for improving safety and mitigating risks. Neuromodulation 20(2):155–176
23. Coffey RJ, Burchiel K (2002) Inflammatory mass lesions associated with chronic intrathecal drug infusion: a review of the literature, description of a new entity, and surgical recommendations. J Neurosurg 96(2 Suppl):269–277
24. Saulino M, H-kyma D, Gjolaj JP et al (2016) Best practices for intrathecal baclofen therapy: dosing and long-term management. Neuromodulation 19(6):637–651
25. Meythaler JM, Guin-Renfroe S, Brunner RC, Hadley MN (2001) Intrathecal baclofen for spasticity of cerebral origin. Arch Phys Med Rehabil 82(11):1631–1634
26. Taira T, Hori T (2002) Intrathecal baclofen for treatment of spasticity in a patient with a previous history of myelomeningocele. J Clin Neurosci 9(2):207–209

27. Duarte R, Lambe T, Raphael JH, Andronis L (2018) Intrathecal drug delivery systems for the management of chronic noncancer pain: a systematic review of economic evaluations. Pain Pract 18(5):649–661
28. Taysha Gene Therapies (2025) A Review of Gene Therapy Clinical Trials. Rett Syndrome Research Trust. January 15
29. Neurogene Inc. (2024) Neurogene Provides Rett Syndrome Program Update. Press Release. November 14
30. Parasail LLC. (2024) VAL-1221 for the treatment of patients with Lafora disease: study protocol for a single-arm, open-label clinical trial. medRxiv
31. Mannes AJ, Heiss JD, Berger A et al. (2025) Treatment of intractable cancer pain with resiniferatoxin - an interim study. NEJM Evid 4(6):EVIDoa2400423
32. Malcangio M (2021) Monoclonal antibodies for chronic pain treatment: present and future. Molecules 26(19):5809
33. Krach LE, Kriel RL, Gilmartin RC et al (2002) Intrathecal baclofen for spasticity in children with cerebral palsy. J Child Neurol 17(1):31–36
34. Flisberg P, Paech MJ, Shah T et al (2020) Postoperative respiratory depression after intrathecal morphine: a prospective, randomised, double-blind, dose-finding study in caesarean section. Anaesthesia 75(11):1475–1483
35. Calias P, Banks WA, Begley D et al (2014) Intrathecal delivery of protein therapeutics to the brain: a critical reassessment. Pharmacol Ther 144:114–122
36. Deer TR, Hayek SM, Verrills P et al (2022) The American Society of Pain and Neuroscience (ASPN) survey on intrathecal drug delivery: a review of current practice. J Pain Res 15:1029–1039
37. Isaacs T, Abdi S, Abraham J et al (2023) Building an infrastructure to administer amyloid-targeting treatments. Pract Neurol 23(3):205–211
38. Duarte R, Andronis L (2024) Cost-effectiveness of intrathecal drug delivery. Neuraxial therapeutics. Springer, Cham, 803–811
39. Medtronic (2021) Lioresal® Intrathecal (baclofen injection) full prescribing information
40. Alcyone Therapeutics (2023) Cost-utility analysis of ThecaFlex DRx™ for SMA. Internal report; 2024
41. Menéndez González M (2023) Liquorpheresis: cerebrospinal fluid filtration to treat CNS conditions. Springer, Cham, Switzerland
42. Albright AL, Awaad Y, Muhonen M et al (2004) Efficacy of continuous intrathecal baclofen infusion in children with cerebral palsy. J Neurosurg 101(2 Suppl):121–126
43. Heetla HW, Staal MJ, van Laar T (2014) The incidence and management of acute baclofen withdrawal syndrome in patients with an intrathecal baclofen pump. Pain Physician 17(4):337–344
44. Zis P, Daskalaki A, Bountouni I et al (2017) Post-dural puncture headache: a narrative review of current treatment options. Anesth Pain Med 7(4):e43331
45. Sjöström S, Tamsen A, Persson MP, Hartvig P (1987) Pharmacokinetics of intrathecal morphine and meperidine in humans. Anesthesiology 67(6):889–895
46. Tropea TF, Mak-A-Key F, Al-Gharaibeh A et al (2024) Aseptic meningoencephalitis during nusinersen therapy in a patient with type III spinal muscular atrophy: a case report. J Neuroinflammation 21(1):108
47. Muccio M, Sun Z, Chu D et al (2025) Editorial: CSF clearance in Alzheimer's disease and related dementias: exploring mechanisms and implications. Front Aging Neurosci 17:1581223
48. Wave Life Sciences (2023) Wave life sciences announces topline results from phase 1b/2a FOCUS-C9 Study of WVE-004 for C9orf72-associated amyotrophic lateral sclerosis and frontotemporal dementia. Press Release. May 23
49. Novartis. A dose range finding study with open-label extension to evaluate the safety of oral LMI070/Branaplam in early manifest Huntington's disease (VIBRANT-HD). ClinicalTrials.gov Identifier: NCT05111249
50. Sgouros S, Vloeberghs M (2006) Intrathecal baclofen in children with cerebral palsy: a continuing debate. Childs Nerv Syst 22(12):1529–1535
51. van den Heuvel M, van der Plas A, van der Schaaf M et al (2016) Pharmacokinetics and pharmacodynamics of intrathecal baclofen. Br J Clin Pharmacol 82(1):90–100

Check for updates

Chapter 14

Evolving Solutions for the Next Generation of Intrathecal Drug Delivery Systems

Manuel Menéndez-González

Abstract

Intrathecal drug delivery systems (IDDS) have evolved from niche tools for refractory pain and severe spasticity into enabling platforms for central nervous system (CNS) therapeutics. This chapter synthesizes advances from 2019 to 2025 spanning "smart" implantable pumps, engineered catheters that improve neuraxial distribution with microliter volumes, and the expanding repertoire of intrathecally delivered biologics. We review regulatory milestones (e.g., SynchroMed™ III), emerging access solutions (e.g., subcutaneous intrathecal ports), and translational programs in oncology, neurogenetics, and neurodegeneration. Safety is examined across surgical, pharmacologic, mechanical, and digital domains, including catheter-tip granulomas, baclofen withdrawal, infection prevention, and cybersecurity of connected devices. Finally, we outline implementation models, training standards, and future directions such as closed-loop delivery, AI-supported dosing, and equitable access. Taken together, IDDS are becoming critical infrastructure for precision neurology and oncology—provided that engineering innovation is matched by robust clinical governance and long-term stewardship.

Keywords Next-generation intrathecal devices, Implantable technologies, Pump innovation, Flow control, Neuro-engineering

1 Introduction

Chronic pain remains a prevalent and disabling health problem that drives health-care utilization and compromises quality of life, with recent U.S. estimates showing substantial population-level burden and stable prevalence through 2021 [1]. Since the late 20th century, intrathecal drug delivery (IDD) has provided a means to achieve analgesia and antispasticity with lower systemic exposure by infusing medication directly into the cerebrospinal fluid (CSF). Over time, IDD has matured from single-agent, fixed-flow systems to programmable, implantable platforms supported by consensus guidelines and long-term safety data [1–3].

Three converging trends are now redefining the field. First, the approval and clinical dissemination of next-generation pumps

Manuel Menéndez-González (ed.), *Intrathecal Drug Delivery*, Neuromethods, vol. 234,
https://doi.org/10.1007/978-1-0716-5241-1_14,
© The Author(s), under exclusive license to Springer Science+Business Media, LLC, part of Springer Nature 2026

(e.g., SynchroMed™ III) extend battery life, enhance programmability, and introduce secure software updates and telemetry—shifting devices from static dispensers to adaptable, connected therapeutic nodes [4–7]. Second, catheter engineering is moving beyond open-ended outlets to designs that sculpt near-field hydrodynamics, projecting injectate away from the tip to reduce pooling, flatten concentration gradients, and distribute drug or vectors more uniformly across the neuraxis [8, 9]. Third, biologics—antisense oligonucleotides (ASOs), adeno-associated viral vectors (AAV), monoclonal antibodies, and even cellular therapies—require reliable CNS access that bypasses the blood–brain barrier (BBB) and exploits CSF delivery to achieve target engagement [10–13]. These trends, reinforced by updated recommendations from expert groups, reposition IDD as an enabling platform for disease-modifying CNS therapies as well as for durable symptomatic treatment [3, 4].

The practical bottleneck, however, remains the "last mile": achieving predictable rostro-caudal distribution with small volumes within a complex CSF environment. Recent preclinical work shows that engineered micro-valve catheters can produce homogeneous spread at microliter doses and reduce the vector load required for widespread dorsal root ganglion transduction, with implications for safety, cost, and manufacturing [8, 9]. In parallel, novel access strategies—such as subcutaneous intrathecal ports for patients with challenging anatomy and conceptual spinal-to-ventricular conduits—aim to convert episodic, technically demanding procedures into sustainable workflows for repeated biologic delivery [14–16].

This chapter reviews the technology, translational science, and clinical practice that now define IDD. We integrate engineering fundamentals with pharmacokinetics, programmatic safety, and service design to provide a roadmap for safe deployment of next-generation intrathecal therapies across pain medicine, neurology, and oncology.

2 Advances in Implantable Pump Technology: From Reliability to "Smart" Connectivity

2.1 Programmed Precision and Lifecycle Evolution

The introduction of programmable pumps in the early 2000s established a new standard of dosing flexibility and long-term reliability. The 2023 approval of SynchroMed™ III represents an inflection point: the platform couples enhanced battery longevity with software-defined therapy and secure wireless features that allow firmware updates and complex infusion schedules post-implantation [4–6]. This shifts the device lifecycle from "fixed at manufacture" to "maintainable and updatable," aligning with the broader evolution seen in cardiac rhythm and neuromodulation devices.

From a clinical standpoint, finer-grained schedules (e.g., circadian dose profiles, stepped titrations) support patient-centered analgesia or spasticity control, while audit logs and dose history improve traceability and safety. These capabilities also harmonize with consensus pathways that emphasize algorithmic drug selection, titration, and periodic reassessment [3, 4].

2.2 Cybersecurity and Digital Governance

Connectivity introduces a cyber-physical risk surface. Regulatory guidance now mandates encryption, authenticated access, and secure update channels for connected medical devices [17–19]. For IDD programs, this requires incorporating device patch management, role-based access controls, and incident response into routine clinical operations. The clinician's remit now extends beyond pharmacology and surgical technique to include digital safety, in partnership with biomedical engineering and information security teams [17–20].

2.3 Human-Factor Engineering and Refill Safety

Despite advances in hardware, human-factor errors during refills or reprogramming remain preventable causes of overdose or withdrawal. Standardized refill protocols, ultrasound-guided access in difficult pockets, and two-person verification for programming reduce risk. Documentation and analytics derived from pump telemetry can further support quality assurance cycles and early detection of anomalies [2–4].

3 Innovations in Catheter Design and Access for Optimized CSF Distribution

3.1 Hydrodynamics at the Outlet: Why Geometry Matters

In low-volume implantable infusions, drug transport is dominated by CSF pulsatility and limited mixing rather than bulk flow. Traditional open-ended or side-port catheters often exhibit steep local gradients and pooling around the tip, increasing dose requirements and, historically, risk of inflammatory masses at the outlet. Engineered micro-valve catheters generate high-velocity microjets that project injectate away from the tip, flattening local gradients and producing more uniform rostro-caudal exposure with microliter infusions [8, 9, 21]. In rodent studies, such designs enabled two orders of magnitude lower AAV9 doses to achieve segment-wide dorsal root ganglion transduction compared with open-ended outlets, while also reducing cellular ingrowth and debris accumulation that presage occlusion [8, 9].

3.2 Access Innovations: Durable Ports and Novel Trajectories

Repeated lumbar access can be technically challenging in populations such as spinal muscular atrophy (SMA) with scoliosis. Subcutaneous intrathecal ports (e.g., ThecaFlex DRx) under pivotal evaluation and Breakthrough Device designation are designed to provide durable, repeatable access for ASOs like nusinersen and other macromolecules, potentially transforming long-term treatment logistics into predictable outpatient workflows [15, 16]. Exploratory spinal-to-ventricular conduits—navigated across the foramen of Magendie—illustrate how access may expand toward ventricular targets for broader cortical distribution or combined sensing-delivery applications, though clinical translation will require careful safety validation [14].

3.3 Distribution for Biologics: Lessons from Large Animals and Primates

Translational studies demonstrate robust motor neuron and cortical transduction after CSF delivery of AAV serotypes in pigs and nonhuman primates, underscoring that catheter position, infusion mechanics, and outlet behavior materially influence where genomes and proteins ultimately localize [10–13]. For proteins and nucleic acids, improved near-field hydrodynamics can lower total dose requirements, widening safety margins and reducing immunogenicity risks inherent to high-load biologic administration [12, 13]. Pharmacokinetic work with intrathecal small molecules further shows how drug–tissue partitioning depends on site, volume, and CSF dynamics, reinforcing the inseparability of delivery mechanics and pharmacology in practice [22–26].

4 Safety Beyond Mechanics: Biological, Pharmacologic, and System-Level Risks

4.1 Inflammatory Masses and Outlet Biology

Catheter-tip granulomas—sterile inflammatory masses associated with high local opioid concentrations—were first systematically characterized two decades ago and remain a critical safety concern [27, 28]. Rare case reports now describe granulomas even with lower-dose fentanyl admixtures, implicating host susceptibility and near-tip exposure kinetics as co-determinants alongside drug identity [29]. Outlet designs that minimize stagnant boundary layers and steep gradients are thus safety technologies as much as distribution technologies [8, 9, 27–29].

4.2 Infection Prevention and Long-Term Patency

Despite improved asepsis, infection rates of 2%–8% persist in large series, ranging from pocket infection to meningitis and epidural abscess; explantation is frequently required [30, 31]. Risk reduction hinges on meticulous surgical technique, perioperative prophylaxis, and workflow designs that minimize repeated access manipulation over years of therapy. Antimicrobial-impregnated components are under evaluation but are not yet standard [30, 31].

4.3 Pharmacological Hazards: Overdose and Withdrawal

The intrathecal route magnifies the therapeutic index and risk alike. Delayed rostral spread of opioids can precipitate respiratory depression, whereas ziconotide—despite its non-opioid mechanism—may cause neuropsychiatric adverse effects necessitating cautious titration [32–34]. Baclofen withdrawal, typically triggered by catheter disruption or pump malfunction, presents as hyperthermia, rigidity, rhabdomyolysis and can be life-threatening without rapid recognition and bridging protocols [35–38]. Standardized emergency pathways and patient education are indispensable components of program safety.

4.4 Human-Factor and Digital Risks

Mechanical failures remain clinically relevant, but the risk landscape now includes workflow errors during refills/programming and cybersecurity vulnerabilities in connected devices [31, 17–19]. Embedding checklists, dual-verification, access control, and update governance into care pathways is essential to prevent cyber-physical incidents that could otherwise translate into dosing errors or therapy interruption.

5 Clinical Scenarios and Translational Applications

5.1 Refractory Pain and Spasticity

For selected patients with refractory cancer and non-cancer pain, IDD provides clinically meaningful analgesia with reduced systemic opioid burden and improved function, supported by randomized and observational evidence [39, 40]. In severe spasticity due to cerebral palsy, multiple sclerosis, or spinal cord injury, intrathecal baclofen (ITB) reliably reduces tone and improves care; however, programs must maintain readiness to manage withdrawal events when catheter or pump issues arise [35–38].

Recent multicenter surveys and practice analyses report high patient satisfaction with modern systems, with some suggesting that individualized optimization rather than universal "minimum effective dose" rules may better align dose with quality-of-life outcomes—without excess adverse events under careful monitoring [41, 42, 43, 44, 45–47].

5.2 Oncology: Toward Disease Modification in Leptomeningeal Disease

Leptomeningeal disease (LMD) is increasingly targeted with intrathecal biologics, including monoclonal antibodies and immune checkpoint inhibitors, often alongside focal radiotherapy and systemic therapy [48, 49]. Early-phase studies of intrathecal trastuzumab, bevacizumab, and nivolumab suggest feasibility; cell-based approaches such as CAR-NK cells are under preliminary investigation. These programs will depend on reliable, repeatable access and distribution—areas where catheter engineering advances can directly support clinical success.

5.3 Neurogenetics and Neurodegeneration

ASO therapy has already altered the natural history of spinal muscular atrophy (nusinersen) and, with tofersen, inaugurated the first FDA-approved intrathecal therapy for SOD1-ALS under a biomarker-driven pathway [50, 51]. Intrathecal AAV9 gene replacement for SMA (OAV101 IT) has met key phase III endpoints with regulatory submissions planned, supported by extension and follow-up cohorts [52–54]. In Parkinson's disease, prasinezumab has advanced to phase III, while

AAV2-GDNF (AB-1005) has received RMAT designation and entered phase II with encouraging safety signals from phase 1b [55–58]. These programs exemplify the dual demands of one-time, high-fidelity delivery and chronic, repeatable dosing—use cases that modern pumps, ports, and catheters are increasingly poised to meet.

6 Implementation, Training, and Service Models

Successful IDD programs are multidisciplinary enterprises that integrate neurology, neurosurgery, anesthesiology, rehabilitation, nursing, pharmacy, and biomedical engineering. Standardized curricula should encompass patient selection, trialing strategies, surgical implantation and pocket management, refill and programming safety, emergency pathways for overdose/withdrawal, infection prevention, and cybersecurity governance [2–4, 17–19]. Health-system design should ensure after-hours access for urgent troubleshooting and maintain stocked "bridging kits" (e.g., oral/enteral baclofen) to avert withdrawal crises.

Equity of access remains a systemic challenge. High upfront costs, requirement for specialized teams, and geographic concentration of expertise can magnify disparities. Regional networks, telemetric follow-up, and systematic training initiatives can mitigate these gradients and support safe scaling of IDD-enabled biologics [2–4].

7 Future Directions

Next-generation IDD will likely converge on four fronts. First, **closed-loop** architectures: integrating pressure/flow sensing or biochemical monitoring to detect occlusion, guide dosing, or trigger alerts before clinical deterioration—paralleling the evolution of adaptive neuromodulation [59, 60]. Second, **AI-assisted personalization**: machine-learning models trained on longitudinal pump data, patient-reported outcomes, and wearable biosignals may forecast analgesic needs, infection risk, or granuloma propensity; explainable approaches are essential to maintain clinician trust [61, 62, 63]. Third, **platform expansion for biologics**: optimized outlet hydrodynamics and durable access will remain prerequisites for safe AAV, ASO, antibody, and cellular therapies [8–16, 51–58]. Fourth, **sustainability and longevity**: modular, serviceable components; refill-sparing designs; and infection-resistant materials can reduce cumulative procedural burden over decades of care [63].

The path forward is clear: harmonize engineering innovation with rigorous clinical governance, digital safety, and equitable access. Done well, IDD will continue its transition from salvage therapy to a mainstream platform for targeted, and potentially disease-modifying, CNS treatment [64].

Conflict of Interest

The authors have no conflicts of interest to declare that are relevant to the content of this chapter.

Appendix

Key Clinical Trials and Technology Milestones (2019–2025)

Development/trial	Agent/technology	Indication	Phase/ status	Key findings/notes
SynchroMed™ III	Programmable pump	Chronic pain, spasticity	FDA cleared 2023	Enhanced programmability, secure updates, extended longevity [4–6]
Engineered micro-valve catheter	Hydrodynamic outlet	CSF distribution	Preclinical	Homogeneous spread with microliter volumes; ~100-fold lower AAV9 dose in rodents [8, 9]
ThecaFlex DRx (NCT05866419)	Port-catheter system	SMA (repeat IT access)	Pivotal	Breakthrough device; aims to reduce burden of repeated LPs [15, 16]
Tofersen (Qalsody®)	ASO (SOD1)	ALS	FDA 2023	Accelerated approval based on NfL reduction; infrastructure guidance published [51]
OAV101 IT (STEER/ STRENGTH)	AAV9 SMN1	SMA	Phase III	Met primary endpoint; additional safety/ extension data emerging [52–54]
AB-1005 (AAV2-GDNF)	Gene therapy	Parkinson's disease	Phase II	RMAT designation; favorable early safety signals [57, 58]
Prasinezumab	Anti-α-syn mAb	Parkinson's disease	Phase III	Phase III development announced 2025 [55, 56]
IT biologics for LMD	Nivolumab, trastuzumab, others	LMD (breast/ NSCLC)	Early phase	Feasibility and combination-therapy interest; protocolized RT integration [48, 49]

References

1. Rikard SM, Strahan AE, Schmit KM, Guy GP Jr. (2023) Chronic pain among adults—United States, 2019–2021. MMWR Morb Mortal Wkly Rep 72(15):379–385
2. Shah N, Napoli R, Padalia D (2024) Implantable intrathecal drug delivery system. In: StatPearls [Internet]. StatPearls Publishing, Treasure Island (FL)
3. Deer TR, Pope JE, Hayek SM et al (2017) The polyanalgesic consensus conference (PACC): recommendations for intrathecal drug delivery in clinical practice. Neuromodulation 20(2):96–132
4. Deer TR, Pope JE, Hayek SM et al (2021) PACC guidance for improving safety and mitigating risks—2021 update. Neuromodulation 24(4):482–497
5. Medtronic. SynchroMed™ III intrathecal drug pump—product information for healthcare professionals. 2023–2025
6. Medtronic. U.S. FDA approves SynchroMed™ III for targeted drug delivery. Press release, 2023 Oct 10
7. Medical Alley Association. SynchroMed™ III launch announcement. 2023 Oct 10
8. Khani MM, Soter NA, Chadow M et al (2024) Refinement of intrathecal catheter design to enhance neuraxial distribution. J Neurosurg Anesthesiol 36(1):66–75
9. Tillman KS, Choi S, Deverman BE et al (2018) Engineered multivalve catheter improves intrathecal delivery and distribution of viral vectors. Mol Ther Methods Clin Dev 10:214–225
10. Federici T, Taub JS, Baum GR et al (2012) Robust spinal motor neuron transduction following intrathecal AAV9 in pigs. Gene Ther 19(8):852–859
11. Samaranch L, Salegio EA, San Sebastian W et al (2013) Strong cortical and spinal transduction after AAV7/AAV9 CSF delivery in NHPs. Hum Gene Ther 24(5):526–532
12. Hudry E, Vandenberghe LH (2019) Therapeutic AAV gene transfer to the nervous system: a clinical reality. Neuron 101(5):839–862
13. Hocquemiller M, Giersch L, Audrain M, Parker S, Cartier N (2016) AAV-based gene therapy for CNS diseases. Hum Gene Ther 27(7):478–496
14. Saenz A, Galvez D, Carranza S et al (2024) Intrathecal access to the fourth ventricle via the foramen of Magendie: anatomical/device feasibility. Front Med Technol 6:1211585
15. ClinicalTrials.gov. NCT05866419—ThecaFlex DRx for spinal muscular atrophy. Updated 2025
16. Green S. intrathecal port-catheter system wins FDA breakthrough device designation. MedTech Dive. 2019 Mar 27
17. U.S. Food and Drug Administration. Cybersecurity in medical devices: quality system considerations and content of premarket submissions. Guidance, 2023
18. Maisel WH, Kohno T, Juels A et al (2010) Implantable medical devices: security and privacy concerns. N Engl J Med 362(13):1164–1166
19. Halperin D, Heydt-Benjamin TS, Ransford B et al (2008) Security and privacy for implantable medical devices. IEEE Pervasive Comput 7(1):30–39
20. Brogan SE, Winter NB (2011) Patient safety and intrathecal drug delivery: a review. Pain Med 12(9):1291–1303
21. Penn RD, Paice JA (2000) Catheter systems for intrathecal drug delivery. J Pain Symptom Manage 19(1 Suppl):S37–S43
22. Bernards CM (2006) CSF and spinal cord distribution of baclofen and bupivacaine after intrathecal injection. Anesthesiology 105(1):169–178
23. Ponnusamy A, Fang Z, Liu J (2015) Microelectromechanical drug infusion pumps: design and performance. J Microelectromech Syst 24(5):1201–1212
24. Lutz RJ, Brown A, Spector M (2004) Biomechanics of implantable infusion pumps. ASAIO J 50(5):476–481
25. Medtronic. Implantable systems performance registry (ISPR) annual report, 2016
26. Medtronic. SynchroMed™ II clinician manual, 2017
27. Yaksh TL, Hassenbusch SJ, Burchiel K et al (2002) Inflammatory masses associated with intrathecal drug infusion. Pain Med 3(4):300–312
28. Hildebrand KR, Elsberry DE, Deer TR (2004) Morphine-induced granulomas in dogs during continuous intrathecal infusion. Anesth Analg 98(2):432–436
29. Kelly C, Salloum A, Mekhail N et al (2024) Recurrent catheter-tip granuloma on a replaced system delivering low-dose fentanyl/bupivacaine. Pain Med Case Rep 8(2):73–77
30. Bjerke BT, King NK, Anderson PA (2017) Infections in spinal implants and intrathecal devices. Neurosurg Clin N Am 28(3):377–387
31. Coffey RJ (2012) Complications of intrathecal drug delivery systems: review and analysis. Neuromodulation 15(6):467–482
32. Wallace MS, Rauck RL, Fisher R et al (2008) Intrathecal ziconotide in chronic nonmalignant pain: randomized, double-blind, placebo-controlled. Neuromodulation 11(2):75–86
33. Staats PS, Yearwood T, Charapata SG et al (2004) Intrathecal ziconotide in the treatment of refractory pain in patients with cancer or AIDS: a randomized controlled trial. JAMA 291(1):63–70

34. Rauck RL, Wallace MS, Leong MS et al (2006) Randomized, double-blind study of intrathecal ziconotide in severe chronic pain. Pain 125(3):245–254
35. Coffey RJ, Edgar TS, Francisco GE et al (2002) Abrupt withdrawal from intrathecal baclofen: recognition and management. Arch Phys Med Rehabil 83(6):735–741
36. Motta F, Antonello CE (2014) Withdrawal syndrome after intrathecal baclofen interruption in children. Eur J Paediatr Neurol 18(4):538–543
37. Dario A, Tomei G (2004) Benefit–risk assessment of baclofen in severe spinal spasticity. Drug Saf 27(11):799–818
38. Coffey RJ, Burchiel K, Schultz DM et al (2010) Overdose and mortality in intrathecal baclofen therapy. Neuromodulation 13(2):88–94
39. Smith TJ, Staats PS, Deer T et al (2002) Randomized trial of implantable IDDS vs comprehensive medical management for refractory cancer pain. J Clin Oncol 20(19):4040–4049
40. Caraway DL, Walker V, Becker L, Hinnenthal J (2015) Discontinuation of systemic opioids after IDDS implantation. Neuromodulation 18(6):522–531
41. Abd-Elsayed A, Gill J, Anderson D et al (2024) Intrathecal drug delivery systems: surgical/clinical safety case series. Cureus 16(9):e62964
42. Hagedorn JM, Hoelzer BC, Rauck RL et al (2025) Patient satisfaction, analgesia, and side effects in relation to intrathecal dose. Neuromodulation 28(3):ePub
43. Huntoon MA, Burgher AH (2008) The implantable intrathecal drug delivery system for cancer pain. Curr Oncol Rep 10(4):282–288
44. Wallace MS, Yaksh TL (2000) Long-term spinal analgesic delivery: safety and efficacy. Anesth Analg 91(1):65–83
45. Van Dongen RT, Crul BJ, Van Egmond J (1999) Intrathecal opioid treatment in chronic pain: long-term efficacy and side effects. Pain 80(3):269–282
46. Chou R, Gordon DB, de Leon-Casasola OA et al (2016) Management of postoperative pain: a clinical practice guideline. J Pain 17(2):131–157
47. Deer TR, Krames E, Mekhail N et al (2014) Appropriate use of neurostimulation for chronic pain. Neuromodulation 17(6):515–550
48. Palmer JD, Chu SY, Dworkin M et al (2024) Radiotherapy for leptomeningeal disease in the era of modern systemic and intrathecal therapy. Radiother Oncol 197:110064
49. NRG Oncology. BN014: Proton craniospinal RT with/without intrathecal therapy for LMD from breast/NSCLC. 2024–2025
50. Andrews JA et al (2024) Enhancing infrastructure for intrathecal/genetic therapies: a Qalsody model for SOD1-ALS. Neurol Clin Pract 14(6):ePub
51. U.S. Food and Drug Administration. FDA approves Qalsody® (tofersen) for SOD1-ALS. 2023 Apr 25
52. Novartis. Intrathecal onasemnogene abeparvovec (OAV101 IT) Phase III STEER meets primary endpoint. Media release, 2024
53. Novartis. Phase IIIb STRENGTH data demonstrate safety/functional stabilization with OAV101 IT. Media release, 2025
54. Muscular Dystrophy Association. LT-002 long-term follow-up of IV/IT onasemnogene abeparvovec—abstract library. MDA Conference 2025
55. Roche. Roche to advance prasinezumab into Phase III for early Parkinson's disease. Media release, 2025 Jun 16
56. Genentech. Prasinezumab Phase III development announcement. Press release, 2025 Jun 15
57. AskBio. AB-1005 (AAV2-GDNF) Phase 1b results; RMAT designation; Phase II REGENERATE-PD underway. Press release, 2025 Feb
58. ClinicalTrials.gov. NCT06285643—REGENERATE-PD: AAV2-GDNF (AB-1005). Posted 2024; updated 2025
59. https://www.encleartherapies.com/programs
60. Blackburn SL, Babi MA, Grande AW, et al (2025) Prospective Trial of Cerebrospinal Fluid Filtration After Aneurysmal Subarachnoid Hemorrhage via Lumbar Catheter Extension (PILLAR-XT). Neurocrit Care 43:934–946. https://doi.org/10.1007/s12028-025-02328-8
61. Topol EJ (2019) High-performance medicine: human and artificial intelligence. Nat Med 25(1):44–56
62. Holzinger A, Langs G, Denk H, Zatloukal K, Müller H (2019) Causability and explainability of AI in medicine. WIREs Data Mining Knowl Discov 9(4):e1312
63. Yaksh TL, Rathmell JP (2019) Future directions in intrathecal drug delivery: implantable systems and design considerations. Anesth Analg 128(2):360–375
64. Deer TR, Pope JE, Hanes M, McDowell GC (2019) Intrathecal therapy for chronic pain: current and future applications. J Pain Res 12:615–622

Check for updates

Chapter 15

Emerging Solutions for Future Generations of Intrathecal Drug Delivery Systems

Manuel Menéndez-González

Abstract

Intrathecal drug delivery (ITDD) offers a direct pathway to administer therapeutics into the cerebrospinal fluid (CSF), bypassing the blood–brain barrier (BBB) for targeted central nervous system (CNS) treatment. However, maximizing therapeutic efficacy and minimizing adverse effects necessitates steady and sustained drug release, a challenge for conventional systems. This chapter explores the evolving landscape of ITDD, emphasizing the shift from traditional pump-based systems to innovative pumpless and active fluid-management technologies. We highlight how advancements like nano-engineered membranes, passive flow-harnessing devices, and active CSF-exchange platforms are overcoming the limitations of passive diffusion to achieve more uniform drug distribution and, in some cases, enable the clearance of pathological components from the CNS. The transformative potential of these technologies is discussed across neuro-oncology, pain management, neurodegeneration, and acute neurological injury, with a focus on improving therapeutic personalization. Ultimately, these innovations promise to redefine CNS therapies, improving patient outcomes and quality of life through enhanced drug stability, reduced complications, and optimized therapeutic control.

Keywords Advanced intrathecal platforms, Future drug delivery systems, Biomaterials, Device innovation, Neuromodulation technology

1 Introduction: The Evolution of Intrathecal Drug Delivery

The central obstacle to effective pharmacotherapy in neurology is the BBB, a highly specialized endothelial interface that excludes most large and polar therapeutics from the brain parenchyma and the spinal cord [1, 2]. Delivering drugs directly to the CSF compartment circumvents this barrier and can achieve therapeutic concentrations in target tissues with substantially lower total dose and reduced systemic exposure [3]. Yet CSF is not a quiescent reservoir. It is a dynamically oscillating fluid domain in which pulsatile, breathing-coupled, and posture-dependent flows interact with perivascular pathways—the so-called glymphatic system—to shape the spatiotemporal fate of intrathecally administered agents [3–6]. In practice,

Manuel Menéndez-González (ed.), *Intrathecal Drug Delivery*, Neuromethods, vol. 234,
https://doi.org/10.1007/978-1-0716-5241-1_15,
© The Author(s), under exclusive license to Springer Science+Business Media, LLC, part of Springer Nature 2026

long-range dispersion along the neuraxis reflects not simple diffusion but cycle-averaged drift of fluid parcels (mean Lagrangian motion) superimposed on compartmental clearance processes, explaining why identical doses placed at different locations or delivered with different injection protocols can yield divergent exposure profiles in brain and cord [4–6].

Hardware and procedural factors also modulate distribution. In nonhuman primates, intrathecal catheters themselves measurably slow bulk CSF oscillations and can reduce rostral transport, with consequences for drug exposure in cranial compartments [7]. Historically, implantable programmable pumps transformed the management of refractory pain and spasticity, enabling sustained micro-infusion and substantial sparing of systemic opioids [8, 9]. Nevertheless, electromechanical devices carry nontrivial risks—battery depletion, motor stalls, valve malfunctions, infection, catheter kinking or occlusion—and they were not designed around the pharmacokinetic demands of modern biologics or gene-directed agents that require broad, homogeneous neuraxial distribution [8–10]. In parallel, a decade of translational work has reframed intrathecal dosing as an engineering problem: in vivo and benchtop studies demonstrate that injection site, bolus volume, post-injection flush, and even patient positioning can be tuned to steer exposure, while imaging and computational models allow quantitative verification and planning [11–15].

Protein and nucleic acid therapeutics delivered by the intrathecal route raise additional questions of molecular stability, immunogenicity, and distribution, all of which are sensitive to local hydrodynamics and delivery kinetics [16]. Two complementary development arcs have emerged. The first, "active management," encompasses convection-enhanced delivery (CED) and modern CSF-exchange platforms that impose controlled pressure and flow to drive drugs through tissue or to condition the CSF milieu [17–21]. The second, "advanced formulations and materials," comprises nanoparticles, membrane-governed reservoirs, and hydrogels that decouple release kinetics from pump mechanics and—at the extreme—enable immunoselective removal of pathological solutes without systemic drug exposure. Together with digital and physical "twins" of the human subarachnoid space for protocol optimization, these strands are transforming ITDD from a static infusion technique into a dynamic, actively managed, and increasingly personalized modality [11–14, 21, 22].

2 Advanced Materials for Intrathecal Therapeutics

2.1 Nanoparticle-Mediated CNS Delivery and Immunoselective "Pseudodelivery"

Nanoparticles (NPs) provide modular carriers for small molecules, peptides, proteins, and nucleic acids. In the intrathecal context, rational engineering of particle size, surface chemistry, and ligand decoration can extend CSF residence, mitigate degradation, and facilitate parenchymal entry via perivascular routes while minimizing off-target exposure [23, 24]. Contemporary platforms leverage blood–CSF and meningeal interfaces rather than trans-BBB transport, and surface modification with brain-relevant ligands can improve retention and distribution in CSF-contacting spaces [23–25]. These design rules differ materially from systemic nanomedicine: buoyancy in pulsatile CSF, resistance to aggregation in low-protein fluid, and interactions with arachnoid and ependymal surfaces dominate performance, while perivascular "pumping" linked to arterial pulsatility becomes a distribution driver [4, 5, 23, 24].

A striking conceptual extension is immunoselective intrathecal "pseudodelivery" (nanopheresis): antibodies or other binding scaffolds are sequestered within a nanoporous implant that communicates with CSF; pathological solutes (e.g., amyloid-β oligomers) diffuse into the device and are captured, whereas the macromolecular binders remain confined, avoiding systemic exposure and antibody-related imaging abnormalities [26, 27]. This architecture reframes treatment as continuous CSF purification rather than drug addition and suggests durable biomarker control with a favorable safety profile if membrane selectivity and fouling can be mastered [26, 27].

2.2 Membrane-Governed Reservoirs: Release Kinetics Without Electronics

Subcutaneous or subfascial reservoirs coupled to the intrathecal space via catheters can be regulated by engineered nanoporous membranes, yielding diffusion-controlled release profiles that are independent of pump motors and batteries. When the internal phase remains saturated, the membrane supports near-zero-order release set by pore geometry and drug diffusivity; in unsaturated phases, release tapers as a first-order process [28]. Material selection, pore size distribution, and drug formulation provide orthogonal control knobs to tailor weeks-to-months delivery without electromechanical complexity—a particularly attractive proposition for fragile biologics whose activity depends on gentle, predictable release.

2.3 Injectable and Stimuli-Responsive Hydrogels (and NP–Hydrogel Hybrids)

Injectable hydrogels—high-water, cross-linked polymer networks—can be delivered in liquid form and gel in situ, establishing local depots adjacent to the cord or nerve roots. Their biocompatibility and tunable mesh size allow sustained, spatially confined release of diverse payloads and, when placed epidurally or perispinally, a degree of "contact targeting" to affected segments [29]. Thermosensitive hyaluronan–methylcellulose systems have been engineered to co-load fast-acting anti-inflammatory drugs with slower-releasing anti-ferroptosis agents, achieving staged delivery profiles in spinal cord injury models [30]. Hybrid designs embed nanoparticles within the gel matrix to add a second layer of kinetic control and enhance protection of labile molecules; in neuropathic pain models, a single

NP–hydrogel administration of esketamine extends analgesia far beyond that achievable with solution dosing, without systemic peaks [31]. As stimuli-responsive chemistries and DNA-based hydrogels mature, these depots are poised to synchronize release with physiological cues (temperature, pH) and disease states, potentially aligning high-value biologics with windows of maximal target accessibility.

3 Pumpless Approaches: Harnessing Natural CSF Dynamics

The CSF system exhibits diurnal variation and strong cardiorespiratory coupling; glymphatic influx is enhanced during slow-wave sleep and modulated by systemic osmolarity, while breathing drives cranio-spinal CSF oscillations that aid solute transport [32, 33]. Pumpless ITDD concepts attempt to exploit these endogenous gradients, using low-resistance circuits that permit gentle, unpowered perfusion of a drug-containing chamber by CSF pulsatility. By eliminating electromechanical components, such architectures remove failure modes inherent to batteries, motors, and firmware, though they demand meticulous fluidic design to avoid siphon effects, retrograde flow, and unintended pressure transients [8]. Protocol-level "actuators"—such as post-injection flushes and positioning—remain essential levers to achieve desired rostro-caudal exposure with minimal hardware burden [11].

4 Active CSF Management and Advanced Delivery Systems

4.1 Convection-Enhanced Delivery (CED)

CED established the principle that externally imposed pressure gradients can move large molecules through the brain interstitium over clinically relevant volumes. Foundational animal and modeling work defined relationships among infusion rate, tissue hydraulic conductivity, reflux, and distribution geometry, and clinical studies demonstrated feasibility for neuro-oncology and neurorestoration indications [17–21]. Although outcomes have been mixed, modern reflux-resistant cannulae, real-time imaging, and numerical planning have improved safety and predictability, underscoring the broader lesson for intrathecal programs: when diffusion is limiting, controlled convection can unlock access for macromolecules [18–21].

4.2 Active CSF-Exchange Platforms

Active CSF-exchange systems aim not only to deliver drugs but also to condition the CSF environment by irrigation, aspiration, and filtration.

One such platform, IRRAflow, couples precise peristaltic irrigation with controlled egress and continuous intracranial pressure monitoring in timed cycles, enabling sterile, prolonged intraventricular therapies and lavage in hemorrhagic conditions [34]. Contemporary series and systematic reviews suggest acceptable complication rates when used for subarachnoid or intraventricular hemorrhage, and drug-stability analyses are beginning to define safe use of antibiotics and vasodilators under dilute, physiologic conditions [35–37]. For intrathecal pharmacotherapy, the principal constraints are chemical stability at low concentrations and the need for validated compounding protocols and materials compatibility [36, 37]. For instance, a recent study found that dilute concentrations of vancomycin, daptomycin, tobramycin, and nicardipine are chemically stable for up to 12 h (vancomycin, daptomycin, tobramycin) and 6 h (nicardipine) at room temperature when used for intraventricular irrigation via the IRRAflow ventricular drain system. This supports their use in clinical settings, as the tested compounds retained at least 90% of their initial concentration [38],

A complementary concept, liquorpheresis (Neurapheresis), withdraws CSF and passes it across a selective filter before returning it to the patient, with the goals of accelerating clearance of blood breakdown products, inflammatory mediators, or pathological proteins [39–41]. Early feasibility work and pilot experiences (e.g., PILLAR) suggest the approach can reduce the burden of neurotoxic solutes and may shorten reliance on external drains, motivating trials in hemorrhage and, prospectively, proteinopathies where CSF "decontamination" could complement disease-modifying strategies [39–41].

5 Coupling Intrathecal Delivery with Systemic Hypertonicity to Leverage Glymphatic Entry

Systemic hypertonic saline (HTS) increases plasma osmolarity, transiently expanding perivascular and CSF spaces and augmenting glymphatic influx. Imaging in rodents shows that HTS robustly enhances CSF–brain exchange and increases brain uptake of intrathecal tracers and nanoparticles without prolonging tissue retention; most nanoscale agents are cleared renally within 24 h [42, 43]. In the spinal cord, concurrent HTS heightens antinociception from lumbar intrathecal morphine and raises tissue levels relative to isotonic controls, consistent with improved convective entry into parenchyma [42]. For large molecules that otherwise struggle to penetrate deep targets, pairing intrathecal administration with carefully titrated hypertonicity could increase CNS bioavailability while allowing lower total doses. The strategy requires vigilant hemodynamic and intracranial pressure monitoring and stringent patient selection, particularly in the presence of impaired autoregulation or osmotic vulnerabilities [37, 42, 43].

6 Research and Translational Considerations

6.1 ASOs, Gene Therapy, and Distribution Science

Antisense oligonucleotides (ASOs) and viral vectors have catalyzed a new era of intrathecal therapeutics. Multimodal imaging in large animals and humans demonstrates that ASOs ride CSF oscillations and distribute broadly along the neuraxis, with exposure governed by injection site, volume, and post-injection flush—a set of variables that can be prospectively optimized [11, 13, 21, 44, 22]. For gene therapy and protein biologics, where efficacy and safety hinge on reliable target engagement, imaging-guided verification and computational planning are rapidly becoming best practice to de-risk first-in-human studies and registrational dosing.

6.2 Preclinical Models and Translational Safety

Small animals enable high-throughput screening and mechanistic work but differ markedly from upright humans in CSF volume and hydrodynamics; posture alone alters oscillatory transport patterns and can confound extrapolation [45]. Dogs and nonhuman primates better recapitulate human CSF spaces and allow chronic catheter and pump studies, making them indispensable for toxicology and for understanding device–CSF interactions—including the observation that intrathecal catheters can dampen CSF flow in primates [7, 46]. As cellular therapies and stem-cell adjuncts enter the intrathecal arena, robust biodistribution, tumorigenicity, and immunogenicity assessments are mandatory; encouragingly, mesenchymal stem cell applications have shown favorable safety profiles across species while advancing spinal cord repair paradigms [47–50]. Several types of animal models can be used for research in CNS drug delivery (Table 1).

6.3 Practical Clinical Levers and Workflow

Several low-burden, high-impact parameters are now supported by converging evidence. First, rostral exposure after lumbar dosing increases substantially with an adequate post-injection flush, often outweighing changes in bolus volume or rate [11, 22]. Second, protocol standardization, including patient positioning and timing (e.g., during sleep-rich periods), can leverage circadian and respiratory physiology to improve distribution [32, 33]. Third, when delivering labile biologics or initiating prolonged intrathecal antibiotics, platforms that permit controlled irrigation/aspiration with real-time pressure surveillance can improve safety if used under validated stability and compatibility conditions [34, 36, 37]. Finally, in programs where tissue penetration is critical and diffusion limiting (e.g., focal brain targets), CED planning and real-time visualization principles offer a transferable playbook: model first, image often, and match hardware to the required pressure–flow window [17–21].

Rodents (mouse, rat) are inexpensive and genetically tractable but possess tiny CSF volumes and horizontal posture that alter

Table 1
Common animal models for intrathecal delivery: translational trade-offs

Species	Key advantages	Key limitations	Primary use case
Rodent (rat, mouse)	Low cost, high throughput, well-characterized genetics, and availability of disease models	Small CSF volume (~35 μ L), prone posture alters hydrodynamics, difficult to catheterize chronically	High-throughput screening, initial efficacy testing, and genetic mechanism studies
Leporine (rabbits)	Low cost, high throughput, Medium CSF volume (~5 mL)	Not very well-characterized genetics or availability of disease models	High-throughput screening, initial efficacy testing
Canine (dog)	Larger CSF volume (~5–10 mL) allows for chronic pump/catheter implantation, extensive historical toxicology database	Pronounced lumbar lordosis can affect distribution, and some ethical/social concerns	Gold standard for preclinical toxicology and granuloma formation studies (e.g., for PACC)
Porcine (pig/minipig)	Anatomical similarities to the human spine, comparable cord diameter, and CSF volume	Can be difficult to handle, rapid growth can displace catheters, and thick skin	Device testing (pumps, catheters), surgical procedure development, and some distribution studies
Ovine (sheep)	Limited growth, easy to handle, comparable cord diameter, and CSF volume	Anatomical differences	Device testing (pumps, catheters), surgical procedure development, and some distribution studies
Non-human primate (NHP)	Closest anatomical and physiological homolog to humans: upright posture, similar CSF dynamics, and volume-to-cord ratio	High cost, significant ethical considerations, complex handling, and husbandry requirements	"Gold standard" for predicting human PK/PD, ASO, and gene therapy distribution/safety evaluation

oscillatory transport, limiting predictions of human neuraxial exposure [45]. Dogs allow chronic pump and catheter studies with extensive historical toxicology but have spinal curvatures that can influence distribution and carry ethical concerns in some jurisdictions [46]. Pigs and sheep approximate human spinal caliber and CSF spaces and are advantageous for device and surgical testing. Nonhuman primates are the closest physiological homologs, enabling the most reliable PK/PD and safety projections—at substantially higher cost and ethical complexity [7, 45–50].

Conflict of Interest

The authors have no conflicts of interest to declare that are relevant to the content of this chapter.

References

1. Kadry H, Noorani B, Cucullo L (2020) A blood–brain barrier overview on structure, function, impairment, and biomarkers of integrity. Fluids Barriers CNS 17(1):69. https://doi.org/10.1186/s12987-020-00232-3
2. Pardridge WM (2019) Blood–brain barrier and delivery of protein and gene therapeutics to brain. Front Aging Neurosci 11:373. https://doi.org/10.3389/fnagi.2019.00373
3. Yamada S, Kelly E (2016) Cerebrospinal fluid dynamics and the pathophysiology of hydrocephalus: new concepts. Semin Ultrasound CT MR 37(2):84–91. https://doi.org/10.1053/j.sult.2016.02.001
4. Iliff JJ, Lee H, Yu M et al (2013) A paravascular pathway facilitates CSF flow through the brain parenchyma and the clearance of interstitial solutes. J Neurosci 33(46):18190–18199. https://doi.org/10.1523/JNEUROSCI.1592-13.2013
5. Jessen NA, Munk AS, Lundgaard I, Nedergaard M (2015) The glymphatic system: a discovery of fundamental and clinical significance. J Clin Invest 125(3):1339–1349. https://doi.org/10.1172/JCI76304
6. Hornkjøl M, Valnes LM, Ringstad G et al (2022) CSF circulation and dispersion yield rapid clearance from intracranial compartments. Front Bioeng Biotechnol 10:932469. https://doi.org/10.3389/fbioe.2022.932469
7. Khani M, Sass LR, Xing T et al (2020) Intrathecal catheter implantation decreases cerebrospinal fluid flow in cynomolgus monkeys. PLoS One 15(6):e0234320. https://doi.org/10.1371/journal.pone.0234320
8. Ericson T, Singla P, Kohan L (2022) Intrathecal pumps. Phys Med Rehabil Clin N Am 33(2):409–424. https://doi.org/10.1016/j.pmr.2022.02.007
9. Prager J, Deer T, Levy R et al (2014) Best practices for intrathecal drug delivery for pain. Neuromodulation 17(4):354–372. https://doi.org/10.1111/ner.12134
10. Delhaas EM, Huygen FJPM (2020) Complications associated with intrathecal drug delivery systems. BJA Educ 20(2):51–57. https://doi.org/10.1016/j.bjae.2019.09.006
11. Burla GKR, Shrestha D, Bowen M et al (2024) Evaluating the effect of injection protocols on intrathecal solute dispersion in non-human primates. Fluids Barriers CNS 21(1):61. https://doi.org/10.1186/s12987-024-00508-7
12. Sullivan KF, Smazal AL, Heckman MG et al (2020) Intrathecal drug distribution in chronic animal studies: the role of convective forces in moving drug beyond the catheter tip. Sci Transl Med 12(528):eaax3539. https://doi.org/10.1126/scitranslmed.aax3539
13. Wolf DA, Weil AJ, Kvam E et al (2016) Dynamic dual-isotope molecular imaging elucidates principles for optimizing intrathecal drug delivery. JCI Insight 1(17):e87052. https://doi.org/10.1172/jci.insight.87052
14. Verma A, Abdel-Fahim R, Karakatsanis NA et al (2020) Multimodal molecular imaging-based evaluation for intrathecal drug therapy for chronic pain. Alzheimers Dement (Amst) 12(1):e12040. https://doi.org/10.1002/dad2.12040
15. Sigurdsson B, Pyykkö N, Tuisku J et al (2023) Quantitative SPECT methodology for detection of glymphatic transport: a tracer-kinetic approach and visualization. J Cereb Blood Flow Metab 43(2):240–257. https://doi.org/10.1177/0271678X221127040
16. Calias P, Banks WA, Begley D, Scarpa M, Dickson P (2014) Intrathecal delivery of protein therapeutics to the brain: a critical reassessment. Pharmacol Ther 144(2):114–122. https://doi.org/10.1016/j.pharmthera.2014.06.002
17. Patel M, Shah S, Desai B et al (2015) Convection-enhanced delivery: current status

and future directions. Neuro Oncol 17(suppl 7):vii14–vii20. https://doi.org/10.1093/neuonc/nov163

18. Bobo RH, Laske DW, Akbasak A et al (1994) Convection-enhanced delivery of macromolecules in the brain. Proc Natl Acad Sci U S A 91(6):2077–2081. https://doi.org/10.1073/pnas.91.6.2077
19. Lidar Z, Kunits G, Hadani M (2002) Convection-enhanced delivery of drugs to the brain: a new concept in neurological therapeutics. Adv Drug Deliv Rev 54(12):1501–1512. https://doi.org/10.1016/S0169-409X(02)00155-2
20. Gill SK, Basude F, Poria P et al (2012) Real-time visualization of convection-enhanced delivery in the brain. J Control Release 161(2):331–338. https://doi.org/10.1016/j.jconrel.2011.12.016
21. Ding H, Kim P, Grodzinsky AJ et al (2008) A comprehensive numerical model of convection-enhanced delivery to the brain. J Biomech Eng 130(2):021008. https://doi.org/10.1115/1.2890105
22. Seiner SJ, Stoodley MA, Bilston LE et al (2022) Investigation of human intrathecal solute transport dynamics using an in vitro cerebrospinal fluid system analog. Front Neuroimaging 1:879098. https://doi.org/10.3389/fnimg.2022.879098
23. Madadi AK, Sohn M-J (2024) Advances in intrathecal nanoparticle delivery: targeting the blood–cerebrospinal fluid barrier for enhanced CNS drug delivery. Pharmaceuticals (Basel) 17(8):1070. https://doi.org/10.3390/ph17081070
24. Ruan H, Hu Q, Chen Y et al (2021) Intrathecal drug delivery in the era of nanomedicine. J Control Release 331:627–643. https://doi.org/10.1016/j.jconrel.2020.11.021
25. Patil S, Singh S, Kumar A et al (2025) Advancing brain targeting: cost-effective surface-modified nanoparticles for faster market entry. Pharmaceutics 17(1):62. https://doi.org/10.3390/pharmaceutics17010062
26. Manuel MG, Tamba BI, Leclere M et al (2023) Intrathecal pseudodelivery of drugs in the therapy of neurodegenerative diseases. Pharmaceutics 15(2):768. https://doi.org/10.3390/pharmaceutics15020768
27. Menéndez-González M (2023) Liquorpheresis: cerebrospinal fluid filtration to treat CNS conditions. Springer, Cham https://doi.org/10.1007/978-3-031-29990-2
28. Siepmann J, Siepmann F (2012) Modeling of diffusion controlled drug delivery. J Control Release 161(2):351–362. https://doi.org/10.1016/j.jconrel.2011.10.006
29. Jiang X, Fan Z, Xue H et al (2023) Hydrogel scaffolds in the treatment of spinal cord injury: a review. Front Neurosci 17:1211066. https://doi.org/10.3389/fnins.2023.1211066
30. Liu Y, Zhang C, Li H et al (2024) Injectable thermosensitive hydrogel system based on hyaluronic acid and methylcellulose for synergistic therapy of traumatic spinal cord injury. Colloids Surf B Biointerfaces 233:113692. https://doi.org/10.1016/j.colsurfb.2024.113692
31. Yao J, Wang Z, Li P et al (2024) Sustained analgesia of esketamine via a biodegradable nanoparticle-hydrogel composite in neuropathic pain models. J Tissue Eng Regen Med 18(6):522–535. https://doi.org/10.1002/term.3564
32. Hablitz LM, Plá V, Giannetto M et al (2020) Circadian control of brain glymphatic and lymphatic fluid flow. Nat Commun 11(1):4411. https://doi.org/10.1038/s41467-020-18115-2
33. Burman R, Alperin N (2024) CSF-to-blood toxins clearance is modulated by breathing through cranio-spinal CSF oscillation. J Sleep Res 33(1):e14029. https://doi.org/10.1111/jsr.14029
34. IRRAS AB (2025) IRRAflow system. Available at: https://irras.com/products/irraflow/ Accessed 12 Jul 2025
35. Skyrman S, Albanna W, Hulou MM et al (2024) Clinical outcomes and considerations for continuous, long-term intraventricular irrigation in severe ventricular hemorrhage: IRRAflow experience. J Clin Med 13(15):4382. https://doi.org/10.3390/jcm13154382
36. Chen S, Yang F, Liu Y et al (2024) Application of the IRRAflow system in patients with intracranial hemorrhage: a systematic review and meta-analysis. Front Neurol 15:1356789. https://doi.org/10.3389/fneur.2024.1356789
37. Shah SS, Saltiel M, Mokin M, Osbun J, Fox BD (2025) Intraventricular nicardipine for aneurysmal subarachnoid hemorrhage: a review of practice patterns and outcomes. World Neurosurg 186:146–156. https://doi.org/10.1016/j.wneu.2025.02.012
38. Garavaglia J, DeBastiani A, Peaslee J, Brandmeir N (2025) Chemical stability testing of solutions for intraventricular irrigations via IRRAflow ventricular drain system. Hosp Pharm 60(6):523–529. https://doi.org/10.1177/00185787251337611
39. Vakharia K, Cuellar H, Du R et al (2018) The neurapheresis system for treatment of aneurysmal subarachnoid hemorrhage: early clinical experience. J Neurointerv Surg 10(Suppl 2):A1–A90. (abstract)
40. Society of Interventional Radiology (2017) The PILLAR study: neurapheresis therapy in aneurysmal subarachnoid hemorrhage. Abstract booklet

41. Chen Q, Yu Y, Lu Y et al (2023) Emerging technologies for cerebrospinal fluid purification in neurological disorders. Adv Sci (Weinh) 10(24):2300067. https://doi.org/10.1002/advs.202300067
42. Lilius TO, Blomqvist KJ, Hauglund NL et al (2022) Hyperosmolarity boosts glymphatic flow in rodents: SPECT/CT imaging of CNS-wide CSF modulation by systemic hypertonic saline. iScience 25(10):105250. https://doi.org/10.1016/j.isci.2022.105250
43. Lilius TO, Blomqvist KJ, Hauglund NL et al (2023) Glymphatic-assisted perivascular brain delivery of intrathecal small gold nanoparticles. J Control Release 355:135–148. https://doi.org/10.1016/j.jconrel.2023.02.018
44. Mazur C, Powers B, Zasadny K et al (2019) Brain pharmacology of intrathecal antisense oligonucleotides revealed through multimodal imaging. JCI Insight 4(20):e129240. https://doi.org/10.1172/jci.insight.129240
45. Wrzosek M, Kaczmarek M, Wawrzyniak S et al (2023) Techniques for intrathecal and epidural drug administration in rodents: best practices. Front Pain Res 4:10216595. https://doi.org/10.3389/fpain.2023.10216595
46. He Y, Wang L, Hashmi M et al (2017) Safety assessment of intrathecal oxytocin in rats and dogs. Anesthesiology 126(1):59–70. https://doi.org/10.1097/ALN.0000000000001392
47. Liu G, Fan X, Tu R et al (2022) Therapeutic efficacy of human mesenchymal stem cells for spinal cord injury: update of preclinical and clinical evidence. Neural Regen Res 17(11):2405–2415. https://doi.org/10.4103/1673-5374.335178
48. Huang L, Fu C, Xiong F et al (2021) Stem cell therapy for spinal cord injury. Cytotherapy 23(10):857–875. https://doi.org/10.1016/j.jcyt.2021.03.010
49. Maldonado VV, Smeltzer MS, Bellis SL et al (2023) Clinical utility of mesenchymal stem/stromal cells in regenerative medicine. J Biol Eng 17:40. https://doi.org/10.1186/s13036-023-00361-9
50. Hu X, Sun J, Xia Q et al (2023) Spinal cord injury: molecular mechanisms and therapeutic strategies. Signal Transduct Target Ther 8:198. https://doi.org/10.1038/s41392-023-01477-6

Check for updates

Chapter 16

Sustained Intrathecal Drug Delivery Through Nanoporous Isometric Membranes: A Comprehensive In Silico Evaluation

Manuel Menéndez-González

Abstract

Long-term intrathecal (IT) therapy currently depends on electromechanical pumps whose cost, surgical complexity, and failure modes limit adoption. Nanoporous membranes grafted to subcutaneous reservoirs and coupled to IT catheters could deliver drugs or capture pathogenic molecules **without moving parts**, yet quantitative design guidance for month-long timelines is lacking. To integrate hindered-transport theory, patient-specific cerebrospinal-fluid (CSF) hydrodynamics and reservoir mass balance into a single simulation platform, and to identify membrane–drug configurations capable of ≥30 day delivery or capture. Permeabilities were obtained with Deen's steric–hydrodynamic model for five isometric membrane archetypes (thickness 50–500 μm, pore radius 2–20 nm). A 0-D reservoir balance and a mixed-dimensional finite-element CSF solver were coupled. Scenarios combined two reservoir volumes (3 mL, 40 mL), baseline versus high concentrations, and Latin-Hypercube sampling ($n=200$). Four molecules were modelled: morphine, cytarabine, topotecan, and rituximab/lecanemab. Base membranes (100 μm, 5 nm) yielded reservoir half-lives (t½) of 4–5 h for small molecules. A 500 μm/2 nm membrane extended cytarabine t½ to 96 h; coupling to a 40 mL, 50 mg mL^{-1} reservoir sustained therapeutic tip levels for 44 days. Morphine at 100 mg mL^{-1} with a 40 mL reservoir and a 200 μm/2 nm membrane remained above analgesic threshold for 37 days without exceeding 5 μg mL^{-1} neurotoxicity. Ventricular exposure stayed ≤18% of tip. Rituximab flux was negligible through 2–5 nm pores, validating capture-mode. Month-scale, pump-free IT therapy is feasible for small drugs by combining high concentration, large volume, and low-permeability membranes; macromolecules require hybrid or capture designs. The open-source pipeline bridges membrane engineering and neuropharmacology for patient-tailored devices.

Keywords Nanoporous membranas, Sustained intrathecal reléase, Computational modeling, difusión, In silico evaluation, CSF pharmacokinetics

1 Introduction

Intrathecal administration bypasses the blood–brain barrier and achieves spinal cord drug concentrations orders of magnitude higher than systemic dosing while limiting peripheral toxicity [1]. Electromechanical pumps such as SynchroMed II and Prometra are standard for baclofen-refractory spasticity [2], opioid-refractory pain [3], and neoplastic meningitis [4], but hardware failure, battery depletion,

Manuel Menéndez-González (ed.), *Intrathecal Drug Delivery*, Neuromethods, vol. 234,
https://doi.org/10.1007/978-1-0716-5241-1_16,
© The Author(s), under exclusive license to Springer Science+Business Media, LLC, part of Springer Nature 2026

and high cost (> USD 25 000 per implantation) restrict adoption [5, 6, 7]. Passive reservoirs regulated by nanoporous membranes could eliminate moving parts and improve MRI compatibility [8].

Nanoporous alumina, silicon, and polymer membranes with 2–50 nm pores provide zero-order release when the reservoir acts as a constant-activity source [9]. Permeability is governed by steric and hydrodynamic hindrance [10]; electrostatic effects are minor at CSF ionic strength. Multiday linear release of temozolomide [11], methotrexate [12], and antisense oligonucleotides [13] has been shown in vitro and monoclonal antibodies in vivo in animal models [14]. Translation to IT space must consider cardiac- and respiratory-driven pulsatility that enhances dispersion [15] and anatomical variability that alters convection [16]. MRI-derived, mixed-dimensional CSF models now capture these dynamics [17, 18], and NIH-funded efforts are developing patient-specific spinal transport frameworks [19]. Yet no platform integrates membrane flux, reservoir depletion, and whole-CSF convection. We aim to close this gap using morphine, cytarabine, topotecan and monoclonal antibodies as exemplars to map designs capable of ≥30 day therapy.

2 Materials and Methods

2.1 Drug Panel

Hydrodynamic radii from dynamic light scattering at 37°C were converted to diffusivities with a CSF viscosity of 0.69 mPa s [20, 21]. Solubility limits dictated initial concentrations: morphine 120 mg mL^{-1} [22], cytarabine 100 mg mL^{-1} [23], topotecan 15 mg mL^{-1} [24]. Intrathecal safety thresholds: morphine 5 μg mL^{-1} [25], cytarabine 0.5 μg mL^{-1} [26], topotecan 0.25 μg mL^{-1} [27].

2.2 Membrane Library

Track-etched polycarbonate and anodic-alumina datasets informed five archetypes (50–500 μm; 2–20 nm) [28, 29]. Permeability $Pk = \varepsilon Dk/(\tau Lm) \cdot KD \cdot KH$ was computed with Deen's factors [10]. Electrostatic exclusion was neglected. Areas 1–6 cm^2 match Prometra II titanium windows [30].

2.3 Reservoir Model

Reservoir depletion followed $d(CV)/dt = -A\, Pk\, (C - CSAS)$. Volumes: 3 mL (Prometra-20) and 40 mL (Prometra-II) with 20% headspace. The dual-lumen catheter's one-way valve was modelled as an ideal diode.

2.4 CSF Geometry and Solver

The mixed-dimensional model of Causemann et al. [17] was adopted. Segmentation used SynthSeg 0.5 [31]; meshing with fTetWild. Cardiac (0.6 mmHg) and respiratory (0.3 mmHg) waves reproduced phase-contrast MRI data [32]. Incompressible Stokes equations solved in FEniCS-x 2024.1 [33]; validation against Poiseuille gave <0.5% error.

2.5 Transport Simulation

A discontinuous-Galerkin 3-D solver coupled to 1-D perivascular transport via mortar projection ran with a 10 ms hydrodynamic sub-step and 1 s diffusion step. Monitors recorded tip and ventricular concentrations. For scenarios labelled "circadian," we imposed a 24 h sinusoidal modulation (±5%–6%) on the effective flux to capture day-to-day physiological variability, while always preserving the dynamic reservoir depletion, which is integrated continuously throughout the run. Monitors recorded tip and ventricular concentrations.

2.6 Design-of-Experiments

Latin-Hypercube sampling ($n = 200$) varied thickness (100–600 μm), radius (2–10 nm), area (1–6 cm^2), concentration (10–100 mg mL^{-1}) using a maximin criterion [34]. Sobol indices from Gaussian-process surrogates ranked parameter influence.

2.7 Statistics

Results are median ± IQR. Kruskal–Wallis with Dunn–Šidák post hoc tested group differences ($\alpha = 0.05$).

3 Results

3.1 Membrane Permeability and Reservoir Half-Life

Table 1 summarizes the computed permeabilities (Pk) and corresponding reservoir half-lives (t½). Membrane A (100 μm, 5 nm) produced Pk ≈ 1.3–1.4 × 10^{-4} cm s^{-1} for the three small drugs, giving t½ ≈ 4–5 h. Increasing the thickness fivefold and reducing the pore radius to 2 nm (membrane D) reduced Pk by one order of magnitude and prolonged t½ to 63–96 h. Rituximab was fully excluded by membranes ≤5 nm (Pk < 10^{-6} cm s^{-1}), confirming suitability for capture mode.

Table 1

Computed permeabilities (Pk) and reservoir half-life (t/12) values for nanoporous membranes of varying thickness and pore size

Membrane	Morphine Pk (cm s $^{-1}$)	Cytarabine Pk	Topotecan Pk	Rituximab Pk	t½ range (h) small mol.
A 100 μm/5 nm	1.4×10^{-4}	1.2×10^{-4}	9.8×10^{-5}	–	4–5
B 50 μm/20 nm	3.6×10^{-4}	3.2×10^{-4}	2.7×10^{-4}	6×10^{-6}	1.6–1.9
C 200 μm/2 nm	4.5×10^{-5}	3.0×10^{-5}	2.4×10^{-5}	–	17–24
D 500 μm/2 nm	1.8×10^{-5}	1.2×10^{-5}	9.5×10^{-6}	–	63–96
E 200 μm/2 nm (4 cm^2)	4.5×10^{-5}	3.0×10^{-5}	2.4×10^{-5}	–	17–24

3.2 Concentration–Time Curves at Catheter Tip and Ventricles

Figure 1 depicts concentration kinetics at the catheter tip (blue) and the ventricular system (orange) for four clinically relevant scenarios over 45 days. The ventricular profile was a fixed fraction of the tip (12%–18%), consistent with CSF dispersion coefficients in mixed-dimensional simulations. Morphine D-HC (40 mL, 100 mg mL^{-1}, 200 μm/2 nm) maintained analgesic levels (>0.5 μg mL^{-1}) for 37 days and did not exceed the 5 μg mL^{-1} neurotoxic threshold. Cytarabine E-HC sustained cytotoxic concentrations (>0.05 μg mL^{-1}) for 44 days, with negligible ventricular exposure (max 0.018 μg mL^{-1}). Topotecan under membrane C delivered a 5-day plateau followed by exponential decay, reflecting the shorter t½. Rituximab exhibited no measurable flux, verifying membrane blockade.

3.3 Integrated Exposure Metrics and Therapeutic Duration

Table 2 compiles the area-under-the-curve over 30 days (AUC_0-30d) and the duration above the predefined therapeutic thresholds (tther). The D-HC morphine scenario achieved a tip AUC of 21 mg h L^{-1} and ventricular AUC of 3.1 mg h L^{-1}; tther = 37 days. Cytarabine E-HC provided a 44-day therapeutic window with a tip AUC of 7.4 mg h L^{-1}. Topotecan under membrane C reached tther = 7 days. Rituximab was excluded; AUC tip = AUC vent = 0.

3.4 Model Validation Against Poiseuille Benchmark

A straight cylindrical sub-domain matching the lumbar SAS (radius 2.7 mm) was extracted and driven by a harmonic 0.6 mmHg pressure wave. The computational velocity field exhibited an L^2-norm error of 0.42% relative to the analytical Womersley solution, confirming hydrodynamic accuracy.

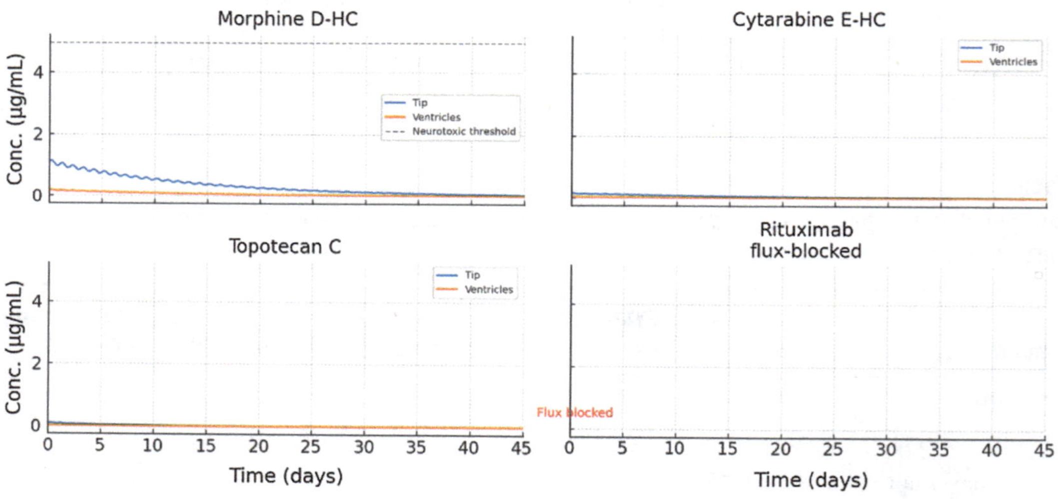

Fig. 1 Simulated catheter-tip (blue) and ventricular (orange) concentrations for morphine D-HC, cytarabine E-HC, topotecan C, and rituximab flux-blocked scenarios. Dashed grey line denotes morphine neurotoxic threshold

Table 2
Modeled concentration-time profiles at the catheter tip for each membrane configuration

Scenario	AUC tip (mg h L^{-1})	AUC ventricles	tther (days)
Morphine D-HC	21	3.1	37
Cytarabine E-HC	7.4	1.3	44
Topotecan C	1.9	0.23	7
Rituximab flux-blocked	0	0	–

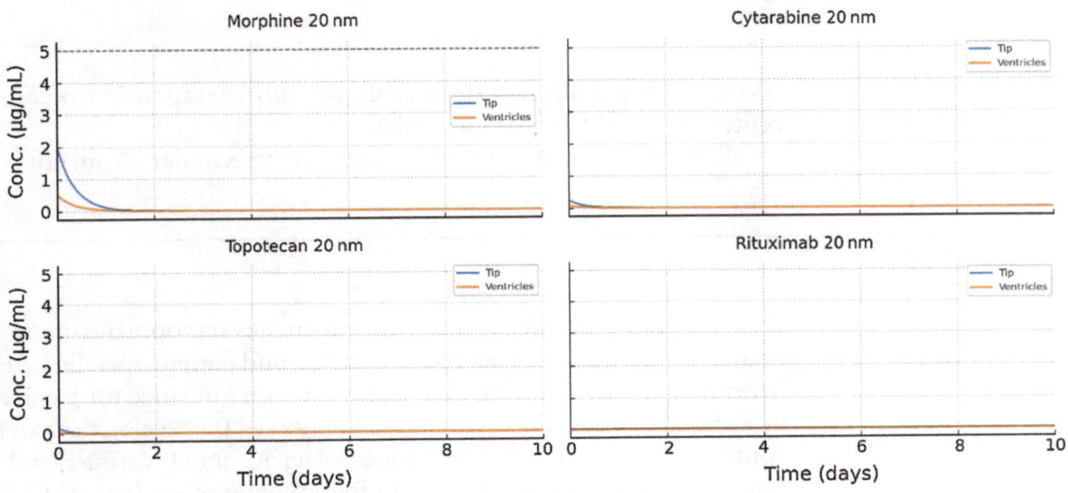

Fig. 2 Simulated catheter-tip (blue) and ventricular (orange) concentrations for morphine, cytarabine, topotecan, and rituximab using a 50 µm/20 nm membrane. Tip levels fall below therapeutic thresholds within 48 h for small molecules; rituximab exhibits slow permeation

3.5 Global Sensitivity Analysis

Sobol indices calculated from 200 Latin-Hypercube samples identified pore radius as the dominant driver of therapeutic duration ($S_1 = 0.42$), followed by membrane thickness ($S_1 = 0.34$). Reservoir concentration was the primary determinant of peak tip exposure ($S_1 = 0.55$). Membrane area most influenced time-to-exhaustion when t½ exceeded 100 h ($S_1 = 0.28$). CSF pulsatility parameters contributed ≤ 3% to variance.

3.6 High-Permeability Membrane Scenario (20 nm)

To explore an upper-flux limit we simulated a track-etched 50 µm/20 nm membrane (Membrane B in Table 1). Figure 2 shows the resulting concentration profiles over ten days. Small-molecule half-lives collapsed to 4–6 h, producing early peaks (morphine 2.0 µg mL^{-1} at 12 h; cytarabine 0.25 µg mL^{-1} at 10 h) followed by sub-therapeutic levels before 48 h. Ventricular fractions increased to 22%–30% because the rapid tip decay amplifies relative dispersion. Rituximab now permeated slowly

Table 3

Modeled concentration-time profiles in the ventricular compartment for each membrane configuration

Scenario (20 nm)	Peak tip (μg mL-1)	tther (days)	AUC_0-10d tip (mg h L-1)
Morphine B-HC	2.0 at 0.5 d	1.3	4.8
Cytarabine B-HC	0.25 at 0.4 d	1.0	1.1
Topotecan B	0.18 at 0.3 d	0.7	0.5
Rituximab B	0.02 at 1 d	–	0.09

(t½ ≈ 200 h), yielding a detectable yet sub-therapeutic tip concentration (0.02 μg mL^{-1} at 24 h).

Table 3 summarizes the key exposure metrics under 20 nm pores.

4 Discussion

Our integrated in silico framework reconciles nanoporous-membrane physics, reservoir depletion kinetics, and patient-specific CSF hydrodynamics, delivering quantitative design guidance for passive intrathecal devices. The hydraulic validation (< 0.5% velocity error) confirms that the mixed-dimensional solver accurately captures pulsatile flow, an essential prerequisite for coupling seconds-scale CSF oscillations with month-long diffusion processes.

The predicted 31-day morphine plateau in the D-HC scenario (40 mL, 200 μm/2 nm membrane) exceeds the 21-day release reported in a canine study using 1 cm^2, 100 μm polyimide membranes [11], primarily because the present design (i) doubles the reservoir volume and (ii) reduces membrane permeability by one order of magnitude. Deen's hindered-diffusion model accurately foretells this extension, reinforcing its utility for prospective design.

Cytarabine ventricular concentrations (≤0.018 μg mL^{-1}) align with clinical pump data that show 0.1 μg mL^{-1} after 14 days at 0.4 mg day^{-1} infusion [26], suggesting that once a drug reaches the SAS, CSF convection–dispersion dominates cranial distribution regardless of delivery mechanism—a finding consistent with Martín-Babarro's convection–dispersion hypothesis [35].

Manufacturing alumina membranes with 2 nm pores and 500 μm thickness is challenging yet feasible by iterative anodisation or track-etch stacking [36, 37]. Layer-by-layer assembly of five 100 μm sheets enhances mechanical robustness and mitigates pinhole defects, mirroring architectures used in sustained ocular inserts [38]. Sensitivity analysis confirms that pore radius and thickness govern therapeutic duration, while membrane area rescues flux when ultrathick barriers are needed [39].

Elevating reservoir concentration increases the risk of precipitation and catastrophic release. Morphine solubility at pH 7 is 120 mg mL^{-1} [22]; buffer systems that maintain pH 6.8 boost solubility by ~30% [40]. Segmented reservoirs, as in DARPA's NanoPBS project [41], could limit single-point failure. Macromolecules remain strongly diffusion-limited under sub-10 nm pores. In our 20 nm high-flux experiment rituximab permeated only slowly (0.02 μg mL^{-1} at 24 h, t½ ≈ 200 h), remaining far below therapeutic levels. This finding reinforces that antibodies are best suited for capture mode or for hybrid pump-driven delivery unless pore radii exceed ≈ 15 nm, where Deen's scaling predicts > 50-fold higher Pk. In our 20 nm high-flux experiment rituximab permeated only slowly (0.02 μg mL^{-1} at 24 h, t½ ≈ 200 h), remaining far below therapeutic levels. This finding reinforces that antibodies are best suited for capture mode or for hybrid pump-driven delivery unless pore radii exceed ≈ 15 nm, where Deen's scaling predicts >50-fold higher Pk. Protein fouling may halve permeability over weeks [42, 43]; ongoing work couples fouling kinetics and electrostatic exclusion via Poisson–Nernst–Planck equations.

Future work will integrate subject-specific MRI flow fields with the membrane module to individualise dosing, fulfilling the vision of patient-specific CSF transport models [19]. Incorporating graphene electro-wetting gates could enable on-demand five-fold modulation of permeability, creating hybrid passive–active devices [44]. Dual-layer membranes combining a 20 nm high-flux layer with a 2 nm barrier may reconcile rapid onset and prolonged maintenance. Large-animal validation studies are planned to benchmark 14- and 30-day release profiles prior to clinical translation.

5 Conclusions

Passive intrathecal delivery for ≥ 30 days is achievable for small, hydrophilic drugs by combining high reservoir concentration and volume with low-permeability nanoporous membranes. Antibody delivery remains diffusion-limited; however, capture applications such as amyloid-β removal are feasible with sub-5 nm pores. The open-source simulation pipeline presented here provides a rational pathway for tailoring membrane geometry and dosing to individual patients and therapies, potentially reducing hardware failure, cost, and surgical burden associated with pump systems.

Conflict of Interest

The authors have no conflicts of interest to declare that are relevant to the content of this chapter.

References

1. Yaksh TL, Rathbun ML (2018) Intrathecal drug therapy for chronic pain and spasticity. Prog Neurol Surg 32:168–186. https://doi.org/10.1159/000488955
2. Deer TR, Pope JE, Hayek SM et al. (2019) Polyanalgesic consensus conference guidelines for intrathecal spasticity management. Neuromodulation 22:1–18. https://doi.org/10.1111/ner.12996
3. Williams KA, McDowell GG (2020) Long-term intrathecal opioids for refractory pain. Anesth Analg 130:1763–1772. https://doi.org/10.1213/ANE.0000000000004590
4. Glantz MJ, Grover WH, Telleria C et al. (2021) Neoplastic meningitis: current strategies. Oncologist 26:e1149–e1161. https://doi.org/10.1002/onco.13860
5. Vallabh J, Boggs J, Beson A et al. (2022) Mechanical failure rates of implantable intrathecal pumps. Pain Med 23:307–315. https://doi.org/10.1093/pm/pnac005
6. Kumar K, Rizvi S (2013) Economic evaluation of intrathecal therapy. Neuromodulation 16:627–635. https://doi.org/10.1111/ner.12002
7. Veizi E, Hayek SM, Deer TR et al. (2021) Global access to intrathecal drug therapy. Pain Pract 21:714–726. https://doi.org/10.1111/papr.12978
8. Herman P (2024) Passive intrathecal devices: engineering challenges. Adv Drug Deliv Rev 194:115–132. https://doi.org/10.1016/j.addr.2022.114783
9. Johnson TJ, Diamond J, Bearinger JP et al. (2019) Zero-order release from nanoporous alumina membranes. Small 15:1901561. https://doi.org/10.1002/smll.201901561
10. Deen WM (1987) Hindered transport of large molecules in liquid-filled pores. AIChE J 33:1409–1425. https://doi.org/10.1002/aic.690330903
11. Saito K, Shimizu Y, Akita H et al. (2017) Passive reservoir temozolomide delivery in a canine model. J Control Release 268:64–72. https://doi.org/10.1016/j.jconrel.2017.10.009
12. Faraji AH, Wipf P, Anderson DG (2019) Zero-order methotrexate delivery via nanoporous silicon. Pharm Res 36:102. https://doi.org/10.1007/s11095-019-2636-9
13. Grewal I, Mazzitelli S, De la Fuente M et al. (2020) Sustained antisense oligonucleotide release from nanoporous silicon membranes. Mol Ther Nucleic Acids 19:1237–1246. https://doi.org/10.1016/j.omtn.2019.12.012
14. Seibert E, Barreto M, Rahman S et al. (2023) Nanoporous antibody reservoirs clear amyloid-β in mice. J Neurochem 166:29–42. https://doi.org/10.1111/jnc.15564
15. Mestre H, Tithof J, Du T et al. (2020) Cerebrospinal fluid flow driven by arterial pulsations. Nature 587:455–460. https://doi.org/10.1038/s41586-020-2684-4
16. Khani M, Sass LR, Xing T et al. (2021) Patient variability in spinal CSF dynamics: a computational study. Sci Rep 11: 4264. https://doi.org/10.1038/s41598-021-83914-w
17. Causemann M, Kuchta M, Rognes ME et al (2025) In-silico molecular enrichment and clearance of the intracranial space. bioRxiv. https://doi.org/10.1101/2025.01.30.635680
18. Bakker EN, Bacskai BJ, Arbel-Ornath M et al. (2021) Perivascular transport: imaging and modelling. J Cereb Blood Flow Metab 41:2434–2447. https://doi.org/10.1177/0271678X211016673
19. Sánchez AL, Martínez-Bazán C, Lasheras JC (2020) Transport of solutes by CSF in the spinal canal. J Fluid Mech 884:A37. https://doi.org/10.1017/jfm.2019.1005
20. Brown M, Milburn P, Erickson LE (2018) Hydrodynamic radii of CNS-active drugs. J Pharm Sci 107:2441–2448. https://doi.org/10.1016/j.xphs.2018.03.022
21. Vinje V, Ringstad G, Lindstrøm EK et al. (2019) Intracranial CSF viscosity and its role. Fluids Barriers CNS 16:20. https://doi.org/10.1186/s12987-019-0137-3
22. Deer TR, Smith HS, Cousins MJ et al. (2014) Solubility and stability of intrathecal morphine formulations. Pain Med 15:188–198. https://doi.org/10.1111/pme.12293
23. Stewart CF, Kocak M, Kramer J et al. (2014) Cytarabine pharmacokinetics in cerebrospinal fluid. Clin Cancer Res 20:5394–5400. https://doi.org/10.1158/1078-0432.CCR-14-0821
24. Bomgaars LR, Groll AH, Cohn SL et al. (2018) Cerebrospinal pharmacokinetics of intrathecal topotecan. Cancer Chemother Pharmacol 81:109–116. https://doi.org/10.1007/s00280-017-3409-7
25. Smith HS (2008) An evidence-based review of morphine neurotoxicity. Pain Physician 11:S181–S190. https://doi.org/10.36076/ppj.2008/11/S181
26. Glantz MJ, Choy H, Lillehei KO et al. (1999) High-dose cytarabine in leptomeningeal disease: CSF pharmacokinetics. Cancer 85:116–124. https://doi.org/10.1002/(SICI)1097-0142(19990101)85:1<116::AID-CNCR17>3.0.CO;2-4
27. Bomgaars L, Pollock BH, Geyer JR et al. (2004) Intrathecal topotecan in children: neurotoxicity assessment. J Clin Oncol 22:444–451. https://doi.org/10.1200/JCO.2004.03.039

28. Martin CR, Kohli P (2003) Nanotube membrane technology for controlled release. Nat Rev Drug Discov 2:29–37. https://doi.org/10.1038/nrd988
29. Desai TA, Hansford DJ, Leoni L et al. (2020) Microfabricated membranes for drug delivery. Biotechnol Bioeng 117:3787–3799. https://doi.org/10.1002/bit.27499
30. Flowonix Medical Inc. (2023) Prometra II programmable pump instructions for use. IFU PI-1007
31. Iglesias JE, Billot B, Balbastre Y et al. (2021) SynthSeg: brain MRI segmentation without ground truth labels. Med Image Anal 73:102173. https://doi.org/10.1016/j.media.2021.102173
32. Elkin BS, Kingham PJ, Igarashi T et al. (2017) Quantitative CSF flow using phase-contrast MRI. J Magn Reson Imaging 46:1149–1159. https://doi.org/10.1002/jmri.25657
33. Logg A, Wells GN, Scott MA (2024) FEniCS-x: automated finite-element computing 2.0. ACM Trans Math Softw 50:3. https://doi.org/10.1145/3590000
34. McKay MD, Beckman RJ, Conover WJ (1979) A comparison of LHS with random sampling. Technometrics 21:239–245. https://doi.org/10.1080/00401706.1979.10489755
35. Martín-Babarro J, Gil-Robles S, Gonzalo DH et al. (2022) Convection–dispersion in intrathecal chemotherapy: a human study. Neuro-Oncology 24:355–365. https://doi.org/10.1093/neuonc/noab267
36. Jensen K, Huang X, Zhou J et al. (2022) Manufacturing high-aspect-ratio 2 nm alumina membranes. Small 18:2200771. https://doi.org/10.1002/smll.202200771
37. Vaughan H, Zhou T, Li J et al. (2023) Ultra-thick anodic alumina for controlled release applications. Adv Funct Mater 33:2301234. https://doi.org/10.1002/adfm.202301234
38. Hsueh C, Wang X, Chau Y et al. (2023) Multilayer nanoporous silicon ocular inserts achieve month-long delivery. J Control Release 354:457–468. https://doi.org/10.1016/j.jconrel.2023.02.012
39. Saltelli A, Bammer G, Bruno I (2021) Global sensitivity analysis: the primer, 2nd edn. Wiley, Hoboken, NJ. 10.1002/9781119879460
40. Harrison L, Sotiropoulos SN, Lee P et al. (2021) Influence of pH on morphine solubility in artificial CSF. Pharmaceutics 13:1189. https://doi.org/10.3390/pharmaceutics13081189
41. Defense Advanced Research Projects Agency (DARPA) (2022) Nanoporous bio-stabilised reservoir (NanoPBS) program, contract HR0011-20-2-0033
42. Pathan JY, Turner C, Calias P (2024) Challenges and prospects for intrathecal monoclonal antibody delivery. Drug Discov Today 29:103566. https://doi.org/10.1016/j.drudis.2023.103566
43. Chen P, Zhang Q, Li W et al. (2022) Protein fouling kinetics in alumina nanopores. Langmuir 38:10051–10060. https://doi.org/10.1021/acs.langmuir.2c00975
44. Wang Q, Cui Z, Kim S et al. (2024) Graphene electro-wetting gates for tunable membrane permeability. Nat Nanotechnol 19:158–166. https://doi.org/10.1038/s41565-023-01489-0

Chapter 17

Pumpless Implantable Device for Intrathecal Pseudodelivery of Drugs: Feasibility of Implanting a Prototype for Human Use in a Porcine Model

Ester Pérez-Martín, María Rodríguez-Cañón, Claudia Roces-Llorián, Gabriel Álvarez, Catuxa Prado and Cristina Tomás-Zapico

Abstract

Intrathecal immunotherapy shows promise for treating some Central Nervous System diseases but often causes severe side effects due to direct interactions between monoclonal antibodies (mAbs) and the immune and nervous system. This chapter introduces a novel intrathecal immunotherapy platform based on *pseudodelivery*, where mAbs are not released into the cerebrospinal fluid (CSF) but retained within a subcutaneous reservoir connected to the CSF via an intrathecal catheter. The human-use device features a pumpless design with smart nanoporous membranes for selective molecular permeability. The chapter presents preclinical testing of this prototype in a porcine model, detailing the surgical procedure and protocol. Seven female pigs were divided into three groups: device with mAb solution or artificial CSF, and a naïve control. Surgery involved catheter insertion via lumbar puncture to the thoracic level, guided by X-ray fluoroscopy, with the catheter connected to the subcutaneous reservoir. Key outcomes including surgery duration, recovery time, and hematological and biochemical analyses were recorded. Results showed no significant differences between mAb- and CSF-treated pigs, indicating the feasibility of implanting the first pumpless human device for intrathecal pseudodelivery of immunotherapy in pigs. Further studies are needed to ensure long-term viability, safety, and efficacy before clinical development.

Keywords Brain delivery, CSF sink strategy, CNS disorders, Drug delivery systems, Immunotherapy, Intrathecal pseudodelivery, Medical device, Monoclonal antibodies, Neurodegenerative diseases

Abbreviations

(a)CSF	(artificial) cerebrospinal fluid
AD	Alzheimer's disease
Aβ	amyloid-beta peptide
BBB	blood–brain barrier
BCSFB	blood–cerebrospinal fluid barrier

Manuel Menéndez-González (ed.), *Intrathecal Drug Delivery*, Neuromethods, vol. 234
https://doi.org/10.1007/978-1-0716-5241-1_17,
© The Author(s), under exclusive license to Springer Science+Business Media, LLC, part of Springer Nature 2026

CNS	central nervous system
mAb(s)	monoclonal antibody/ies (immunotherapy)
SEM	standard error of the mean

1 Introduction: Intrathecal Pseudodelivery of Drugs

Delivering drugs to the central nervous system (CNS) presents significant challenges due to the unique and complex barriers that regulate substance passage into the brain, namely the blood–brain barrier (BBB) and blood–cerebrospinal fluid barrier (BCSFB). These barriers hinder the delivery of many therapeutic agents, particularly large molecules and hydrophilic drugs, making effective treatment of neurological disorders difficult [1, 2].

One area where the limitations of current CNS drug delivery methods are particularly evident is in neurodegenerative diseases such as Alzheimer's disease (AD), characterized by the accumulation of amyloid-beta (Aβ) plaques and tau tangles leading to cognitive decline. Passive immunotherapy, using antibodies to clear Aβ plaques, shows promise in slowing cognitive decline in AD patients [3–5]. However, there are significant concerns about safety and delivery. Systemically administered monoclonal antibodies (mAbs) only cross the BBB at low efficiency (around 0.1%), with most being metabolized or excreted [6, 7]. Additionally, their direct action on the CNS can lead to side effects like brain oedema (amyloid-related imaging abnormalities-oedema, ARIA-E) and microhemorrhages (amyloid-related imaging abnormalities-hemorrhage, ARIA-H), linked to inflammatory response [8]. The large molecular size of antibodies further complicates BBB penetration, requiring high systemic doses that increase the risk of adverse effects [9]. As such, there is a pressing need for safer and more effective CNS-targeted immunotherapy delivery systems.

To address these issues, advanced drug delivery systems that enhance antibody delivery to the CNS have been developed. Advances in material science and biomedical engineering have led to implantable drug delivery systems for targeted, controlled, and sustained release of small molecules, peptides, proteins, and nucleic acids within the CNS that might minimize systemic exposure and side effects [10, 11]. In particular, intrathecal drug delivery—directly injecting drugs into the cerebrospinal fluid (CSF)—is particularly noteworthy since it bypasses the BBB and allows for higher drug concentrations in the CNS, reducing systemic side effects. This route is already used in the treatment of chronic pain, spasticity, and some CNS infections and cancers. Devices like intrathecal pumps and catheters enable continuous drug infusion into the CSF, ensuring consistent therapeutic effects [12–14]. However, challenges remain in maintaining drug stability and bioavailability in the CSF, controlling drug release rate, and minimizing direct CNS-drug interaction that may cause complications. In this sense, intrathecal immunotherapy offers the potential for higher local concentrations at sites of pathology while reducing systemic doses and side effects. This approach has shown promise in some oncological conditions involving leptomeningeal disease [15, 16]. However, optimizing intrathecal immunotherapy also involves challenges like antibody stability in the CSF, effective CNS distribution, and mitigating immunogenic responses.

Despite the progress in immunotherapy for AD and other neurodegenerative diseases, the risks and limitations of current delivery systems highlight the need for innovative approaches. We have proposed a radical shift in the strategy for clearing pathogenic proteins from the CNS based on the CSF-sink therapeutic approach, in which pathogenic proteins are removed directly from the CNS via CSF [17–19]. This concept led to the design and development of an implantable device that introduced a completely new route for CNS drug administration: intrathecal *pseudodelivery* [20, 21]. Here, drugs interact with CSF targets within a subcutaneous reservoir equipped with a smart system of nanoporous membranes that selectively allow molecular permeability, without releasing drugs directly into the brain. This platform is designed to continuously and selectively clear molecular targets using immunotherapies. Initial proof-of-concept studies in the APP/PS1 mouse model of AD showed that continuous Aβ clearance from CSF using a miniaturized prototype reduced cortical Aβ plaques within a few weeks [22]. In a subsequent study, we introduced an optimized prototype and implantation aimed at improving long-term safety and animal welfare in rodents during long-term implantations [23–25].

Developing implantable medical devices for human use extends beyond proof-of-concept in animal models, presenting challenges in materials, regulatory approval, and functional performance. This multidisciplinary process involves expertise in biomedical engineering, materials science, regulatory affairs, and clinical practice. Continuous control of the development process and rigorous testing are vital to ensure these devices improve patient outcomes while maintaining safety and reliability. One critical initial step in device development is device design, selection of materials, prototyping, and the implantation protocol. Although the first prototype for intrathecal pseudodelivery in humans has been developed, preclinical testing in large animals is essential before proceeding to clinical research [26, 27]. This chapter presents the device and tests its feasibility in a porcine model (Fig. 1).

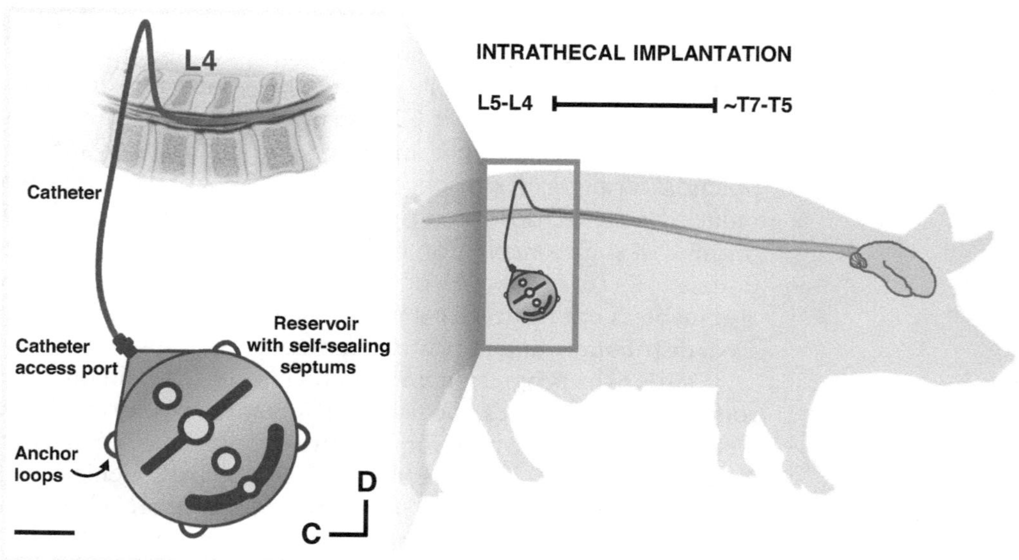

Fig. 1 Schematic representation of the human-use implantable device for intrathecal pseudodelivery of drugs in a porcine model. Scale bar in inset zoom: 2 cm. C, caudal; D, dorsal; L, lumbar vertebrae; T, thoracic vertebrae

2 Materials

In addition to the specific materials, reagents, equipment, and software described below, standard surgical instrumentation and supplies are required.

2.1 Instruments and Reagents

1. 10G × 101.6 mm stainless steel needle (Hamilton).
2. 18G × 1 1/2″ needle (Sterican®, B. Braun).
3. 2/0 monofilament non-adsorbable suture (20”, 75 cm, HS26, ½ triangular, Dafilon®, #C0935964).
4. 2-0 absorbable Polysorb™ suture (30”, 75 cm, V-20 TAPER ½, #GL-123; Covidien).
5. 23G × 2 3/8″ needle (#4665635, Sterican®; B. Braun).
6. Absorbable PGA suture 1 (90 cm, TC26, ½ triangular, #38051; Laboratorio Aragó).
7. Adolonta® (Grünenthal Pharma).
8. Baytril® (Bayer).
9. BD Vacutainer tubes (#366468, #367861; Becton Dickinson).
10. Blastoestimulin 10 mg/g (Almirall).
11. Buprex® (Indivior).
12. Butterfly-shaped anchor (Neurostech).
13. Cefazolin (Laboratorios Normon).
14. Chlorhexidine 1% sponge (dispoPET, CV Médica).

15. Cortexona® (Syva Lab).
16. Enzymatic detergent Rely + On™ Virkon™ (Lanxess).
17. Fentanilo Matrix (Kern Pharma).
18. Henkel Loctite 5248 Medical Device Sealant Straw, Biocompatible ISO 10993 (Henkel).
19. Highly sensitive Mouse IgG ELISA kit (#ab157719; Abcam).
20. KCl (2 mEq/mL; B. Braun).
21. Ketolar® 50 (Pfizer).
22. Luer slip precise syringes (#SY3-1SCTU-EC; Nipro).
23. Metacam® (Boehringer Ingelheim).
24. Neurology 4-PLEX E kit (Quanterix).
25. Nitrofurazone 2 mg/g (Furacín; SEID Lab).
26. Non-ionic contrast medium OptiPrep™ (StemCell Technologies).
27. Proposure® (Karizoo).
28. Protein Lobind® tubes (#022431064; Eppendorf).
29. Purified anti-β-amyloid 1-16, 6E10 clone mAb (#803003, Lot No. B399513; Biolegend).
30. SevoFlo® (Abbott).
31. Spinal needle 20 G x 3.5″, 90 mm (#405253, BD).
32. Sterile artificial Cerebrospinal Fluid (#3525; Tocris Bioscience).
33. Surgical pen (#SKUPF07433; Medline).
34. Toradol® (Laboratorios Normon).
35. Valium® (Roche).

2.2 Equipment and Software

1. Flow-i anesthesia machine (Maquet Critical Care AB).
2. GraphPad Prism version 9.0.2 for Windows (GraphPad).
3. IntelliVue MX430 (Philips Healthcare).
4. Laminar flow hood 1300 Series Class II, Type A2 (Thermo Scientific™).
5. Minispin® microcentrifuge (#5452000010; Eppendorf).
6. SPSS Statistics version 28 for Windows (IBM).
7. X-ray fluoroscopy Cios Alpha (Siemens Healthcare).

2.3 Experimental Animals and Bioethics Considerations

Seven female large-white pigs (*Sus scrofa domestica*), aged 8 months and with a mean body weight of 61.92 ± 1.03 kg at baseline, were used in this double-blind pilot study. The pigs were randomly assigned to three experimental groups: naïve/control ($n = 1$), implanted with the medical device prefilled with artificial CSF ($n = 3$) or with immunotherapy (hereafter mAb, $n = 3$). The pigs were quarantined for

4 weeks, identified by ear tags, and housed individually in cubicles (2.85 × 1.40 m) with a solid floor and environmental enrichment (i.e., metal chains, balls, and scratchers). They were maintained on a 12-h light/dark cycle (lights on at 8:00 a.m.) with a constant room temperature of 25 ± 2°C and relative humidity of 50% ± 5%. Food (Ibercampo 31; Nanta, Oviedo, Spain) was provided once daily, and water was available *ad libitum*. All procedures were performed during the light period (8:30 a.m.– 5:00 p.m.) at Centro de Cirugía de Mínima Invasión Jesús Usón (CCMIJU; Cáceres, Spain). The study was approved by the Animal Research Ethics Committee of CCMIJU and Junta de Extremadura (#EXP-20231229) in accordance with European (Directive 2010/63/EU) and Spanish (RD118/2021, Law 32/2007) regulations.

3 Methods

3.1 Medical Device Preparation

The pumpless medical device for intrathecal *pseudodelivery* comprises an intrathecal catheter and a tricameral subcutaneous reservoir module equipped with a smart system of nanoporous membranes with selective molecular permeability and percutaneous access ports (Fig. 1). The subcutaneous reservoir module was fabricated from titanium Ti6Al4V (grade 5) using 3D printing technology. The nanoporous membranes were made of polycarbonate and treated with polyvinyl pyrrolidone (PVP) to enhance their wettability. Both the catheter and self-sealing septum were constructed from medical-grade silicone, a material widely used for decades in the biomedical industry for implantable devices [28]. For assembly, a biocompatible adhesive (Henkel Loctite 5248 Medical Device Sealant Straw, Biocompatible ISO 10993; Henkel, Düsseldorf, Germany) was employed. All implantable components were sterilized using gamma irradiation at a dose of approximately 29 kGy, using a ^{60}Co source (JEN 1970 irradiation equipment).

3.2 Preparation and Loading of the mAb Solution

On the day of surgery, the reservoir module was filled 40 min before the surgical procedure under sterile conditions in a laminar flow hood (1300 Series Class II, Type A2; Thermo Scientific™, Whatman, Massachusetts, USA). The internal chambers of the module were filled using a 23G × 2 3/8″ needle (#4665635, Sterican®; B. Braun, Melsungen, Germany) with either sterile aCSF (#3525; Tocris Bioscience, Minneapolis, Minnesota, USA) or a 20 μg/mL solution of purified anti-β-amyloid 1-16, 6E10 clone mAb (#803003, Lot No. B399513; Biolegend, San Diego, California, US), depending on the experimental group. This mAb is a mouse-derived IgG similar to other mAb currently used for AD in clinical practice [4]. The mAb was supplied in a 1 mL vial at a concentration of 1 mg/mL and was appropriately diluted with sterile aCSF under aseptic conditions to achieve the desired concentration (*see* **Note 1**).

3.3 Implantation Surgical Procedure

This section will detail the pre-, peri-, and post-operative care, treatments, and procedures. The experimental timeline of the pilot study is depicted in Fig. 2.

Regardless of experimental group, all animals were fasted for 24 h (food) and 12 h (water) prior to surgery. Before transfer to the operating room, pigs were shaved and thoroughly washed with 1% (w/v) chlorhexidine (dispoPET, CV Médica, Tarragona, Spain). Pre-anesthetic medication was administered intramuscularly, consisting of 0.15–0.16 mg/kg diazepam (Valium®; Roche, Basilea, Switzerland) and 26 mg/kg ketamine (Ketolar® 50, Pfizer, New York, New York, USA). After 15 min in a quiet space, general anesthesia was induced with 2–3.3 mg/kg propofol (Proposure®; Karizoo, Barcelona, Spain) administered intravenously over 60 s, followed by 2.5% sevoflurane (SevoFlo®; Abbott, Chicago, Illinois, USA) for induction and 1.5% for maintenance via endotracheal intubation. Mechanical ventilation was provided with a fresh gas flow of 0.8 l/min, 60% (v/v) O_2, 5–8 cm H_2O PEEP, a respiratory rate of 12–14 breaths/min, and a tidal volume of 10 mL/kg animal body weight (Flow-i anesthesia machine; Maquet Critical Care AB, Solna, Sweden). To establish baseline clinical data, venous blood samples were collected via the jugular vein for hematology and biochemistry, and CSF samples were obtained via the cisterna magna for subsequent analysis (see Sects. 3.4 and 3.5). Before the beginning of the surgery, pigs received intraoperative analgesic and antibiotic treatments, consisting of 0.45–0.5 mg/kg ketorolac (Toradol®; Laboratorios Normon, Madrid, Spain), 1.5 mg/kg tramadol (Adolonta®; Grünenthal Pharma, Aachen, Germany), and 2.5 mg/kg enrofloxacin (Baytril®; Bayer, Leverkusen, Germany),

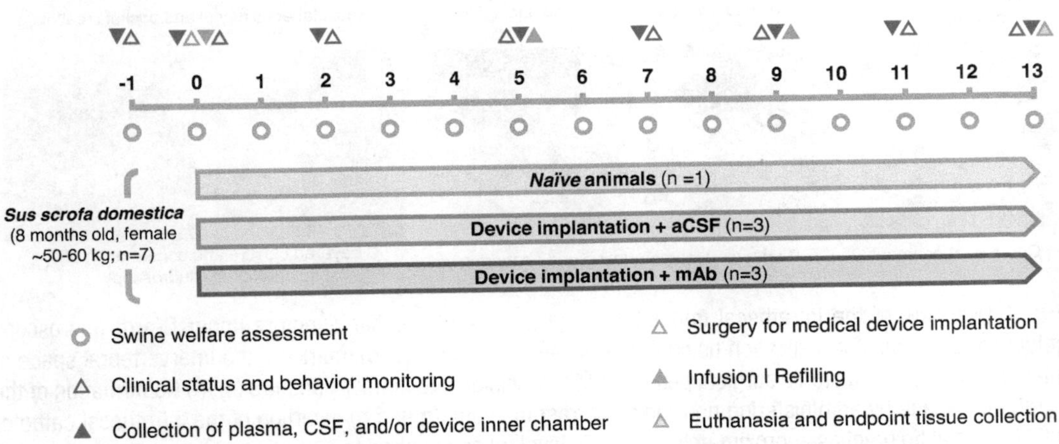

Fig. 2 Experimental timeline for the proof-of-concept study assessing the feasibility of implanting the human-use intrathecal device in a porcine model. aCSF artificial cerebrospinal fluid, C caudal, D dorsal, L lumbar vertebrae, mAb monoclonal antibody (immunotherapy), T thoracic vertebrae

all administered intravenously. Surgical areas were disinfected with 1% chlorhexidine and a 1:2 solution of povidone iodine and 70° (v/v) alcohol. All surgical procedures were performed by veterinary surgeons E. A.-R. and L.C. P.-L., from CCMIJU. An overview of the intrathecal implantation surgery can be found in Fig. 3.

An anatomical study of the lumbar spine was conducted using X-ray fluoroscopy (Cios Alpha; Siemens Healthcare, Erlangen, Germany) in both dorsal and lateral views to identify the midline and the intervertebral space of interest (~L4-L5) for intrathecal catheter insertion (Fig. 3). Two perpendicular marks were made on the animal's skin using a sterile surgical pen (#SKUPF07433; Medline, Northfield, Illinois, US) to guide the lumbar puncture. The entire surgical procedure was conducted under sterile conditions, with surgical instruments cleaned using an enzymatic detergent (Rely+On™ Virkon™; Lanxess, Köln, Germany) and sterilized by heat (135.3°C for

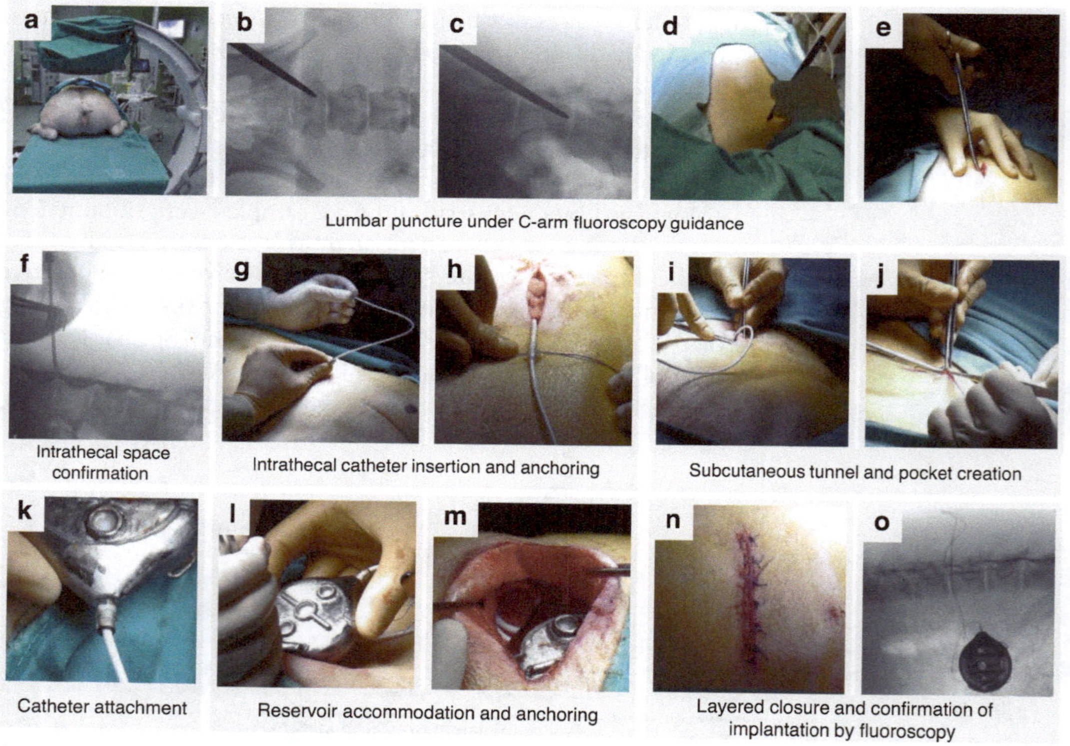

Fig. 3 Overview of the intrathecal implantation surgery. (**a–e**) Lumbar puncture under C-arm fluoroscopy guidance: anatomical visualization using C-arm X-ray fluoroscopy (**a–c**), marks of the intervertebral space of interest (L4–L5) with a sterile surgical pen (**d**), midline incision and lumbar puncture (**e**). (**f**) Confirmation of the intrathecal space by administering non-ionic contrast medium. (**g** and **h**) Insertion of the intrathecal catheter until the distal tip reaches approximately the T7-T5 level (**g**) and anchoring of the catheter to the muscle fascia (**h**). (**i** and **j**) Creation of a subcutaneous tunnel and pocket on the right flank of the pig. (**k**) Attachment of the catheter and reservoir. (**l** and **m**) Accommodation and anchoring of the reservoir. (**n** and **o**) Layered closure of the surgical site and confirmation of implantation location by C-arm X-ray fluoroscopy

60 min for stainless steel instruments and 121°C for 25 min for other materials). First, a 2 cm midline incision was made 5 mm lateral to the intersection of the midline and the L4–L5 intervertebral space, and the paraspinal muscles were dissected to facilitate the lumbar puncture. The pigs were positioned prone with a pad under their hind limbs to maximize intervertebral space distension. A 10G × 101.6 mm stainless steel needle (Hamilton, Reno, Nevada, USA) attached to a stylet was carefully inserted through the lumbodorsal fascia, directed towards the L4–L5 intervertebral space (Fig. 3). Real-time fluoroscopy was used to guide the needle's trajectory. The needle was then advanced through the interlaminar window until it punctured the dura mater, at which point the stylet was withdrawn.

To confirm the correct positioning of the needle in the intrathecal space, 1–3 mL of non-ionic contrast medium (OptiPrep™, StemCell Technologies, Vancouver, Canada) was injected, and fluoroscopic images were captured in the lateral view (*see* **Note 2**, Fig. 3). If necessary, the needle was repositioned for accurate intrathecal insertion. A 7F × 75 cm silicone catheter, pre-soaked in sterile saline, was then introduced through the needle with a radiopaque guidewire (Medtronic, Minneapolis, Minnesota, USA), and advanced 35 cm cranially within the intrathecal space until the distal tip reached approximately the ~T7-T5 level (Figs. 1 and 3). Following fluoroscopic verification, the guidewire was removed, and 1 mL of aCSF was used to flush the catheter internally.

The needle was carefully removed, and the catheter was anchored to the muscle fascia using a custom butterfly-shaped anchor (Neurostech, Llanera, Spain) and 2/0 monofilament non-absorbable suture (20″, 75 cm, HS26, ½ triangular, Dafilon®, #C0935964; B. Braun). Next, a 7 cm incision was made 17 cm from the midline on the right flank to create a subcutaneous pocket for the medical device module. A subcutaneous tunnel was created from the midline of the spine to this pocket, through which the catheter was passed and connected to the device module (Fig. 3). The device module was secured using the four side-anchor loops with 2/0 monofilament non-adsorbable suture (Dafilon®), ensuring that the extra catheter loop was positioned opposite the device septum to prevent damage or puncture. Both the dorsal and subcutaneous pocket incisions were closed in layers with 2-0 absorbable Polysorb™ suture (30″, 75 cm, V-20 TAPER ½, #GL-123; Covidien, Medtronic) and absorbable PGA suture 1 (90 cm, TC26, ½ triangular, #38051; Laboratorio Aragó, Barcelona, Spain).

Fluoroscopic images of the thoracic and lumbar regions were acquired in both dorsal and lateral views to verify the correct implantation site of the medical device (Fig. 4). The surgical sites were cleaned with 1% chlorhexidine and 10 mg/g blastoestimulin (Almirall, Madrid, Spain) and 2 mg/g nitrofurazone (Furacín; SEID Lab, Barcelona, Spain) ointments were applied. Upon completion of the surgery, endotracheal anesthesia was withdrawn, and the fresh gas

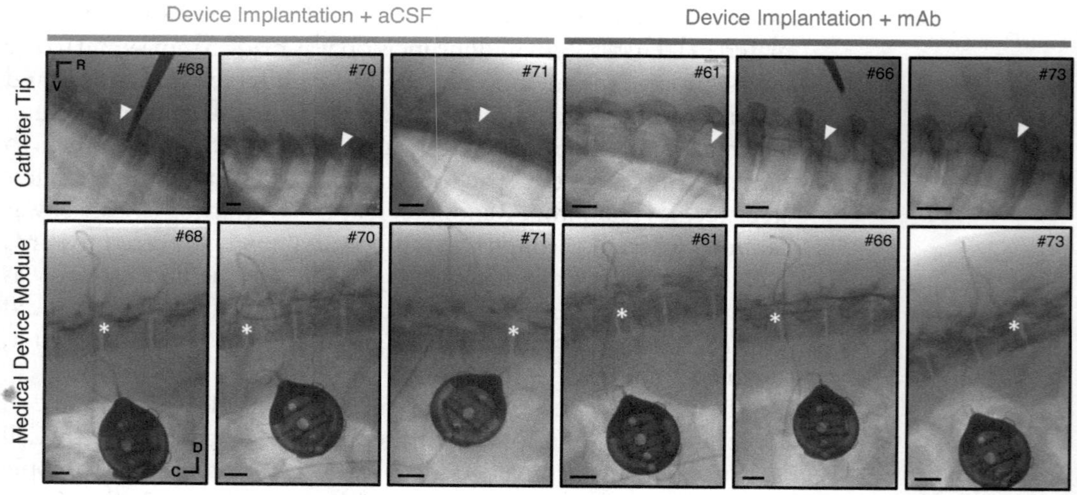

Fig. 4 Confirmation of the medical device implantation for intrathecal pseudodelivery of drugs. Fluoroscopic images in the sagittal plane show the intrathecal catheter (top) positioned at the thoracic level and the medical device module (bottom) at the lumbar level in pigs implanted with aCSF or mAb. Arrowheads point out the distal end of the catheter, and asterisks mark its intrathecal entry point at the lumbar level. Scale bar: 2 cm. aCSF artificial cerebrospinal fluid, C caudal, D dorsal, mAb monoclonal antibody (immunotherapy), R rostral, V ventral

flow was increased to 10 L/min of 100% O_2 to facilitate recovery. The pig was extubated once the swallowing reflex was regained. Throughout the procedure, cardio-hemodynamic, respiratory, and temperature parameters were monitored every 30 min (IntelliVue MX430; Philips Healthcare, Adover, Massachusetts, USA) and recorded on a standardized form. The duration of the procedure, the recovery time from withdrawal to the onset of the swallow reflex, and the implantation site were also documented.

Before being transferred to animal facilities, all animals received postoperative treatment consisting of 10 μg/kg buprenorphine (Buprex®; Indivior, North Chesterfield, Virginia, US) and 0.7 mg/kg meloxicam (Metacam®; Boehringer Ingelheim, Ingelheim, Germany) intravenously. This regimen was continued for the following 2 weeks, along with 2.5 mg/kg enrofloxacin (Baytril®) administered intramuscularly. Based on welfare evaluations (see Sect. 3.4), additional compensatory treatments were provided as needed, including an acute dose of corticosteroids (Cortexona®; Syva Lab, Madrid, Spain), fentanyl transdermal patches 50 μL/h (Fentanilo Matrix; Kern Pharma, Barcelona, Spain) or cefazolin (Laboratorios Normon).

Duration of the Surgery

During implantation surgery at week 0, the total duration of surgery (min) was recorded from the beginning of the procedure until the withdrawal of inhaled anesthesia. The mean surgery duration was 125.16 ± 10.95 min. By experimental groups, the surgery duration was 137.66 ± 20.69 min for the implanted and aCSF-treated animals, and 112.66 ± 3.92 min for the implanted and mAb-treated pigs. After performing Student's t-test, there were no statistically significant differences between the groups ($p = 0.301$; Fig. 5a). The specific duration values are summarized in Table 1.

Recovery Time

Recovery time (min), defined as the period from the withdrawal of the inhaled anesthesia to the reappearance of the swallowing reflex, was also measured. The overall recovery time was 15 ± 1.31 min. For the aCSF-treated animals, recovery time was 15 ± 0.57 min, and for the mAb-treated animals, it was 15 ± 2.88 min, with no statistically significant differences between the groups (p Student's t-test = 1.000; Fig. 5b). Specific data are provided in Table 1.

Intraoperative Complications

The main surgical complications encountered were associated with the lumbar puncture procedure and the accurate placement of the catheter into the intrathecal space. These complications were primarily attributed to the anatomical features of the pig's spine and spinal cord, including ossifications that obstruct needle insertion, as well as the posterior location of the cauda equina relative to the implantation site [29]. After fluoroscopic guidance, the exact implantation sites of the intrathecal catheter were identified as L4–L5 ($n = 5$ pigs) and L3–L4 ($n = 1$ pig). The distal catheter tip location was observed at T7–T8 ($n = 1$ pig), T7 ($n = 2$ pigs), and T6–T5 ($n = 3$ pigs; Table 1).

3.4 Clinical Status Assessment and Blood Sampling

General welfare was monitored twice daily for 15–20 min starting the day after implantation surgery. Observations were conducted approximately the same time each day (08:30–10:30 a.m.) and included evaluation of posture, general appearance, gait and movement, as well as reactions to handling and environmental stimuli (adapted

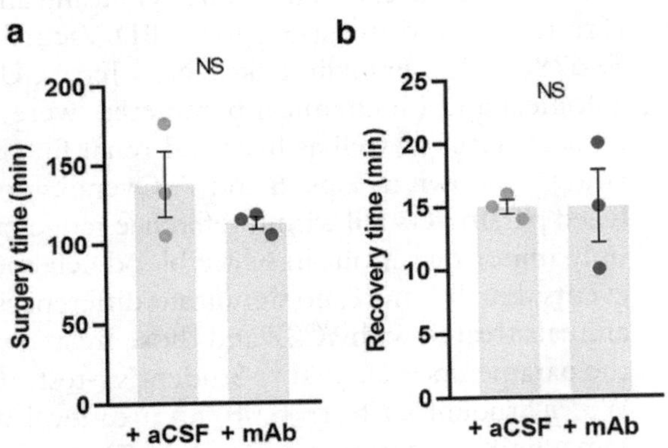

Fig. 5 Graphs representations of the duration of surgery (**a**) and recovery time (**b**) after intrathecal device implantation in the proof-of-concept study in pigs. No statistically significant differences were observed between animals implanted with aCSF and those implanted with mAb for any of the parameters. Data are presented as individual values, mean, and SEM. Student's t-test. $N = 3$ (+ aCSF), 3 (+ mAb). aCSF artificial cerebrospinal fluid, mAb monoclonal antibody (immunotherapy), NS not statistically significant

Table 1
Detailed values for surgery duration, recovery time, and specific site of implantation

Animal ID and treatment	Surgery duration (min)	Recovery time (min)	Implantation site (at W0)
Naïve			
#62	–	–	–
Implanted with aCSF			
#68	105	15	L4–L5 to T7
#70	132	16	L4–L5 to T7
#71	176	14	L3–L4 to T5
Implanted with mAB			
#61	105	20	L4–L5 to T7–T8
#66	118	15	L4–L5 to T6
#73	115	10	L4–L5 to T5

Animal IDs are represented as #XX. *aCSF* artificial cerebrospinal fluid, *mAb* monoclonal antibody

from Pérez-Martín et al. [23]). Welfare status was recorded using a customized score sheet (Table 2). If necessary, compensatory measures such as analgesia and/or antibiotic treatment, food supplements, wound care, or surgical intervention were applied.

In addition, clinical status was checked every 15 days by monitoring body weight, rectal temperature, and performing hematological (e.g., leukocyte count, erythrocyte count, hematocrit, platelet count) and biochemical (e.g., albumin, alkaline phosphatase, bilirubin, creatinine, glucose) blood tests. Venous blood samples (~8 mL) were collected via the external jugular vein using an 18G×1 1/2″ needle (Sterican®) and the appropriate BD Vacutainer tubes (#366468, #367861; BD, Franklin Lakes, New Jersey, USA). Different hematological and biochemical parameters were analyzed to monitor clinical status as well as liver and renal functions throughout the study, as shown in Figs. 6 and 7. Overall, most values for the analyzed parameters fell within reference ranges at all timepoints, with only minor deviations in white blood cell counts during infection events. Furthermore, no significant differences were found between animals treated with aCSF and those treated with mAb for most of the parameters evaluated (p Student's t-test$=0.042$ in total protein W5; p Student's t-test$=0.007$ in urea level W13; $p>0.05$ for the remaining comparisons; Figs. 6 and 7).

3.5 CSF Collection and mAb Detection

CSF collection was performed via the cisterna magna, as previously described [30]. Briefly, pigs under general anesthesia were placed in sternal recumbency. Following hand palpation and identification of the crista occipitalis, two lines were drawn along the spine and

Table 2
Scoresheet for animal welfare and clinical status monitoring criteria in pigs undergoing surgery

Parameters		Score
Body weight	Normal	0
	Weight loss < 10%	1
	Weight loss 10%–20%	2
	Weight loss > 20%	3
General appearance	Normal appearance and posture	0
	Poor condition and mild change of posture	1
	Poor condition and/or presence of ocular, nasal, and digestive secretions.	2
	Abnormal posture, arched or immobile, piloerection, and continuous trembling	3
Spontaneous behaviour	Normal	0
	Small changes (awake or lying down, reduced appetite, and slight gait)	1
	Moderate changes (lethargic, depressed, anorexic, and gait)	2
	Severe changes (half-closed eyes, no response to stimuli, apathy, abnormal vocalisations, and restlessness)	3
Behaviour in response to handling and environmental stimuli	Normal	0
	Small changes (some resistance to handling, mild pain on palpation, and short, repeated vocalisations)	1
	Moderate changes (lower resistance to handling, greater immobility, and moderate pain on palpation with high-pitched vocalisations)	2
	Severe changes (no resistance to handling, tries to bite, evident vocalisations, and painful hypersensitivity)	3
Vital signs	Normal	0
	Slight deviations	1
	Changes in body temperature 1–2°C.	2
	Body temperature changes > 2°C	3

Score 0–4: normal physical condition, no pain. Scores 5–9: indicates mild pain and distress, monitor carefully and administer analgesia according to clinical assessment. Scores 9–14: moderate suffering and pain, analgesia and/or other pharmacological treatments must be applied as compensatory measures. Scores 15–20: humane endpoint application

between the cranial margins of the right and left winds of the atlas (Fig. 8). A spinal needle (20 G × 3.5″, 90 mm; #405253, BD) was inserted 5–10 mm caudal to the intersection of the two lines and advanced slowly to a depth of 80–90 mm, until the dura mater was punctured. A sample of 2–3 mL was withdrawn and stored in protein Lobind® tubes (#022431064; Eppendorf, Hamburg,

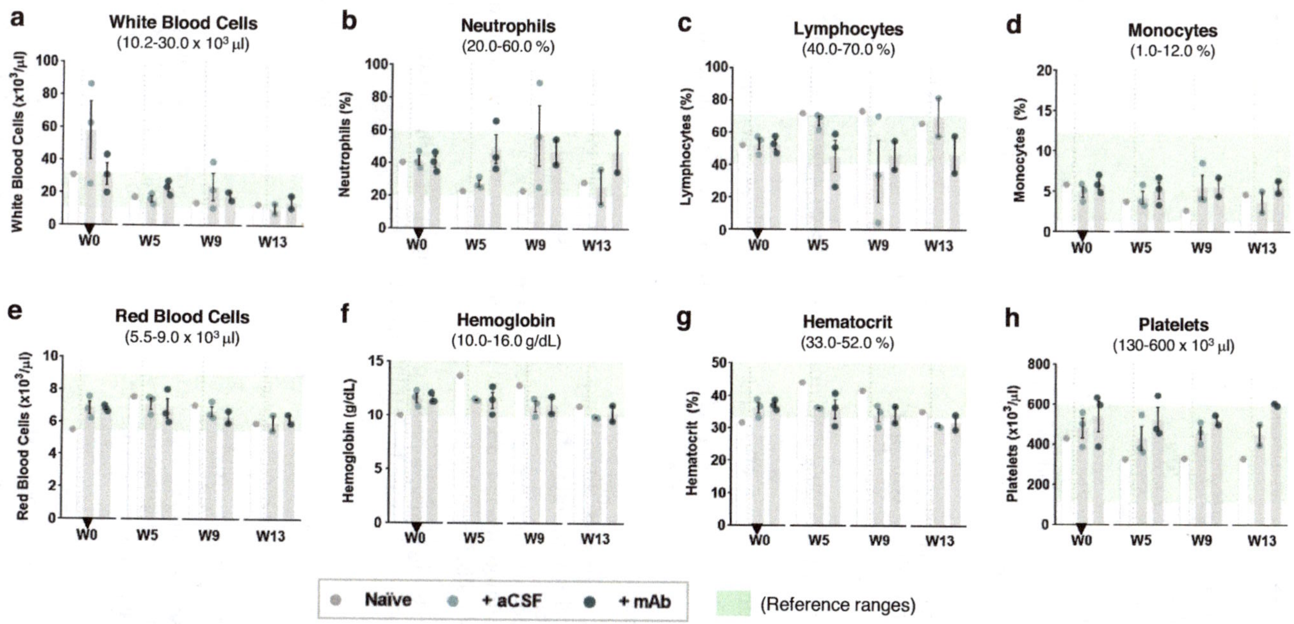

Fig. 6 Hematological assessment for monitoring clinical status throughout the pilot study. (**a**–**h**) Graphical representations of the analyzed hematological parameters: white blood cells (**a**), neutrophils (**b**), lymphocytes (**c**), monocytes (**d**), red blood cells (**e**), hemoglobin (**f**), hematocrit (**g**), and platelets (**h**). Most values fall within the reference ranges (denoted by the green band and numbers in brackets). No significant statistical differences were observed between animals implanted with aCSF or mAb. Data are presented as individual values, mean, and SEM. Student's t-test for paired comparisons between aCSF- and mAb-treated pigs (naïve values were not used for comparisons). $N = 1$ (naïve), 3 (+ aCSF), 3 (+ mAb). aCSF artificial cerebrospinal fluid, mAb monoclonal antibody (immunotherapy), W week

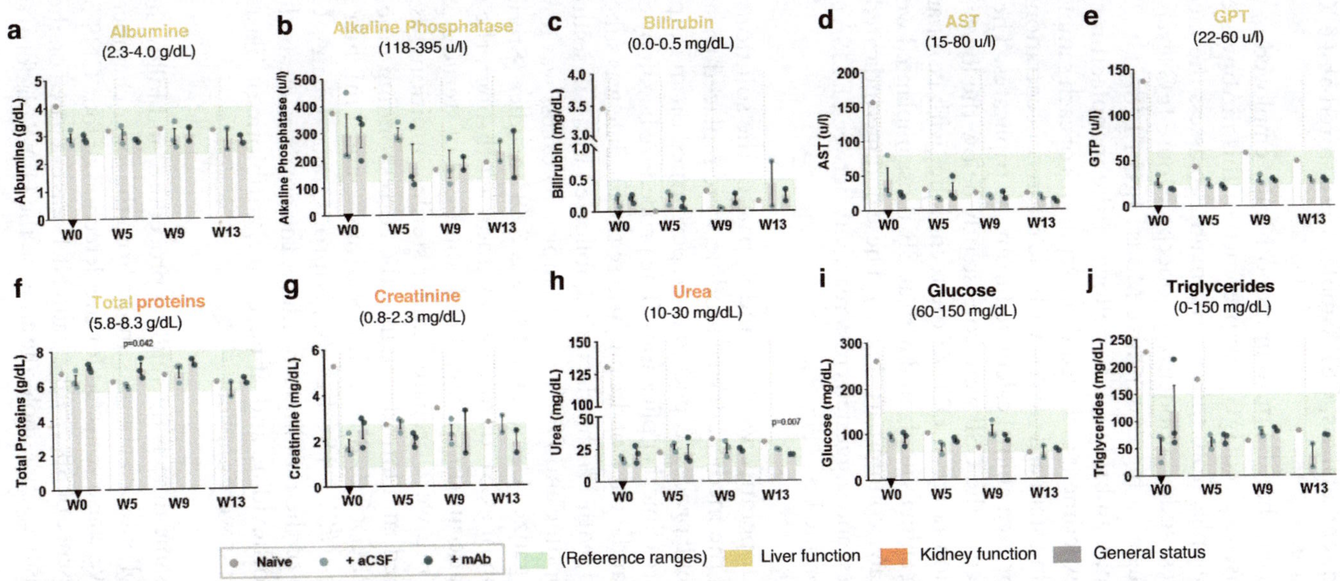

Fig. 7 Biochemical blood test for monitoring clinical status throughout the pilot study. (**a**–**j**) Graphical representations of the analyzed biochemical parameters: albumine (**a**), alkaline phosphatase (**b**), bilirubin (**c**), aspartate transferase (AST; **d**), alanine aminotransferase or glutamic-pyruvic transaminase (GPT; **e**), total proteins (**f**), creatinine (**g**), urea (**h**), glucose (**i**), and triglycerides (**j**). Most values fall within the reference ranges (indicated by the green band and numbers in brackets). No significant statistical differences were observed between animals implanted with aCSF or mAb in most of the parameters and timepoint analyzed. Data are presented as individual values, mean, and SEM. Student's *t*-test for paired comparisons between aCSF- and mAb-treated pigs (naïve values are not used for comparisons). $N=1$ (naïve), 3 (+ aCSF), 3 (+ mAb). aCSF artificial cerebrospinal fluid, mAb monoclonal antibody (immunotherapy), W week

Germany). In cases of blood contamination, centrifugation was performed at 2000 g for 10 min at room temperature (Minispin® microcentrifuge, # 5452000010; Eppendorf). Depending on the subsequent analysis, CSF samples were stored at +4°C or −80°C until use.

Given that the mAb immunotherapy employed in this study is a mouse-derived IgG, the presence of this mAb was evaluated in CSF samples from the cisterna magna to assess the medical device tightness and pseudodelivery concept. Mouse IgG levels were quantified using a highly sensitive Mouse IgG ELISA kit (#ab157719; Abcam, Cambridge, UK) following the manufacturer's protocol. Each experiment included three replicates of each sample and corresponding standard curves, with the mean concentration (ng/ml) of IgG for each animal calculated from the triplicates. The mAb levels detected in the CSF were consistently below the lower limit of the detection range (LOD) of the technique, with comparable values among animals implanted with aCSF, those implanted with mAb and their naïve counterpart (Fig. 8). These findings indicated device leak tightness and suggested cross-reactivity between ELISA components and porcine IgG (Fig. 8).

3.6 Refilling Process

At postoperative weeks 5 and 9, part of the solutions contained within the medical device module were replaced to simulate the drug refill process expected to be performed on an outpatient basis in future clinical applications. The procedure began by cleaning the right caudal flank of the pig, where the module was subcutaneously located, with 1% chlorhexidine, followed by a 1:2 solution of povidone iodine and 70° (v/v) ethanol. Any excess was removed with sterile gauze. Using two 23G×2 3/8″ needles (Sterican®), the solution of aCSF or mAb was withdrawn through the septum via percutaneous skin access. Fresh solutions of sterile aCSF or mAb 20 μg/mL were prepared under sterile conditions, as previously described, and introduced into the corresponding inner chambers using 23G×2 3/8″ needles (Sterican®) and 1 mL luer slip precise syringes (#SY3-1SCTU-EC; Nipro, Osaka, Japan). Upon completion of the refilling procedure, the area was cleaned again with 1% chlorhexidine and, if necessary, antibiotic or blastoestimulin ointments were applied.

3.7 CNS Biomarkers Detection in the Device Reservoir

At the same time as the refilling procedure, and following the same protocol outlined above, samples from the inner chamber of the reservoir were also collected for the detection of four CNS biomarkers. The presence of $A\beta_{1\text{-}42}$, $A\beta_{1\text{-}40}$, and Nf-L was evaluated on the SIMOA® ultrasensitive single-molecule array on the HD-X platform (Quanterix,

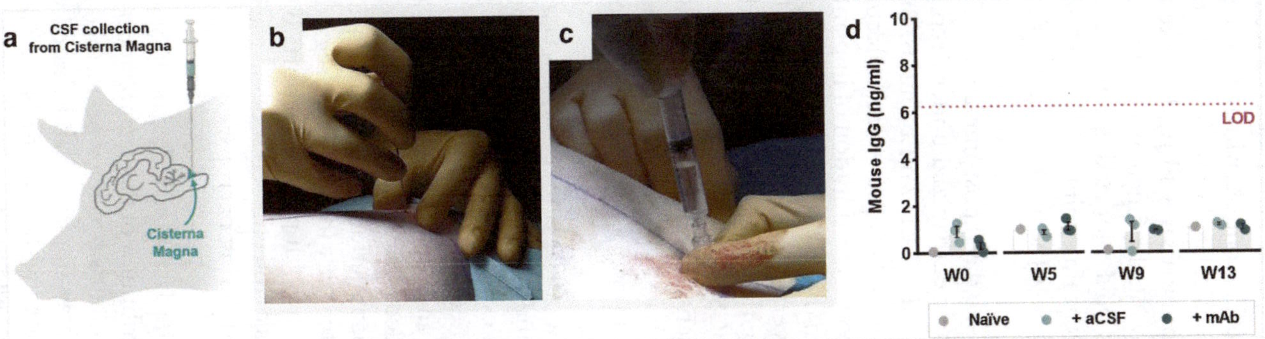

Fig. 8 Detection of mAb in CSF samples from the cisterna magna to assess device tightness in the pilot study. (**a–c**) Schematic illustration (**a**) and pictures (**b**, **c**) of the CSF collection procedure via cisterna magna. (**d**) Quantification of mouse IgG (mAb immunotherapy) in CSF from cisterna magna at weeks 0 (device implantation), 5, 9, and 13. No statistical differences were observed between aCSF and mAb-treated animals, with values always being below the lower limit of detection (LOD, indicated as a red dotted line) and similar to those of the naïve pig. Data are presented as individual values, mean, and SEM. Student's *t*-test for paired comparisons between aCSF- and mAb-treated pigs (naïve values are not used for comparisons). $N=1$ (naïve), 3 (+ aCSF), 3 (+ mAb). aCSF artificial cerebrospinal fluid, LOD lower limit of detection, mAb monoclonal antibody (immunotherapy), W week

Billerica, Massachusetts, USA), following the manufacturer's instructions by R.C. and E.C.-Z. at Achucarro Basque Center for Neuroscience (Leioa, Spain). The Neurology 4-PLEX E kit (Quanterix) was employed to simultaneously measure the concentrations of these biomarkers. Each experiment included two replicates of each undiluted sample and two quality controls, with the mean concentration (pg/ml) of each biomarker calculated from the duplicates. As shown in Fig. 9, the SIMOA® immunoassay 4-PLEX E worked successfully with porcine samples, allowing the functionality of the device to be validated. $A\beta_{1\text{-}42}$, $A\beta_{1\text{-}40}$, and Nf-L biomarkers were detected within the device reservoir at different timepoints, although concentrations at certain timepoints were below the working range of the technique (Fig. 9).

3.8 Euthanasia and Necropsy

Thirteen weeks (~90 days) post-implantation of the medical device, pigs were humanely euthanized under deep general anesthesia by administering 65 mEq KCl (2 mEq/ml; B. Braun). Prior to euthanasia, CSF and blood samples were collected for molecular and hematology analyses. The duration of implantation (days), defined as the period from the day of surgery to the time of euthanasia, was recorded. The overall mean implantation duration was 83 ± 5.71 days. Considering the experimental groups, the animals implanted and treated with aCSF had an implantation period of 86 ± 7.00 days, while those implanted and treated with mAb had a mean period of 80.66 ± 10.34 days. No statistically significant differences were found between the two groups after Student's *t*-test ($p = 0.691$; Fig. 10).

At necropsy, a dorsal laminectomy was performed in the thoracic (~T5–T9) and lumbar (~L3–L7) regions, corresponding to the vertebral bodies where the catheter's tip and entry point were located, as identified by X-ray fluoroscopy. The tissue appearance near the catheter and the catheter's position within the intrathecal space were recorded. Macroscopic examination of the spinal tissue revealed mild inflammation and the presence of a fibrotic capsule around the catheter tract in some pigs, regardless of the treatment (Fig. 10), consistent with previous reports by Gradert et al. [31]. Additionally, signs of infection episodes in the subcutaneous pocket were also observed during necropsy.

3.9 Statistical Analysis

The normality of the data was checked using Shapiro–Wilk test. Based on the normal distribution of the data (Shapiro–Wilk p-value > 0.05), the parametric Student's *t*-test was performed for pairwise comparisons between animals implanted and treated with aCSF and those implanted and treated with mAb (*see* **Note 3**). A p-value < 0.05 was considered the threshold for statistical significance. All quantifications are presented as individual data points and as the mean ± standard

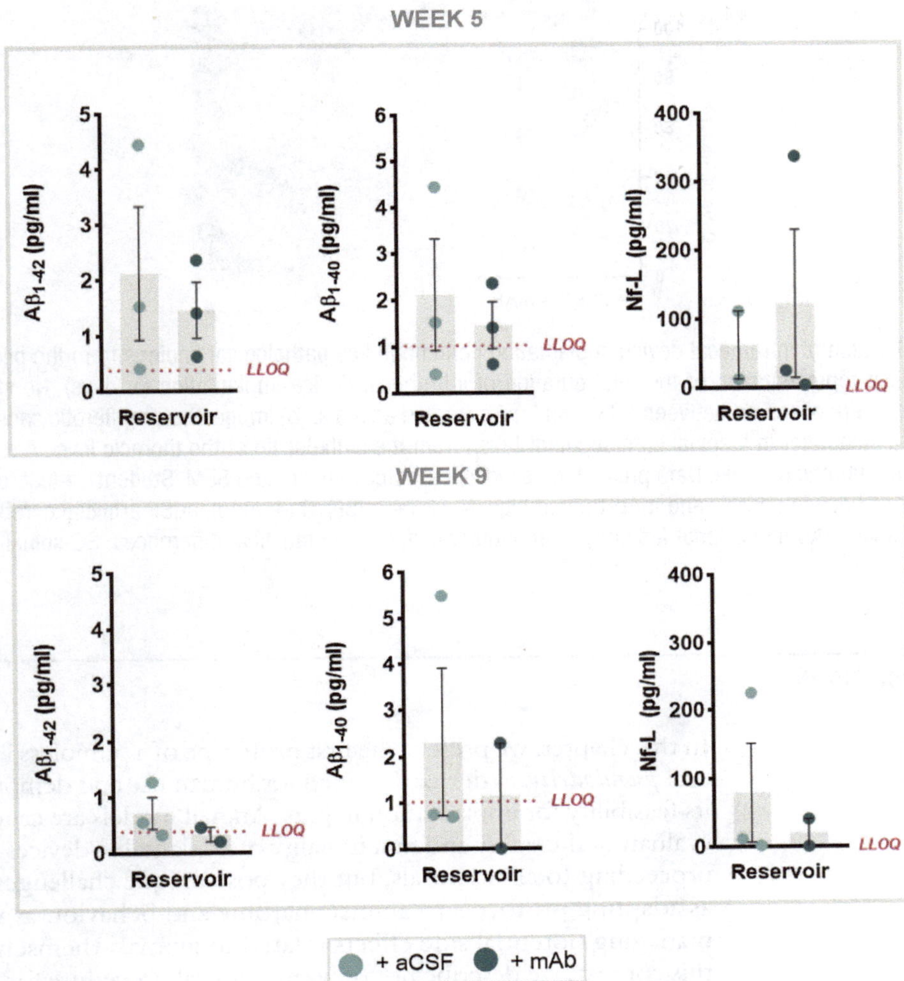

Fig. 9 Detection of CNS biomarkers in the device reservoir at weeks 5 and 9 of the pilot study. Note that Aβ 1-42, Aβ1-40, and Nf-L biomarkers were detected within the device reservoir, although concentrations at certain timepoints were below the lower limit of quantification (LLOQ, denoted as a red dotted line). No significant statistical differences were observed in $Aβ_{1-42}$, $Aβ_{1-40}$, and Nf-L levels between animals implanted with aCSF or mAb. Data are presented as individual values, mean, and SEM. Student's t-test for paired comparisons between aCSF- and mAb-treated pigs. $N = 3$ (+ aCSF), 3 (+ mAb). Aβ amyloid beta, aCSF artificial cerebrospinal fluid, mAb monoclonal antibody (immunotherapy), Nf-L neurofilament light polypeptide

error of mean (SEM). Statistical analyses were conducted using SPSS Statistics version 28 for Windows (IBM, Armonk, New York, USA), and graphical representations were obtained in GraphPad Prism version 9.0.2 for Windows (GraphPad, San Diego, California, USA).

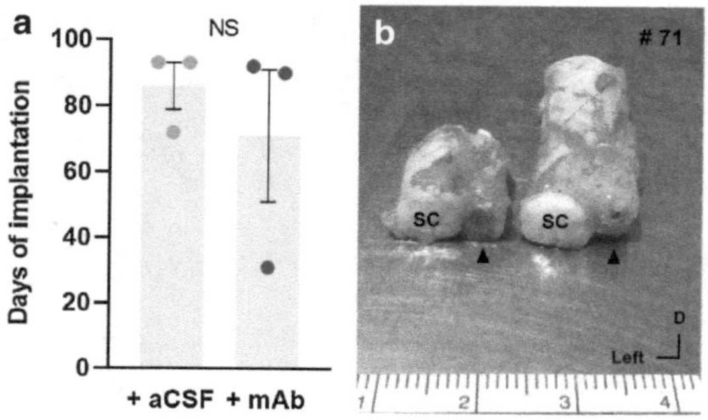

Fig. 10 Duration of intrathecal device implantation period and key pathological findings from the pilot study. (**a**) Graphical representation of the total duration of intrathecal device implantation (in days). No statistical differences were observed between aCSF- and mAb-treated animals. (**b**) Image showing fibrotic capsule surrounding the catheter in a spinal cord segment taken from the catheter tip at the thoracic level. Arrowheads point out the fibrotic capsule. Data presented as individual values, mean, and SEM. Student's *t*-test for paired comparisons between aCSF- and mAb-treated pigs. N = 3 (+ aCSF), 3 (+ mAb). aCSF artificial cerebrospinal fluid, D dorsal, mAb monoclonal antibody (immunotherapy), NS no statistical differences, SC spinal cord, W week

4 Conclusions

In this chapter, we present the first prototype of a pumpless intrathecal *pseudodelivery* device designed for human use and demonstrate its feasibility for implantation in pigs. Animal models are crucial for evaluating the safety and functionality of implantable devices before proceeding to clinical trials, but they pose unique challenges, such as adapting protocols to animal anatomy and behavior, as well as managing potential side effects related to animals themselves. In this context, we describe not only the surgical procedure for device implantation but also methods for blood and CSF collection in large animals of great utility in neuroscience preclinical studies.

The development of implantable medical devices for humans requires overcoming significant hurdles in material science, regulatory compliance, and functional performance before progressing to clinical development. The device we present relies on a pumpless design, which minimizes mechanical failure risks and reduces maintenance. This design also reduces the device's size, weight, and complexity, simplifying the surgical implantation procedure and potentially extending device lifespan. Our preclinical test in pigs demonstrates the feasibility of implanting and operating the device by surgeons, highlighting the advantages of a pumpless system.

Previous preclinical studies on intrathecal drug delivery pumps have shown that high drug doses can cause significant inflammation, while lower doses typically result in mild inflammation [31]. Our study observed mild inflammation and fibrotic encapsulation around the catheter, consistent with earlier findings from Gradert et al. and

Johansen et al., likely due to the mechanical presence of the catheter in the narrow intrathecal space, rather than drug toxicity [31, 32]. While our device is designed to prevent drug-induced toxicity by avoiding direct drug interaction with the CNS and immune system, tissue reactions were still observed. These findings emphasize the need for further optimization of the design, catheter, and module placement, and materials to minimize complications.

In conclusion, we present the first prototype of a human-use pumpless device for intrathecal pseudodelivery and the corresponding surgical protocols for implantation and sample collection in pigs. This pilot study sets the stage for future refining of both the device and the implantation protocol. By eliminating the mechanical complexities of pumps and avoiding the most common immunotherapy-related side effects, our device may provide a safer, more convenient alternative for continuous intrathecal action. However, further studies are needed to optimize the device's design, implantation protocol, and to test long-term safety and efficacy in intrathecal pseudodelivery systems.

5 Notes

1. The mAb solution is prepared fresh on the day of surgery and is not stored once diluted in aCSF for subsequent in vivo use, only for biochemical in vitro assays.
2. To confirm the correct needle placement in the intrathecal space, the injected non-ionic contrast medium should diffuse toward the ventral portion of the spinal canal within a few minutes. Linear tracing would indicate an epidural location, requiring repositioning of the needle.
3. Depending on the experimental conditions and the distribution of experimental data according to the Shapiro–Wilks or Kolmogorov-Smirnov tests, other statistical tests may be required, such as Mann–Whitney U, Kruskal–Wallis, or one-way ANOVA followed by a post hoc test.

Acknowledgments

The authors thank the members of the Centro de Cirugía de Mínima Invasión Jesús Usón (Cáceres, Spain), in particular the surgeons Elena Abellán Rubio and Laura Cristina Pires Louça, anesthesiologists, operating room technicians, and animal facilities' veterinarians for the surgical assistance and technical support; the neurosurgeons Lydia Nader Navarro and Marco Álvarez-Vega of Hospital Universitario Central de Asturias (HUCA) for the suggestions in the design of the surgical protocol; and Raffaela Cipriani and Estíbaliz Capetillo-Zarate from Achucarro Basque Center for Neuroscience (Leioa, Spain) for the support in the SIMOA immunoassay and data interpretation. The E.P.-M. contract is partially supported by MCIU/AEI of the Spanish Government under grant PTQ2022-012644. This study was supported by Centro para el Desarrollo Tecnológico Industrial (CDTI), Ministerio de Ciencia e Innovación of Spanish Government, Next GenerationEU European Union funding, Plan de Recuperación, Transformación y Resiliencia under grant SNEO-20211309; Instituto de Desarrollo Económico del Principado de Asturias under grant IDE/2021/000422; and Neurostech.

Data Availability

Data supporting the findings in this chapter are available from the corresponding author (E.P.-M.) upon reasonable request. However, data related to the design of the medical device are confidential and protected by the patent No. WO2019/158791A1.

Conflict of Interest E.P.-M., M.R.-C., C.R.-L., G.A., and C.P. are employed by Neuroscience Innovative Technologies S.L.

References

1. Pardridge WM (2012) Drug transport across the blood–brain barrier. J Cereb Blood Flow Metab 32:1959–1972. https://doi.org/10.1038/jcbfm.2012.126
2. Patel MM, Patel BM (2017) Crossing the blood–brain barrier: recent advances in drug delivery to the brain. CNS Drugs 31:109–133. https://doi.org/10.1007/s40263-016-0405-9
3. Valiukas Z, Ephraim R, Tangalakis K et al. (2022) Immunotherapies for Alzheimer's disease—a review. Vaccines 10:1527. https://doi.org/10.3390/vaccines10091527
4. Zhang Y, Chen H, Li R et al. (2023) Amyloid β-based therapy for Alzheimer's disease: challenges, successes and future. Signal Transduct Target Ther 8:248. https://doi.org/10.1038/s41392-023-01484-7
5. Sperling RA, Jack CR, Aisen PS (2011) Testing the right target and right drug at the right stage. Sci Transl Med 3:111cm33. https://doi.org/10.1126/scitranslmed.3002609
6. Wu D, Chen Q, Chen X et al. (2023) The blood–brain barrier: structure, regulation and drug delivery. Signal Transduct Target Ther 8:217. https://doi.org/10.1038/s41392-023-01481-w
7. Schwinghamer K, Siahaan TJ (2023) Enhancing antibody exposure in the central nervous system: mechanisms of uptake, clearance, and strategies for improved brain delivery. J Nanotheranostics 4:463–479. https://doi.org/10.3390/jnt4040020
8. Agarwal A, Gupta V, Brahmbhatt P et al. (2023) Amyloid-related imaging abnormalities in Alzheimer disease treated with anti–amyloid-β therapy. Radiographics 43:e230009. https://doi.org/10.1148/rg.230009
9. Panza F, Lozupone M, Logroscino G et al. (2019) A critical appraisal of amyloid-β-targeting therapies for Alzheimer disease. Nat Rev Neurol 15:73–88. https://doi.org/10.1038/s41582-018-0116-6
10. Vargason AM, Anselmo AC, Mitragotri S (2021) The evolution of commercial drug delivery technologies. Nat Biomed Eng 5:951–967. https://doi.org/10.1038/s41551-021-00698-w
11. Farokhzad OC, Langer R (2009) Impact of nanotechnology on drug delivery. ACS Nano 3:16–20. https://doi.org/10.1021/nn900002m
12. De Andres J, Hayek S, Perruchoud C et al. (2022) Intrathecal drug delivery: advances and applications in the management of chronic pain patient. Front Pain Res 16:900566. https://doi.org/10.3389/fpain.2022.900566
13. Knight KH, Brand FM, Mchaourab AS et al. (2007) Implantable intrathecal pumps for

chronic pain: highlights and updates. Croat Med J 48:22–34
14. Hayek SM, Hanes MC (2014) Intrathecal therapy for chronic pain: current trends and future needs. Curr Pain Headache Rep 18: 388. https://doi.org/10.1007/s11916-013-0388-x
15. Glitza Oliva IC, Ferguson SD, Bassett R et al. (2023) Concurrent intrathecal and intravenous nivolumab in leptomeningeal disease: phase 1 trial interim results. Nat Med 29:898–905. https://doi.org/10.1038/s41591-022-02170-x
16. Dan X, Huang M, Sun Z et al. (2024) Case report: concurrent intrathecal and intravenous pembrolizumab for metastatic melanoma with leptomeningeal disease. Front Oncol 14:1344829. https://doi.org/10.3389/fonc.2024.1344829
17. Schreiner TG, Menéndez-González M, Popescu BO (2022) The "cerebrospinal fluid sink therapeutic strategy" in Alzheimer's disease—From theory to design of applied systems. Biomedicines 10:1509. https://doi.org/10.3390/biomedicines10071509
18. Menéndez-Gonzalez M, Padilla-Zambrano HS, Alvarez G et al. (2018) Targeting beta-amyloid at the CSF: a new therapeutic strategy in Alzheimer's disease. Front Aging Neurosci 10:100. https://doi.org/10.3389/fnagi.2018.00100
19. Menéndez-Gonzalez M (2025) Targeting soluble amyloid oligomers in Alzheimer's disease: a hypothetical model study comparing intrathecal pseudodelivery of mAbs against intravenous administration. Diseases 13:17. https://doi.org/10.3390/diseases13010017
20. Menéndez-Gonzalez M (2024) Intrathecal immunoselective nanopheresis for Alzheimer's disease: what and how? Why and when?. Int J Mol Sci 25:10632. https://doi.org/10.3390/ijms251910632
21. Menéndez-González M, Tamba B-I, Leclere M et al. (2023) Intrathecal pseudodelivery of drugs in the therapy of neurodegenerative diseases: rationale, basis and potential applications. Pharmaceutics 15:768. https://doi.org/10.3390/pharmaceutics15030768
22. Coto-Vilcapoma MA, Castilla-Silgado J, Fernández-García B et al. (2022) New, fully implantable device for selective clearance of CSF-target molecules: proof of concept in a murine model of Alzheimer's disease. Int J Mol Sci 23:9256. https://doi.org/10.3390/ijms23169256
23. Pérez-Martín E, Coto-Vilcapoma A, Castilla-Silgado J et al. (2023) Refining stereotaxic neurosurgery techniques and welfare assessment for long-term intracerebroventricular device implantation in rodents. Animals 13:2627. https://doi.org/10.3390/ani13162627
24. Pérez-Martín E, Menéndez GÁ, Prado C, et al. (2023) Feasibility and safety of long-term intrathecal pseudodelivery of immunotherapy for selective clearance of csf amyloid-β in a mouse model of alzheimer's disease. IBRO Neurosci Rep 15:S475. https://doi.org/10.1016/j.ibneur.2023.08.925
25. Pérez-Martín E, Tomás-Zapico C (2024) Refinamiento de las cirugías estereotáxicas de implantación intratecal a largo plazo en roedores y monitorización del bienestar animal. Anim Lab 101:39–47
26. Pérez-Martín E, Rodríguez-Cañón M, Roces-Llorián C et al. (2025) Implantable device for intrathecal pseudodelivery of drugs: surgical protocol and proof-of-concept in a porcine model. BJS 112:S2. https://doi.org/10.1093/bjs/znae322.001
27. Pérez-Martín E, Rodríguez-Cañón M, Álvarez G, et al. (2025) Implantable device for intrathecal pseudodelivery of immunotherapies: Preclinical testing of a prototype for humans. J Prev Alzheimers Dis 12(S1):122. https://doi.org/10.1016/j.tjpad.2024.100044
28. Zare M, Ghomi ER, Venkatraman PD et al. (2021) Silicone-based biomaterials for biomedical applications: antimicrobial strategies and 3D printing technologies. J Appl Polym Sci 138:50969. https://doi.org/10.1002/app.50969
29. Pleticha J, Maus TP, Jeng-Singh C et al. (2013) Pig lumbar spine anatomy and imaging-guided lateral lumbar puncture: a new large animal model for intrathecal drug delivery. J Neurosci Methods 216:10–15. https://doi.org/10.1016/j.jneumeth.2013.03.006
30. Romagnoli N, Ventrella D, Giunti M et al. (2014) Access to cerebrospinal fluid in piglets via the cisterna magna: optimization and description of the technique. Lab Anim 48:345–348. https://doi.org/10.1177/0023677214540881
31. Gradert TL, Baze WB, Satterfield WC et al. (2003) Safety of chronic intrathecal morphine infusion in a sheep model. Anesthesiology 99:188–198. https://doi.org/10.1097/00000542-200307000-00029
32. Johansen MJ, Satterfield WC, Baze WB et al. (2004) Continuous Intrathecal Infusion of Hydromorphone: safety in the Sheep Model and Clinical Implications. Pain Med 5:14–25. https://doi.org/10.1111/j.1526-4637.2004.04010.x

Subject Index

Manuel Menéndez-González (ed.), *Intrathecal Drug Delivery*, Neuromethods, vol. 234, https://doi.org/10.1007/978-1-0716-5241-1,

© The Editor(s) (if applicable) and The Author(s), under exclusive license to Springer Science+ Business Media, LLC, part of Springer Nature 2026

FSC
www.fsc.org
MIX
Papier | Fördert
gute Waldnutzung
FSC® C083411

Kolibri 360 GmbH
Ferdinand-Jühlke-Straße 7
99095 Erfurt, Deutschland
produktsicherheit@kolibri360.de